SEX, LOVE AND DNA

SEX, LOVE AND DNA

What Molecular Biology Teaches

Us About Being Human

PETER SCHATTNER

OLINGO PRESS

Foster City, California

SEX, LOVE AND DNA
Peter Schattner

Editor: Leslie Tilley
Cover Design: Glen M. Edelstein
Book Interior Design: Reality Information Systems
Illustrations: Paul Brittenham
Photography credits are listed in the acknowledgments.

Published in the United States by Olingo Press

Publisher's Cataloging-in-Publication Data

Schattner, Peter, 1948-
    Sex, love and DNA: what molecular biology teaches us about being human / Peter Schattner.
pages cm
    Includes bibliographical references and index.
    ISBN: 978-0-9914225-1-7 (pbk.)
    ISBN: 978-0-9914225-2-4 (e-book)
    1. Human genetics. 2. Molecular biology. 3. Biology—Social aspects. 4. Science—
Popular works.   I. Title.
QH431 .S317 2014
599.93`5—dc23
                                        2014909458

This book is dedicated in loving memory of my father,

Edward Schattner

TABLE OF CONTENTS

PREFACE

Ever since childhood I've wondered why the world is the way that it is. Why are sunsets so colorful, and why do birds return to the same places every spring? Why don't people in the Southern Hemisphere fall off the earth, and why don't we notice that our planet is hurtling through space at hundreds of thousands of miles per hour? Mostly I was curious about people and why we behave in such surprising and unpredictable ways.

Having been raised with a traditional, if not particularly observant, religious upbringing, I was taught that the answer was God. God was this unseen and inscrutable force that causes everything to be the way that it is. Unfortunately, I didn't find this explanation very satisfying. Instead, from the time I was a teenager, I was fascinated by the world of science, which provided answers to questions about nature that I could understand and that I found quite beautiful.

Being curious about people and animals, you might think I'd have been drawn to biology, the science of life. However, when I was going to high school and college, in the 1960s, biology courses were still largely descriptive. Students learned to catalogue and describe animals and plants, as well as their organs, substructures and even their cells. But there was little explanation – and, in fact, still relatively little knowledge – of how the microscopic world of cells affected an entire plant or animal, let alone a human being.

As a result, I found biology unsatisfying. Instead I was attracted to the world of physics. Through physics, I learned how an extraordinary variety of complex natural phenomena and human inventions – from volcanoes and hurricanes to transistors and lasers could be understood from a small number of "fundamental laws" and the logical framework of mathematics. I was instilled with a sense of awe of the natural world's simplicity and beauty when viewed through the lens of modern physics. I decided to study physics, and in particular particle physics, the science of protons, neutrons, electrons and the other fundamental constituents of nature. I completed my PhD in physics and began to do research and teach as a physicist.

Despite my disappointing experiences studying biology in school, my desire to learn more about the scientific basis of life didn't disappear. I maintained the belief that someday understanding living creatures in the systematic manner of physics would be possible. Slowly, I refocused my work life back to biology. I left the world of pure physics and became an applied physicist and engineer, using the techniques and tools of physics to study biology. I worked for several years on the early development of medical ultrasound and magnetic resonance imaging (MRI) scanners, instruments that provide detailed pictures of the inside of the human body.

Although ultrasound and MRI scanners are remarkable tools, they are also frustrating ones. They provide excellent images of the brain and other human organs, but their resolution is too limited to look inside individual human cells. In particular, ultrasound and MRI shed little light on the interplay of proteins and DNA within cells, interactions that play as critical a role in our biological makeup as protons, neutrons and electrons do in the world of physics. Driven by my desire to learn more about how people functioned at the most fundamental level, I began to retrain myself as a molecular biologist, and in 2001 I started working as a researcher in the Biomolecular Engineering Department of the University of California, Santa Cruz.

As I began applying mathematical and computer techniques to questions in biology, I also became aware of how the collective efforts of thousands of biologists were starting to create a coherent picture of the world of proteins and DNA. I started to appreciate how this microscopic universe affects almost every aspect of our lives. I began to share the profound sense of awe that many biologists experienced as they watched their field transform from a descriptive science into an elegant framework in which countless complex phenomena could be explained by a relatively small number of profound scientific principles.

This transformation of biology into a rigorous scientific discipline has long been clear to those working in the field. Yet I realized that my nonbiologist friends, even those who were quite intelligent and intellectually sophisticated, were generally unable to appreciate these developments. The problem appeared to be that while headline-catching snippets of research breakthroughs did reach the mainstream media, these discoveries rarely were tied together and presented in a way that was accessible to the nonscientist. This seemed a great pity, as I believe that the ideas of modern biology are having a profound impact on our understanding of the natural world, and that, if clearly explained, these concepts are well within the grasp of the intellectually curious nonscientist. I realized that I wanted to help nonbiologists appreciate the beauty and profound implications of this exciting new world of molecular biology. And it was this desire that motivated me to write this book.

PART I

PROTEINS AND GENES: THE CONSTITUENTS OF LIFE

1

FROM PROTEINS TO PEOPLE

Can the discoveries of 21st-century molecular biology answer age-old questions about the human experience? Can studying proteins and DNA help us understand how we make our choices in sex and love? How we communicate? Where our emotions come from? Or why we age and die?

In the past such questions have generally been reserved for philosophers or psychologists. Some will argue that this is as it should be. Yet we are biological animals, and by studying biology, and especially the biology of cells and proteins and DNA, we can learn a lot about what it means to be human.

In this book I'll address these questions from the perspective of recent research in molecular biology, using stories and anecdotes that illustrate how the interactions of proteins and DNA shape an animal's life and how they affect the lives of people, as well.

DNA, PROTEINS AND BEHAVIOR

If the idea that proteins and DNA influence who we are seems surprising, consider these examples:*

---

* Some of the terms will be unfamiliar. Don't worry. They'll all be explained when we revisit these examples later.

- Prairie voles and meadow voles are closely related rodents, but male prairie voles are usually devoted monogamous mates, while meadow voles are promiscuous. This different behavior is largely the result of a single gene, AVPR1A, which is involved in the detection of a hormone called oxytocin. In fact, injecting prairie vole AVPR1A DNA into the brains of male meadow voles changes these promiscuous animals into devoted mates that are just as faithful as prairie voles.

- Fucose mutarotase is a protein that links a sugar molecule (called fucose) to other proteins. Although mice that can't produce fucose mutarotase are healthy and appear just like normal mice, the females display typical male sexual behavior. They mount other females and prefer to sniff female urine. Physically, they are 100% female. Yet disabling a single protein somehow switches their brain wiring to become what we might call "homosexual."

- Male zebra finches attract females by singing, with every young male zebra finch learning to sing the same song. But if a zebra finch is deficient in a protein called FOXP2, it is unable to learn how to correctly sing the usual finch song.

- Adding a gene can make a mouse smarter. When scientists administered an extra dose of DNA for a gene called NR2B to mice, they found that the mice solved mazes faster, and remembered the solutions longer, than their otherwise genetically identical litter mates.

- Male mice lacking a protein called MAO-A are more aggressive than other mice. If these mice are given a drug that compensates for their MAO-A deficiency, their aggressive behavior disappears.

It might be tempting to discount such animal experiments as irrelevant to understanding people and as merely confirming that humans are very different from animals. Since scientists can't do the same kinds of experiments on people, they can't directly determine whether similar biological effects occur in people. Nevertheless, indirect evidence increasingly suggests that animal experiments are relevant to understanding human behavior. Consider these recent discoveries:

- Autism is characterized by difficulties in expressing and receiving love and affection. Recent studies have linked some cases of autism to impaired oxytocin signaling, the same brain pathway that affects pair bonding in voles. Some physicians have even started treating autistic children with oxytocin. Although the causes of autism are still unclear, a few autistic individuals have even been found with rare variants of the AVPR1A gene, the same gene involved in oxytocin signaling in voles.

- The adrenal gland uses a protein called 21-OH to produce the hormone cortisol. Approximately 1 out of 1500 women lack 21-OH, and their adrenal glands are unable to synthesize cortisol (a condition called congenital adrenal hyperplasia). Instead their glands synthesize testosterone, a hormone normally produced by men. Although testosterone's effects on women are still poorly understood, a recent study of sexual orientation was able to reach the "overwhelming conclusion" that women with congenital adrenal hyperplasia are more likely to be homosexual or bisexual than other women.

- In the 1990s, British researchers described a family that had severe language difficulties despite having normal intelligence, hearing and vocal chords. Genetic analysis revealed that all the afflicted family members shared a defective variant of the protein FOXP2, the same protein that was later shown to be critical for proper singing in zebra finches.

- Some individuals suffering from intellectual disability have nonfunctional variants of NR2B, the same gene that was inserted into the DNA of the super-smart mice. In addition, genetic studies of people with severe intellectual disability have found mutations in genes that interact closely with NR2B. Recently, researchers have also discovered the first genetic variant linked with IQ scores of healthy people, suggesting that differences in brain biochemistry contribute not only to intellectual disability but also to variations in normal intelligence.

- In 1993, scientists from the Netherlands reported on a family that included eight men, all of whom were violent or impulsively aggressive. Each of these men had a disabling variant in the MAO-A gene, the same gene linked to increased aggression in male mice. This particular genetic variant is extremely rare, but other more common MAO-A variants have since been discovered and also appear to be linked to susceptibility to violent behavior.

These examples illustrate how genes and proteins influence what sort of people we become. Occasionally a single inherited genetic variation can dramatically affect us. Some traits, such as intelligence or height, are influenced by hundreds of genetic variants. In other cases, environmental factors are more important than genetic ones. Yet here as well, scientists are learning that our environment affects us through changes in our DNA and proteins.

ABOUT THIS BOOK

In this book, you'll learn how genetics and the environment interact on the level of cells and how those interactions affect our lives. Each new concept, however elementary, will be explained as it is encountered, so even if you have no background in molecular biology or genetics, you shouldn't have any problem following the explanations in the book. I'll describe biological phenomena using everyday language, rather than with the specialized words and abbreviations used in the scientific and medical literature. When a specialized scientific word does make communicating easier, it will be defined right away or included in the glossary.* And don't worry; you won't be reading a dry biology textbook. You'll be learning biology through stories: stories of people who don't feel pain because of rare genetic variants and children whose DNA enables them to perform unusual feats of strength. Individuals whose genes have given them healthy lives past the age of 100, and people who can't speak or read simply because they lack certain proteins.

To ease our journey into the brave new world of molecular biology, the book is divided into six parts that build upon one another to tell a unified story. We'll begin our explorations by learning about proteins and their central role in life and get an introduction to DNA and genes, the blueprints for building proteins. In Part II of the book, the emphasis will turn more specifically to DNA and the way it provides a window into the past and the future. Part III focuses on genes and introduces RNA, the key intermediary molecule between genes and proteins. The way that our environment affects us is the central topic of Part IV. We'll learn about chromosomes and how the environment affects them, thus altering the genetic recipes that we inherited from our parents. By Part V, we'll be ready to appreciate what modern biology is teaching scientists about the ancient debate between "nature" and "nurture." We'll see how animal experiments, as well as human genetic studies, can disentangle the effects of heredity and environment on our proteins and on our lives. Finally, in Part VI, the various threads of DNA and protein, as well as RNA and chromosomes, will be tied together in a series of chapters illustrating how the world of molecular biology affects our behavior and our emotions.

Although you won't need any scientific background to understand *Sex, Love and DNA*, you should bring an active curiosity and a willingness to

---

* The exception is the abbreviations used for gene and protein names. Because these names are typically assigned long before scientists know a gene's or protein's function, they usually convey little information, and abbreviations work just as well.

stretch your mind. That said, the amazing world of molecular biology that you are about to enter will make the effort worthwhile. Beyond simply enjoying how the collective human mind has unraveled so much of how we function, you will learn to read about the latest scientific breakthroughs with a more critical eye, becoming capable of distinguishing real advances from the sometimes-breathless hype of the popular press. In short, you will be able to share the excitement as the scientific community addresses perhaps the greatest intellectual challenge of all – the challenge that Socrates described more than 2000 years ago as "to know thyself."

2

CAN A PROTEIN SAVE YOU FROM AIDS?

S teve Crohn had more than his share of tragedy and heartbreak. In the
1980s and early 1990s he saw one friend after another, all gay men like
himself, succumb to the deadly scourge that came to be known as acquired
immunodeficiency syndrome (AIDS). By all rights, Crohn should have been
stricken as well. He was certain that, just like his friends, he had been exposed
to the human immunodeficiency virus (HIV), the deadly microorganism
that causes AIDS. Yet Steve Crohn didn't get sick. Like a small number of
other people who had been repeatedly exposed to HIV, Crohn just didn't get
infected with HIV, and he didn't get AIDS.

When it first became apparent that a few people didn't get AIDS
even after repeated exposure to HIV, no one had any idea why. Some
believed alternative, nontraditional AIDS treatments, such as laetrile,
made the difference. Others reasoned that the survivors were simply in
better overall health or had greater mental determination. Some thought
that prayer was the key and that a miracle could cure one of AIDS, if one
believed strongly enough.

But Steve Crohn was not saved from AIDS by a miracle, nor by a
healthy lifestyle or an exotic herb. No, Crohn was saved by a rare quirk in
his genetic makeup. Later in this chapter we'll see how a subtle change in
Crohn's genes prevented him from being infected by HIV. To understand

how this happened we'll first need a basic understanding of AIDS, as well as of genes and DNA, the remarkable molecule that genes are made of. Before that, though, we'll need to explore a class of biological molecules that are the basis for almost everything that happens in our bodies: proteins.

PROTEINS

Proteins are large molecules consisting largely of carbon, hydrogen, nitrogen and oxygen, and are arguably the most important building blocks of life. Approximately 20% of our body is made of proteins. After water, our body's cells, the trillions of microscopic structures that form our tissues, contain more protein than anything else. Our muscles are largely built out of proteins, as are our brain, lungs and other internal organs. Many of the signaling molecules, which our cells use to communicate with each other, and the cell structures that receive these signals, are composed of proteins. The metabolic processes that provide our bodies with energy are controlled by enzymes, which are also proteins. In fact, even the biological molecules that are *not* proteins, such as sugars and carbohydrates and fatty acids, are synthesized by proteins. It is not without reason that the word *protein* comes from the Greek word *protos,* meaning "first."

But proteins are not the *fundamental* biological building blocks, because they themselves are assembled as strings of smaller molecules called amino acids. There are 20 of these amino acids found in living cells.* The amino acids are a family of related molecules, with names such as methionine, lysine, asparagine, and glutamine, which share the same chemical backbone but have differing chemical structures attached to that backbone. Different parts of an amino acid's structure have different charges: positive or negative. Since each amino acid has a different shape, each one also has a different distribution of electrical charges on its surface.

The strings of amino acids that make up proteins are typically quite long, often containing hundreds or even thousands of amino acids. As different proteins consist of varying sequences of amino acids, each with its own distribution of electric charges, each type of protein also has a unique set of positive and negative charges along its length. And because opposite electric charges attract and like charges repel each other, each protein –

---

* Actually 22 different kinds of amino acids have been found in living cells, but 2 of them, selenocysteine and pyrrolysine, are very rare. They are found mainly in microbes and are not incorporated into proteins by the conventional genetic code.

which starts out as a simple, linear string of amino acids – folds itself into a unique three-dimensional shape in a way that puts opposite electrical charges as close to one another as possible while the like charges are as far apart from each other as possible.

PROTEINS FOR BIOLOGICAL SIGNALING

One of the amazing facts of biology is that just by taking on different *shapes* proteins take on different *functions*. Nowhere is this link between a protein's shape and its function more dramatically illustrated than in the body's internal signaling systems. Our bodies are composed of trillions of cells that must function in a coordinated manner for us to be able to move, breathe, see, hear, or do just about anything else. To accomplish such coordinated activity, our bodies use a form of signaling that is both simple in concept and remarkably sophisticated in its details.

The core of our body's signaling system involves two kinds of molecules: ligands and receptors. The ligands are small molecules that cells synthesize and secrete when they need to send a signal to other cells. For communicating with cells that are not located nearby, the transmitting cell will typically transmit the ligand via the bloodstream. Such ligands are called hormones. In contrast, sometimes a cell just needs to send a signal directly to a neighboring cell. This form of communication is particularly important in the brain and nervous system, and ligands used for this form of signaling are known as neurotransmitters. For a ligand to function as a signaling molecule, there needs to be another molecule, called a receptor, to detect the ligand. Receptors are nearly always large molecules composed of multiple proteins with specific three-dimensional shapes. The receptor's shape is said to be complementary to that of its corresponding ligand. This means that the ligand's shape matches the shape of the receptor, in the way that a key matches the shape of a lock. As a result, when a ligand comes close to its matching receptor, there is a strong electrical attraction between the two; in the language of biochemistry, the ligand is *bound* to the receptor. Typically, the force of electrical attraction between the ligand and its receptor causes the shape of the receptor to change ever so slightly. This subtle change in the receptor's shape produces a magnified movement of electrical charge in other parts of the receptor, which initiates some chemical or physical process within the cell. Figure 2.1 illustrates the concept of a ligand and its associated receptor.

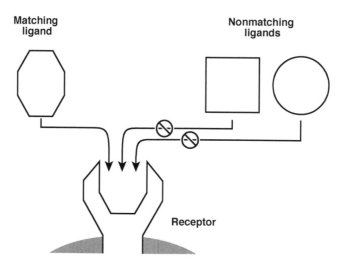

Figure 2.1. Ligands and receptor. The shape of the ligand shown at upper left matches that of the receptor, and consequently it can bind to the receptor. The other ligands do not match the receptor and are unable to bind to it.

The regulation of blood sugar using the secretion of insulin by the pancreas is one example of biological signaling using ligands and receptors. Special cells in the pancreas, called beta cells, detect blood glucose concentration by means of glucose receptors on their surface. When blood sugar levels reach a certain threshold, the beta cells secrete insulin into the blood. The insulin then circulates in the bloodstream until it is detected by insulin receptors in muscle, liver and fat cells. The binding of the insulin to the insulin receptors starts a chain of biochemical reactions causing the muscle, liver and fat cells to take up more glucose from the blood and store it for future use.

All the other functions that proteins carry out, from moving muscles and synthesizing sugars or carbohydrates to extracting energy from food and oxygen, are performed by similar subtle changes in the shapes of proteins. Indeed, a few simple statements summarize much of what scientists have learned about the biological functioning of proteins:

- Proteins are long strings of amino acids in a specific sequence.
- Each type of amino acid has a somewhat different shape and distribution of electrical charge.
- As a result of its unique distribution of electric charges, each type of protein folds itself into a different three-dimensional shape that is precisely determined by the order of the amino acids found in its sequence.

- The shape of a protein determines the possible functions the protein may have in a cell.

That so much biology at the microscopic level can be explained by such a small set of basic principles is truly astonishing. Still, you may be wondering what all this talk of proteins, amino acids, ligands and receptors has to do with Steve Crohn's AIDS immunity. We'll soon see that the connection is very direct.

HIV, AIDS AND T CELL RECEPTORS

As noted above, AIDS is caused by the HIV virus. A virus is a microorganism consisting of a small amount of DNA* or RNA surrounded by a protein envelope. HIV infects people when the virus is transmitted to them from a previously infected individual, by means of some bodily fluid, such as blood, semen or pre-ejaculate, vaginal fluids or breast milk. After HIV enters the bloodstream, it invades a specific type of white blood cells – the so-called helper T cells – and ultimately destroys them. Since helper T cells are a critical part of our immune system (the part of our body that protects us from a wide variety of diseases, especially infectious diseases), HIV-infected individuals are vulnerable to infections and other diseases.

To do its damage, HIV has to enter the person's T cells. But cells have a protective outer membrane. So how does HIV, or any virus, get inside a cell? The answer is that the outside surface of HIV, like the surface of any virus, is made up of proteins. These HIV surface proteins precisely bind to a receptor protein, called CCR5, found in the membranes of T cells, thereby enabling HIV to enter the cell. Of course our CCR5 receptors evolved to bind ligands that our bodies produce naturally, not to bind HIV. But just as a computer virus enters a computer system by exploiting a piece of computer software that was intended for a completely different purpose, so too, HIV enters T cells by exploiting receptors whose purpose is to bind completely different molecules.

What would happen if someone didn't have functioning CCR5 receptors? This is not a hypothetical question. Approximately 1% of people of European ancestry produce CCR5 protein with an altered amino acid

---

* We'll learn about DNA shortly and about RNA in chapter 7. For the moment, it suffices to know they are both large molecules that play key roles in the chemistry of life.

sequence, and this protein has an abnormal shape and is nonfunctional.* HIV can't bind to their CCR5 proteins, so the virus can't enter their T cells. They are immune to HIV infection and can't contract AIDS. In fact, having such nonfunctional CCR5 receptors is what saved Steve Crohn from AIDS.

This still leaves the question of why some people produce CCR5 protein with an altered and nonfunctional shape. Furthermore, we are led to the more basic questions of how a cell "decides" what kinds of proteins to make, and how the cell determines the precise shape of each of those proteins. This time the answers are not found in a cell's proteins but rather in its DNA.

DNA AND CHROMOSOMES

DNA (deoxyribonucleic acid) and its close relative RNA (ribonucleic acid) are not proteins, but nucleic acids. A nucleic acid is a molecule consisting of a long backbone made of carbon, hydrogen, oxygen and phosphorus atoms to which a sequence of small molecules called nucleotides are attached. Along with the proteins, DNA and RNA are the most important molecules of life, and nearly every cell contains both DNA and RNA. While proteins make up most of the structures and tissues of an animal, it's the DNA that contains the cell's "blueprints," which determine precisely what proteins each cell should make.

Like a protein, DNA is a long, linear molecule that is assembled as a precisely ordered sequence of simpler building blocks. However, there are two important differences between the structures of DNA and proteins. First, in contrast to the 20 different amino acids of proteins, DNA only uses four fundamental building blocks, the nucleotides, or bases: adenine, cytosine, guanine and thymine. Often the bases are referred to simply by their initial letters: A, C, G and T.

DNA also differs from protein in that DNA molecules nearly always consist of two paired nucleotide sequences, called DNA strands, which are bound together in the form of a helix. This is the famous double helix initially discovered by James Watson and Francis Crick in 1953. The two strands of the DNA molecule are bound together in a very specific manner in which certain bases are always matched. These matching bases are called complementary pairs; base A is always bound to base T, while C is always bound to G. Using two complementary strands enables a cell to easily duplicate the precise sequence of its DNA, something it needs to do every time it divides into two cells.

---

* Among Asians and Africans, genetic variants leading to nonfunctional CCR5 are even less common.

The number of base pairs* in a DNA molecule is much larger than the number of amino acids in a protein. DNA molecules can be millions or even hundreds of millions of base pairs in length. If the entire DNA in just a single human cell were stretched out, it would be approximately 3 meters (10 feet) long. But DNA is almost never stretched out. Instead each DNA molecule is tightly packed, like a ball of string, and surrounded by proteins that protect it. Such a compact DNA molecule with its protein covering is called a chromosome.

The precise number of chromosomes in each cell depends on the organism. For example, humans have 46 chromosomes in each cell,† as do sable antelopes. In contrast, the kangaroo has only 16 chromosomes per cell, while the dog has 78 chromosomes and the turkey has 80. So if we want to believe that we humans are at least as complex as dogs or turkeys, then apparently the number of chromosomes in a fertilized egg doesn't correlate with how sophisticated an organism the egg will become.

With few exceptions, each of the estimated 50 trillion to 75 trillion cells in our body has copies of the same set of 46 chromosomes with the identical sequences of base pairs. Consequently, essentially every cell – be it a skin cell, muscle cell or white blood cell – contains the same number of chromosomes with all the same genetic information required to construct an entire organism.

GENES

As a result of discoveries by Francis Crick, James Watson and others in the 1950s and 1960s, scientists were eventually able to determine how sequences of nucleotides in DNA serve as blueprints for the strings of amino acids needed to build proteins. A critical initial step was the 1955 discovery of the cell's protein-making machines, called the ribosome, by the Romanian biologist George Palade.

Since there are only four different nucleotides, ribosomes need to use more than one DNA nucleotide to specify which of the 20 amino acids to include at a given position in a protein (see figure 2.2). Through a series of clever and painstaking experiments, scientists learned that precisely three

---

* Since DNA is nearly always double-stranded, DNA lengths are typically given in units of base pairs, or nucleotide pairs, rather than as bases or nucleotides.
† Important exceptions are sperm cells and unfertilized egg cells, which each have 23 chromosomes, half the normal number of chromosomes, so that after the sperm fertilizes the egg, the fertilized egg has the full complement of chromosomes, or 46.

consecutive DNA bases, conventionally called a codon, are used to specify a single amino acid in a protein. The conversion from codons to amino acids uses a set of relatively simple and universal formulas called the genetic code. It's like a cookbook. For example, AAG (meaning two adenine bases followed by one guanine base) means add a lysine amino acid to the protein sequence, while TGG means add a tryptophan. Figure 2.3 illustrates two short, bound DNA strands and the amino acid sequence that the DNA codes for.

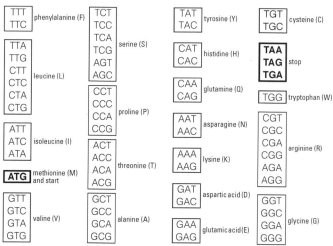

Figure 2.2. Amino acids and their three-letter codes. The 20 fundamental amino acids and their single-letter abbreviations are shown, as well as the three-letter DNA sequences that specify them within the genetic code. A, C, G and T represent the four DNA nucleotides.

As there are only 20 amino acids but 64 ways to form a codon of three nucleotides (that is, there are four possibilities for the first nucleotide, four for the second, and four for the third, and 4 × 4 × 4 = 64), there is some redundancy in the genetic code. For reasons scientists still don't completely understand, some amino acids (such as tryptophan) are specified by only a single codon. Others (such as lysine or glutamine) can be specified by two different codons. While yet other amino acids can be represented by three, four, or even six different codons, as shown in figure 2.2.

The genetic code also needs to have a way of indicating where the blueprint for a new protein starts in the DNA and where it stops, like the capital letter at the beginning of a sentence and the period at the end. To indicate the beginning of a new protein, the genetic code uses a special "start" codon: ATG. Start codons biochemically signal that a sequence of letters that codes for a protein starts at that location. The ATG codon also has a secondary, more conventional function of determining the next amino acid in the protein. ATG specifies the amino acid methionine, and as a result, essentially all proteins begin with a methionine amino acid.

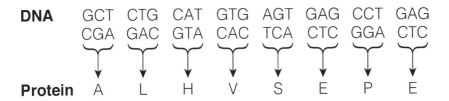

Figure 2.3. Double-stranded DNA and its associated amino acid sequence. The top strand of the double stranded DNA in the upper part of the figure specifies the (protein) amino acid sequence shown in the lower part. A, C, G and T represent the four DNA nucleotides. In the bottom row of the figure, A, L, H, V, S, E and P are abbreviations of 7 of the 20 amino acids. (Note that this picture is very incomplete. The missing pieces will be filled in in chapter 7.)

In a similar manner, the genetic code indicates that a protein blueprint is finished by the presence of one of the three special "stop" codons: TGA, TAA or TAG. Any one of these three codons signals to the ribosome that the protein is complete. That's it. The cell determines what proteins to make by looking for start codons and, once a start codon has been found, by adding additional amino acids to the growing protein according to the sequence of codons that it finds in the DNA. Finally, a stop codon is reached, and the protein is completed.

A section of DNA that is used as a blueprint to build a single protein is called a gene.* It's important to remember that by saying that a gene encodes a protein, we are also saying what a gene does *not* do. Genes do not encode cancer or heart disease or intelligence or athletic ability, or even blue or brown eyes. A gene simply encodes a blueprint for a protein. As we'll see, this concept is central to understanding how genes affect our lives.

One more bit of terminology. Scientists usually give genes and proteins abbreviated names like CCR5 or BRCA1, and often the same name is used for both the gene and the protein it encodes. To distinguish

---

* Some scientists use slightly more complicated definitions of *gene*. A gene is sometimes defined to include the DNA that regulates when a protein is to be produced. In addition, in some cases a cell can synthesize more than one protein from a single section of DNA. Though such reuse of DNA, called alternative splicing, is important, this book will generally discuss only the simpler case, where one section of DNA codes for a single protein.

between the gene and the protein, abbreviated gene names are generally italicized, while protein names aren't. For example, the myostatin gene, *MSTN*, encodes the MSTN protein.

ONE, TWO, MANY HUMAN GENOMES

In June 2000, Francis Collins and Craig Venter, leaders of the two scientific teams sequencing the nucleotides in human DNA, announced the completion of a draft version of the human genome, the entire sequence of human genes and intervening DNA. Although this was an enormous accomplishment, by describing this scientific milestone as the decoding of *the* human genome, the announcement was somewhat misleading. It implied that all people have the same genome, that is, the same sequence of DNA. This is not correct. Each of us has our own unique genome, our own unique DNA sequence. Recent studies have shown that even identical twins* do not have *identical* DNA sequences, though twins' genomes are much more similar to each other than those of any two other people.

In fact, each of our cells has *two* genomes: one that we inherited from each of our parents. The genome that we received from our mother consists of 23 chromosomes and includes a 3 billion base-pair version of all the human genes and intervening DNA sequence. The DNA that we inherited from our father is contained in the 23 remaining chromosomes and also includes a version of every gene and every piece of intervening DNA. Usually, in this book, it will be clear when the term *genome* is referring to the DNA sequence coming from a single parent, or the complete 6 billion base-pair DNA sequence (coming from both parents, with two† versions of each gene) that one actually finds in human cells. If there is a possibility of confusion, I'll use the more precise, scientific term: haploid genome, meaning the DNA sequence from a single parent, or diploid genome, meaning the total DNA sequence from both parents.

---

* Identical twins, also called monozygotic twins, originate from a single fertilized egg cell (called a zygote in scientific parlance). Consequently, monozygotic twins have almost identical DNA sequences. In contrast, nonidentical, or dizygotic, twins come from two separate fertilization events involving distinct sperm and egg cells. As a result, dizygotic twins share only as much DNA sequence as any other pair of nontwin siblings.

† In the cells of males, most genes on the so-called sex chromosomes occur only once. We'll discuss those exceptions in later chapters.

Although each person's genome is unique, the genomes of any two people, even unrelated people, are *very* similar, which is why our bodies mostly synthesize identical proteins and why we all have the same kinds of organs and cells. Indeed, human genomes are quite similar to those of other animals as well, be they dachshunds, lizards or even cockroaches.

How different is your genome from mine? Precisely measuring the DNA difference between two people is not easy, in part because scientists don't always agree on how to count the differences. For example, if there is a location* on a certain human chromosome where my DNA sequence has, say, 10 extra base pairs inserted compared to your DNA sequence, should that be counted as a single difference between our DNA sequences or 10? Until recently, such quibbling about how to measure DNA similarity might have seemed pedantic, but in the past few years, scientists have discovered that the DNA differences between any two people come largely from a small number of long DNA insertions or deletions. In fact most people have at least one DNA insertion or deletion that is more than 100,000 base pairs long, and 1% of us have insertions or deletions that are more than a million base pairs long. Just how important these long, "structural variations" in our DNA are, is still unknown, but there is already evidence suggesting that these large variants are important and may even affect our susceptibility to serious diseases, including cancer and schizophrenia.

When scientists take these differing kinds of DNA variations into account, they estimate that the genomes of unrelated people of the same sex are at least 99% identical. The DNA of unrelated men and women are not quite as similar to each other. Men and women have only 45 of their 46 chromosomes in common; there are two special chromosomes, called the X chromosome and the Y chromosome, and women nearly always have two X chromosomes, while men have one X and one Y chromosome. Consequently, the overall similarity of the genomes of men and women is estimated to be only approximately 98.5%. As an interesting comparison, the current estimate of the level of similarity between the DNA sequences of male humans and male chimpanzees (or between female humans and female chimps) is approximately 96%. So the data indicates that, at least at the DNA sequence level, male humans are more similar to female humans than to male chimpanzees.

---

* Because each individual's DNA sequence is slightly different in length, scientists use an arbitrarily chosen "reference genome" for each chromosome, as a kind of yardstick for location.

GENES, GENE VARIANTS AND MUTATIONS

Overall, you and I differ only at approximately 1% of the locations in our DNA (a little less if you're male, a bit more if you're female). One percent may not seem like much, but with 6 billion letters in our diploid genome, there are millions of differences between your DNA and mine. Even one of those differences can result in producing, or not producing, a protein that can radically affect one's life. For example, having a certain variant at just one out of the six billion locations in your genome can mean you will eventually suffer from Huntington's disease; variations at just two locations cause cystic fibrosis or sickle cell anemia. That said, many other genetic variations seem to have no biological effect whatsoever – or at least none that scientists have been able to figure out.

In this book, I'll often refer to DNA variations as "genetic variants," even though the DNA differences may not be located within a gene.* One kind of genetic variant is particularly important and deserves a special name. A mutation is the genetic variant that has appeared more recently in a family or species than other variants found at that location. Mutations occur at a very low rate, in all kinds of cells, sometimes as the result of errors in duplicating DNA when a cell divides and sometimes from external forces acting on the cell, such as radiation or toxic chemicals.

Sometimes determining whether one genetic variant appeared more recently than another is easy, for example, if the variant is found in an offspring's DNA but not in either of its parents. If a variant is rare, there's also a good chance that it's relatively recent, since there hasn't been much time for it to spread throughout the population. For more common variants, determining which of two variants is newer, and hence is the mutation, can be quite difficult.

Mutations often result in proteins that either have a less functional structure or aren't produced at the optimal time. As a result, mutations often lead to disease or even death. Not all mutations are bad, though. Mutations enable a species to evolve and to have a better chance of surviving in a changing external environment. In some cases a mutation can even save your life, which brings us back to the story of Steve Crohn and AIDS.

---

* Scientists also sometimes call DNA variations "alleles" or "polymorphisms." If the DNA sequences at a location differ by just a single nucleotide, the variants may also be referred to as "single nucleotide polymorphisms" or SNPs. This book won't use this terminology much, but if you want to read the scientific literature, and even some popular articles in genetics, you should be familiar with terms such as allele and SNP.

HIV IMMUNITY AND THE CCR5 RECEPTOR PROTEIN

As we've seen, about 1% of the population, including Steve Crohn, doesn't synthesize functional CCR5 protein. These individuals have inherited two copies of a mutated version of the *CCR5* gene, which is missing 32 base pairs. This genetic deletion leads to a change in the shape of their CCR5 proteins that makes them nonfunctional and prevents the HIV virus from binding to them. As a result, these people are immune to AIDS.

Might it then be possible to treat AIDS by disabling the CCR5 receptors of an AIDS patient who belongs to the 99% of the population that does have functional CCR5 receptors? Timothy Ray Brown, a Seattle, Washington, native in his mid-40s, has shown that it is. In 1995, Brown was a student studying in Berlin, Germany, when he was diagnosed with HIV. Brown was given standard anti-HIV drugs; his life was difficult due to the side effects of these powerful drugs, but he was surviving. Then in 2006, Brown received more bad news; he had also contracted leukemia (cancer of the white blood cells). Brown's doctor, Gero Hütter, gave him standard chemotherapy drugs for his leukemia. These drugs held Brown's leukemia at bay, but by 2007 the side effects of the chemotherapy were severe enough that Brown had to stop taking them.

Hütter then considered a more extreme and more expensive leukemia treatment called bone marrow transplantation (BMT). With BMT, the patient is first irradiated with X rays to kill both the cancerous white blood cells as well as their bone marrow stem cells, the cells that make blood cells. Subsequently the patient receives an infusion of bone marrow from another person, the bone marrow donor, who doesn't have leukemia. If all goes well, the donor's DNA is a close enough genetic match to the patient's that there is no severe immunological reaction from the transplant, and the patient begins to produce cancer-free blood cells from the donated marrow.

Hütter thought that he might be able to use BMT not only to treat Brown's leukemia but his AIDS as well. Hütter reasoned that if he could find a BMT donor with two nonfunctional *CCR5* genes, then Brown's new T cells, produced by the donated bone marrow stem cells, should be immune to infection by HIV. To test this idea, in 2007, Hütter's team at the Charité University Hospital in Berlin performed a BMT on Timothy Brown. In what turned out to be a remarkable stroke of luck, a potential donor was identified who was a good overall genetic match to Brown and who also had two nonfunctional copies of the *CCR5* gene. Immediately after the transplant, Brown had various infections and immunological complications, which often happens after a bone marrow transplant. Eventually these complications were overcome and the transplant was a success. Five years after the transplant, Timothy Brown was alive and well. Not only did his leukemia go into remission, but his HIV infection disappeared as well, with no need for anti-HIV medicine.

In principle, BMT therapy could also be used to treat AIDS patients who don't have leukemia. In fact a company, Stemcyte, was started with the goal of creating a database of potential bone marrow donors with two nonfunctional *CCR5* genes. Unfortunately, this promising-sounding approach faced two serious problems. First, BMT is not only expensive, it's also dangerous. A BMT temporarily destroys the patient's immune system, and the transplanted bone marrow may even attack the recipient's organs, a condition known as graft-versus-host disease. Second, since even among Europeans only about 1% of the population has two nonfunctional *CCR5* genes, finding compatible donors with nonfunctional *CCR5* genes is difficult. To date, Timothy Brown is the only AIDS patient who has been lucky enough to have his AIDS symptoms disappear as a result of a BMT from a compatible donor with two nonfunctional *CCR5* genes.

Because of the challenge of finding suitable BMT donors and the difficulty of the procedure, medical researchers have instead been attempting to develop drugs to block CCR5 receptors. One of these drugs, Maraviroc, has been effective enough in AIDS patients to be approved by the U.S. Food and Drug Administration. So understanding how HIV interacts with CCR5 receptors may yet wind up helping the 99% of the population whose T cells do have CCR5 receptors and therefore are vulnerable to HIV.

HIV AND WEST NILE VIRUS

As promising as CCR5-blocking drugs are, the question remains whether disabling a person's CCR5 receptors might somehow compromise their immune system. For almost a decade after the discovery of the connection between HIV and CCR5 receptors, scientists were unable to find any such negative effect on the immune system caused by disabling CCR5. After all, Steve Crohn and other people who have no CCR5 receptors were healthy; apparently their immune systems were capable of compensating for whatever immune protection CCR5 proteins provided to the rest of the population. More recently, though, scientists have learned that CCR5 does

play an important role in the immune system's defe1
disease: West Nile virus (WNV) infection.

Although WNV and HIV are both viruses, they :
occurs only in humans and other primates. WNV main
it also occurs in other species, ranging from dogs and
people. HIV is spread by the exchange of body fluids, '
the main agent for transmitting WNV. Lastly, whereas
eventually show the symptoms of AIDS, more than 9_.~ ~. μ~~μ.~ ιιιι~~ι~u
with WNV either get only a brief mild fever or don't have any symptoms at all.

Despite the differences between HIV and WNV, experiments on mice
showed that the CCR5 receptor plays important, though different, roles in
these diseases. Cells of mice infected with WNV produce unusually large
amounts of CCR5 protein, and while normal* mice infected by WNV usually
survive, mice lacking functional CCR5 protein often die from WNV infection.
These results motivated researchers to examine the DNA of people who had
been stricken with WNV. They found that 5% of the people severely affected
by the 1999 WNV epidemic in the United States had the nonfunctional *CCR5*
mutation, even though they only represented 1% of the overall population.
The scientists concluded that whereas people with the *CCR5* mutation often
became severely ill from WNV infection, individuals with the common,
functional variant of *CCR5* generally only became slightly ill from WNV, or
were not affected at all.

Another clue suggesting CCR5's relevance to WNV was the observation
that the genetic variant leading to nonfunctional CCR5 receptors is less
common among Asians than among Europeans, and is almost completely
nonexistent among Africans. Although it's difficult to prove cause and effect,
it is tempting to speculate that the reason Africans have functional *CCR5*
genes is that they need them to survive in environments where WNV can
be prevalent. Taken together, the mouse and human data made a strong
case that functional CCR5 receptors are an important component in our
immune system's defense against WNV. So, although nonfunctional CCR5
receptors make you immune to AIDS, they also make you more vulnerable to
consequences of WNV infection. As a result, as promising as the anti-CCR5
drugs are for treating AIDS, they will need to be introduced with caution,
especially in areas at risk for outbreaks of West Nile virus.

---

* In this book and generally in biology, the word *normal* means typical or average.
It does not necessarily imply healthy or desirable.

PART II

DNA: OUR LINK TO THE PAST AND THE FUTURE.

3

WHO ARE OUR FATHERS?

According to the book of Genesis, God promised Abraham that if Abraham and his descendants followed God's laws, God would make them into a great nation. The Bible also tells us that Abraham's grandson, Jacob, had 12 sons who became the leaders of the ancient tribes of Israel, suggesting that God's covenant with Abraham had been fulfilled. In the book of Kings, though, the Bible describes the Assyrian invasion leading to the Assyrian annexation of the northern kingdom of ancient Israel. Only the ancient tribes of Judah and Levi, and parts of the tribes of Simeon and Benjamin, survived intact. Modern day Jews are believed to descend from just these tribes.

The fate of the remaining tribes, which became known as the lost tribes of Israel, is not known. They apparently were assimilated among the ancient Assyrian and neighboring populations until their Jewish roots disappeared. Nevertheless, some believe that remnants of these tribes survived as distinct communities. For example, the Beta Israel community of Ethiopia, the Bnei Menashe of northern India and the Lemba tribe of southern Africa all have oral traditions claiming descent from the ancient Israelites, and in some cases from members of the lost tribes. Typically, these people have rituals that are reminiscent of Jewish traditions. They may light candles on Friday night or circumcise their male infants or have rules against eating pork.* Until recently, verifying such claims of ancient Israelite ancestry has been difficult, if not impossible, but in the last 20 years, tracing family history using DNA has become possible. Scientists can now answer historical questions, such as who are the descendants of the ancient Israelites, by analyzing genetic variations in DNA.

ANCESTRY AND GENETICS

The last chapter introduced DNA as the molecule that contains our genes, the blueprints for our proteins. DNA also plays a central role in connecting us to our past and our future. We inherit our DNA from our parents and pass it along to our children. By examining the DNA variations that we've inherited, we can open a window into the world of our ancestors, and even obtain glimpses of the future lives of our descendants. In this chapter and the next three, we'll explore the implications of DNA's ability to teach us about the past and the future, and the opportunities and challenges that come with this knowledge.

Let's start by recalling that our genome has millions of locations where each person's DNA sequence is slightly different from others'. Half of those genetic variants were inherited from our mother and half from our father. Looking back further in time, we know that we inherited 1/4 of our genetic variants from each of our four grandparents, 1/8 of our variants from each of our great-grandparents, 1/16 of our variants from each of our great-great-grandparents, and so on. So if we go back in time just 10 generations, we see that our DNA sequence is a complicated mixture of genetic variants from over a thousand ancestors.† All of this genetic mixing decreases the occurrence of genetic diseases and is good for us as a species and for us as individuals. But it does make tracing ancestry via genetics more challenging.

Fortunately for geneticists, two special types of DNA are not inherited in such a complicated manner. In these cases, one can know from precisely which of our thousands of ancestors a specific genetic variant has been inherited. One of these exceptions is mitochondrial DNA, which we'll discuss in the next chapter. The second exception is the DNA in the Y chromosome. Since the Y chromosome is passed directly from father to son, without any admixture of DNA from the mother, its DNA sequence remains

---

* Of course, these practices are not unique to Judaism.
† If there has been intermarriage within one's family history, such as first cousin marriages, there will be a smaller number of distinct ancestors.

unchanged between generations. For example, after five generations the DNA sequence of a man's Y chromosome is almost identical* to that of his father's father's father's father's father, rather than being a mixture of DNA sequences from multiple ancestors. We'll call this ancestor the man's fifth-generation extreme paternal ancestor (see figure 3.1). Since Y chromosome DNA is passed directly from one extreme paternal ancestor to the next, studying inheritance from extreme paternal ancestors using the Y chromosome is vastly easier than other forms of genetic ancestry analysis. In fact, until recently, extreme paternal Y chromosome analyses, along with mitochondrial DNA investigations of extreme *maternal* ancestors, were the only forms of genetic ancestry tracing that were feasible.

Perhaps nowhere has the Y chromosome yielded more intriguing information about human ancestry than in the study of the history of the Jews. What kinds of questions of Jewish ancestry might one address by looking at Y chromosomes? Perhaps the most basic question is just that of who (genetically speaking) is a Jew? This simple question has had profound historical implications ranging from determining who was persecuted during the Holocaust to who is eligible for Israeli citizenship.†

To learn more about the genetic identity of Jews, an international research team from the United States, Israel and Great Britain searched for sets of Y chromosome genetic variants (or Y chromosome haplotypes, as such related genetic variants are called in the scientific literature) that are specific to Jewish men – or at least are more frequent among them. The study was partly motivated by the book *The Thirteenth Tribe,* by the well-known Hungarian-British writer, Arthur Koestler (who was also Jewish). Koestler proposed the controversial idea that Jews from Central or Eastern Europe, so-called Ashkenazi Jews, were not descendants of the Middle Eastern Jews of the Bible. Instead, Koestler speculated that Ashkenazi Jews were descendants of the Khazars, a Turkic people, some of whom converted to Judaism and moved to Eastern Europe. Y chromosome DNA analysis would tell the investigators only about the extreme paternal ancestors of the Jews' family tree, yet if it turned out that these ancestors of modern European Jews were not of Middle Eastern origin, it would have been a big deal for historians studying Jewish ancestry.

---

* *Almost* identical because one or more mutations may have occurred as the Y chromosome was transmitted from father to son.

† It's ironic that the Y (male) chromosome has provided deep insights into Jewish history, despite the fact that Jewish law asserts that only a person's maternal ancestry is relevant to determining whether they are truly Jewish.

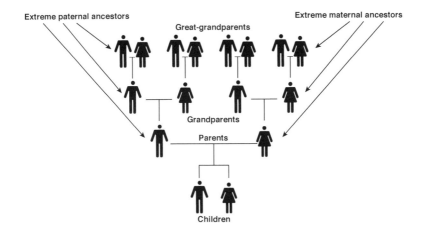

Figure 3.1. Family tree illustrating extreme paternal and maternal ancestors. The great-grandfather on the far left of the top row is the extreme paternal ancestor of that generation. Likewise, the great-grandmother at the far right of the top row is the extreme maternal ancestor of that generation. (Extreme maternal ancestors will be described in more detail in the following chapter.)

The Y chromosome study found that certain haplotypes are common only among Jewish men, and they are found in Jews from Europe, the Middle East or North Africa. These Y-chromosome haplotypes are uncommon among non-Jewish men, and those (rare) non-Jewish men who did have them were nearly all from the Middle East and not from Europe. The Y chromosome evidence showed that, at least with regard to their extreme paternal ancestors, modern Jews have a common ancestry, and that ancestry has Middle Eastern and not European origins. Subsequent studies extended this analysis from just the Y chromosome to all 46 human chromosomes. These more sophisticated genetic tests confirmed that modern Jews are originally from the Middle East, and not from Europe or Central Asia as Koestler had speculated.

Since the initial study of Jewish genetic roots, DNA analysis has taught us far more about Jewish history than simply confirming geographic origins. Using Y chromosome analysis, in 1998 a British and Israeli research team was able to trace the ancestry of the Jewish priestly class called the Cohens. In the Jewish tradition, being a Cohen is an honor that is passed directly from father to son, so being a Cohen should be inherited just like the Y chromosome. Since the Bible also claimed that the Cohens were direct descendants of a single individual (Aaron, the brother of Moses), present-day Cohens should all have extremely similar Y chromosomes. The 3500 years that have passed since the days of Moses and Aaron is too little time for many mutations to have been added into the Cohens' Y chromosomal DNA.

In more recent times, members of the Jewish priestly class can be identified by their surnames, typically Cohen, Kahane, Hakohen or something similar, or by family tradition. Consequently one might expect that Jewish men who claim that they are Cohens, because of their last names or family traditions, will have similar Y chromosome haplotypes. The 1998 British-Israeli study confirmed such similarity among Y-chromosome haplotypes. This class of haplotypes is very rare among non-Jews and is also uncommon among Jewish men who are not self-described Cohens. Since this set of genetic variants is found almost exclusively among Cohens, the scientists called it the Cohen modal haplotype (CMH). The CMH is found among Cohens from all over the world, including Jews from Europe, North Africa and the Middle East. In fact the study found that the Y chromosomes of European Cohens are more similar to those of North African Cohens than they are to the Y chromosomes of European Jews who do not identify as Cohens. So with regard to the origins of the Jewish priestly class, the genetic data once again support history as described in the Bible.

THE LEMBA AND THE Y CHROMOSOME

Besides being fascinating confirmations of Jewish and biblical history, these genetic studies also suggested a way to test the ancestry of those claiming descent from the lost tribes of Israel. One could test the Y chromosome haplotypes of the men and see whether they were similar to those found in male Jews from other parts of the world. There is a serious flaw in this strategy, though. When a group of people moves to a new location, they typically intermix with the local population, at least a little bit. Since Y chromosome analyses only identify extreme paternal ancestral lineage, any non-Jewish male marrying into the group would completely erase the "Jewish" Y chromosome haplotype in all of his descendants.

Despite this problem, there was one group of people who claimed to be descendants of the ancient Israelites for whom Y chromosome ancestry analysis might work: the Lemba tribe of southern Africa. Like others that claimed Jewish ancestry, the Lemba have rituals reminiscent of Jewish ones, such as lighting candles on Friday night and circumcising male babies. With the Lemba, an additional factor made tracing male Jewish ancestry via the Y chromosome feasible: the Lemba have a strong tradition of allowing only *women* to marry into the tribe. (Such rules exist in other tribal societies, but the Lemba seem to have been particularly rigorous in enforcing theirs.) To the extent that this tradition had been rigidly enforced over the hundreds of years since the Lemba's alleged Jewish ancestors migrated from the Middle East, the original Jewish Y chromosome variants should still exist among Lemba men.

In 1996 researchers performed the first genetic analyses of the Y chromosomes of Lemba men. The study compared Y chromosomal haplotypes of Lemba men to Jewish men, as well as to those of men from neighboring Bantu tribes and from non-Jewish Semitic (Middle Eastern) backgrounds. The results showed that more than half of the Lemba men had Y chromosome haplotypes similar to those found in Jewish men from other parts of the world. In contrast, men from other African tribes almost never had "Jewish" Y chromosome variants.

However, while the Lemba's Y chromosome haplotypes resembled those of Jews, they also were similar to those of other Semitic men. So, although the results indicated that the Lemba's paternal ancestry was Middle Eastern, they didn't prove that the Lemba's Semitic ancestors were Jews. Consequently, a second study was carried out in 2000, and it included more detailed Y chromosomal testing. With this higher resolution genetic screening it would also be possible to determine whether any of the Lemba men had the rare Cohen modal haplotype. The results were dramatic.

To appreciate the results, you need to know that the Lemba tribe is divided into clans, one of which is called the Buba. The Buba are the leaders among the Lemba, and being a Buba is passed directly from father to son. In the genetic study, Y chromosome samples were taken from men from six of the Lemba clans, including the Buba. Remarkably, more than 50% of the Buba men had the Cohen modal haplotype. The Bubas were "Cohens"! In contrast, only 3% of the Lemba men who were not from the Buba clan had the Cohen modal haplotype. And not a single one of the men from any of the neighboring tribes did. So the evidence strongly supports the claim that the extreme paternal ancestors of many Lemba men were Jews from the Middle East.

It's worth mentioning that even before these genetic studies, the Lemba had an oral history describing Jewish migration to the Arabian Peninsula at the time of the prophet Jeremiah. According to this history, the Arabian Jews were involved in maritime commerce that took them to southern Africa. It was at least plausible that some of these Jews might have been ancestors of an African tribe. Nevertheless, without the Y chromosomal data, this story of Jewish maritime migration would not be much more than historical speculation.

Figure 3.2. Lemba men wearing traditional Jewish yarmulkes and prayer shawls.

THE INDIAN CASTE SYSTEM AND THE Y CHROMOSOME

The Jewish Cohens aren't the only ancient religious group that passed its cultural identity from father to son. Any paternally inherited tradition should lead to a clustering of a relatively small number of Y chromosomal haplotypes among its present-day male members. One example is the (originally Hindu) Indian caste system. The Indian caste system consists of four official castes, or classes: Brahmins (priests), Kshatriyas (warriors), Vaishyas (farmers and merchants), and Shudras (laborers). In addition, there was a fifth group outside the official caste system, the Dalits – in English often referred to "untouchables." Historically, it was unusual for people to marry outside the caste they were born into, and when people from different castes did marry, it was usually the woman who married "up" and joined her husband's (higher) caste. As a result, Y chromosome haplotypes that were found only in members of a single caste in ancient times would be expected to still be limited to men of that caste today.

To test this idea, several genetic studies were undertaken of Indian men whose families had belonged to different castes. In one such study, in the Indian state of Uttar Pradesh, Brahmin men were found to have Y chromosome haplotypes more similar to those of other Brahmins than to the haplotypes of men from other castes. Kshatriya men, too, were found to have Y chromosome haplotypes more similar to those of other Kshatriyas. However, subsequent studies carried out in other parts of India showed much less genetic clustering on the basis of caste, or no clustering at all.

Why is there so much variety among the Y haplotypes of the Brahmins and Kshatriyas, in contrast to the dramatic clustering of Y haplotypes found among the Jewish Cohens? In part, the explanation is likely to lie in the fact that the Indian caste system is much weaker today than it once was. Shortly after India gained its independence from Britain in 1947, the new government implemented laws intended to end discrimination based on caste and to generally improve the living conditions of those from the lower castes. As a result, marriage between members of different castes has become more common, and many men from lower castes have married into the Brahmin and Kshatriya castes.

Another factor contributing to the large variety of Y haplotypes among higher caste Indian men is that the Brahmins or Kshatriyas are not descended from just one or a very few individuals. Even in ancient times, Brahmin or Kshatriya men had many different Y chromosomes. In contrast, the Jewish Cohens are all claimed to be descended from a single man (Aaron) so, to the extent that this tradition is historically accurate, we shouldn't be surprised that the present-day Cohens have less Y chromosome variation than today's Brahmins and Kshatriyas.

THE GENGHIS KHAN CHROMOSOME

Our last example of how the Y chromosome can illuminate male ancestry is perhaps even more dramatic than that of the Cohens. In 2003, an international team of scientists analyzed the Y chromosomal haplotypes of over 2000 randomly selected men from 16 different Central and East Asian populations between the Pacific and the Caspian Sea. Before this study, no large-scale investigation of male genetic variants had ever found any single haplotype that was much more common than every other haplotype. In this study, though, fully 8% of the men shared a single Y chromosome haplotype. These apparently unrelated men had such extraordinarily similar Y chromosomes that there appeared to be no explanation other than that they were all descendants of a single individual. If one believes that these men were indeed representative of the populations they came from (and there is no reason not to believe so) then one has to conclude that there are millions of men living today who are all descendants of this single person. Could this possibly be the case?

Actually, yes, and that single individual was Genghis Khan, the 13th-century Mongol emperor. Although no remains of Genghis Khan exist, and consequently no source of DNA to directly test this hypothesis, the historical record indicates that Genghis Khan's sons established at least five major Asian dynasties. Since these men were rich and powerful, one would expect that their sons, in turn, would have been similarly rich and powerful, had many wives and concubines, and therefore children (including many sons). Further supporting this idea, 15 of the 16 population groups with the "Genghis Khan" haplotype live within the boundaries of Genghis Khan's former empire. The exception is the Hazara people of Pakistan, but the Hazara are also known to be of Mongol origin, and in fact many of them have family traditions claiming that they are descendants of Genghis Khan. Overall, the genetic evidence makes a compelling case that roughly 1 in 12 men living in Central Asia is a direct descendant of Genghis Khan.

The research on the Lemba tribe and the descendants of Genghis Khan demonstrates that Y chromosome variations can provide powerful insights into human history. That said, Y chromosomal ancestral analysis also has an important limitation. The Y chromosome only provides information about a single ancestral lineage (that of the extreme paternal ancestors) and ignores the genetic contributions of all other ancestors. As a result, Y chromosomal ancestry analysis can lead to confusing conclusions; and, as we will see in the following chapter, nowhere are these misleading results more dramatic than in the study of the roots of African Americans.

4

WHO ARE OUR MOTHERS?

Vy Higginsen is a talented and successful African American woman. Born and raised in Harlem in New York City, her accomplishments have been many: from starting the Gospel for Teens Choir to hosting her own show on WWRL, an AM radio station that was for many years the voice of New York's black community. After her grandmother died in 1978, Higginsen became interested in her family history, as well as in the ancestry of her people. So when commercial DNA ancestry testing became available in the early 2000s, she was eager to take advantage of this new way of learning about one's roots.

Higginsen chose to use Y chromosome DNA testing, the form of genetic ancestry analysis described in the last chapter. (You'll recall that Y chromosome testing can only be performed on men, so Higginsen persuaded her uncle, the Reverend James O. West Jr., to take the test). Shortly after Reverend West submitted his DNA sample, Higginsen received a telephone call from a man named Marion West, announcing that they were "cousins." Though the DNA data did in fact provide strong evidence that West and Higginsen were related, there was a bit of a problem – or at least a surprise – for Vy. Marion West, and as far as he knew the rest of his family, was white. Meanwhile, all of Higginsen's relatives were black. How could this be? Could this proud African American woman from Harlem really be part of a white family?

GENETIC ANCESTRY TESTING AND AFRICAN AMERICANS

Perhaps no people have more reason to be sensitive and curious about their roots than African Americans. Beyond the general interest that many people have about their history, African Americans have special reasons for wanting to trace their roots. The tragedy and injustice of the African slave trade in the Western Hemisphere from the 1500s to the 1800s left most blacks of the Americas uniquely deprived of knowledge of their ancestry. Despite this (or perhaps because of it), African Americans have often been especially motivated to learn about their history. This desire to connect with their past contributed to the excitement that accompanied the 1976 book and subsequent television miniseries *Roots,* describing author Alex Haley's attempt to trace his family history back to Africa. The popularity of DNA ancestry testing among African Americans is also a reflection of the dreams of a violently uprooted people to learn more about their origins.

Unfortunately, Y chromosome ancestry analysis can lead to misleading results. As we saw in the last chapter, Y chromosome tests ignore all of a man's genetic heritage except his extreme paternal ancestry. Ignoring most of one's ancestry can lead to confusion for anyone. In the case of African Americans, the results can be particularly misleading. Just how misleading? In an estimated 30% of cases, when African Americans take a Y chromosome test they, like Vy Higginsen and James O. West, are told that their genetic ancestry is European and not African.

From our knowledge of Y chromosomal testing and of the history of slavery in the Western Hemisphere, these results are not surprising. If at *any* time in the history of an African American family, a slave owner (or other white male) fathered a son with a black slave, then all of his direct male descendants would have genetic variants generally found in European men. Even after many generations, at which point the descendant's overall genetic makeup might be overwhelmingly African, Y chromosomal testing would only identify a single ancestor, the extreme paternal (white) one.

Despite its limitations, paternal ancestry tracing can occasionally be useful to African American families tracing their roots. One such family includes the descendants of Sally Hemings, a slave who belonged to Thomas Jefferson. Hemings' descendants have long claimed that she was Jefferson's mistress and that they are descended from Jefferson. Proving these claims had been difficult before the advent of Y chromosome testing. However, genetic testing in 1998 comparing Y chromosomal DNA of direct male descendants of Eston Hemings Jefferson (Sally Hemings' son) with that of Field Jefferson (a paternal uncle of Thomas Jefferson) provided strong evidence that Thomas Jefferson was in fact Eston Hemings Jefferson's father.[*]

In contrast to the vindication experienced by the descendants of Sally Hemings, most of the 30% of African American men with "European Y chromosomes" feel frustration, confusion or disbelief when they learn the results of their Y chromosome ancestry tests. To understand their ancestry, African Americans need to use a DNA ancestry test that includes more than the Y chromosome. The easiest way to accomplish this (and, until recently, the only way) is by studying the DNA of the mitochondrial chromosome.

THE GENETICS OF MITOCHONDRIA

Mitochondria are small specialized substructures found inside the cells of plants and animals. They are essential to the cell because they convert food molecules such as glucose into the biochemical adenosine triphosphate, the form of chemical energy that cells use. Depending on the type of cell and its energy requirements, there can be anywhere from 100 to 10,000 mitochondria in a single cell.

Geneticists are interested in mitochondria because they contain chromosomes with small amounts of DNA (called mitochondrial DNA, or mtDNA). Each mitochondrial chromosome is only approximately 16,000 base pairs long, in contrast to the nuclear chromosomes, which can be hundreds of millions of base pairs in length. More important for ancestry analysis is that mitochondrial chromosomes are transmitted without modification from the mother to the egg cell. In contrast, for chromosomes other than the mitochondrial chromosome (and the Y chromosome), the DNA from both parents is mixed together before it is packaged in egg and sperm cells for transmission to the next generation. So, after five generations, your mitochondrial DNA is not a mixture from 32 great-great-great-grandparents but is almost identical to the mtDNA of your mother's mother's mother's mother's mother, the single individual whom we are calling the extreme maternal ancestor (see figure 3.1).

---

[*] Whether the Hemings' extreme paternal ancestor was Thomas Jefferson or his younger brother, Randolph, is still unclear. Since the two brothers would have had essentially identical Y chromosomes, there is no way Y chromosomal DNA analysis can resolve this last piece of the puzzle.

Black Americans can generally learn more about their African roots from mtDNA than from Y chromosome DNA, as mtDNA is not affected by the presence of any white males in their family tree. A 2007 study by the Laboratory of Genomic Diversity of the U.S. National Cancer Institute* clearly confirmed this fact. The study showed that while almost 30% of the Y chromosomes of African American men were of European ancestry, less than 10% of the African Americans had mtDNA of European origin.

LIMITATIONS OF MTDNA ANCESTRAL ANALYSIS

To African Americans, simple confirmation of their African ancestry is a bit anticlimactic. Typically, they also want to know which African tribe or ethnic group they are descended from. Although addressing this question by studying mtDNA seems to be straightforward, maternal mtDNA tracing of African American ancestry turns out to have its own challenges.

The limitations of mtDNA analysis became apparent with the first detailed African American mtDNA study, published in 2006 by researchers at the University of South Carolina. The study's strategy was to first build a database of sets of mtDNA genetic variants (in other words, mtDNA haplotypes) from almost 4000 individuals representing a large number of West African tribes. They then collected mtDNA samples from African Americans. Finally, they matched the mtDNA haplotypes of the African Americans to ones found in the African database.

In principle, the strategy was sound, but when the researchers attempted to pinpoint the West African origins of the extreme maternal ancestors of the African Americans, they ran into problems. The scientists discovered that only 10% of the African American samples could be linked to a unique tribe or locale using the African database. In more than half of the cases, the African American mtDNA haplotypes were shared by multiple West African tribes, and 40% of the African American mtDNA haplotypes couldn't be found in the West African database at all.

In part, the researchers' problems were merely technical. If the scientists had tested more Africans, including individuals from additional tribes, then the fraction of African American mtDNA variants that couldn't be found in the West African database would surely have decreased. On the

---

* That the National Cancer Institute does genetic ancestry research might at first be surprising. Since the effectiveness of certain drugs varies among ethnic groups with different common genetic variants, identifying someone's ancestry from their DNA may someday help doctors tailor drug prescriptions for people from different population groups.

other hand, as the researchers themselves noted, as the size of their database increased, so too did the number of African Americans with mtDNA haplotypes matching more than one African tribe. In other words, even as more African mtDNA data accumulated, determining precisely from where in Africa a given mtDNA haplotype originated wasn't getting any easier.

The challenge with African American mtDNA ancestral analysis wasn't that the databases were too small but that mtDNA haplotypes weren't specific enough to trace an individual's ancestry to a specific tribe or geographic region. In West Africa, marriages often occur between individuals from different tribes and locations. Usually, the woman leaves her home to join the family (and tribe and geographic location) of the man. As a result, women with the same or very similar mtDNA haplotypes have migrated to many tribes and regions across West Africa, making the determination of African American ancestry with mtDNA very difficult.

THE STORY OF ANASTASIA

Although maternal mtDNA ancestry analysis has its shortcomings, it has also resolved some dramatic historical puzzles. Arguably, no such puzzle has been more emotional and contentious than that of the fate of Anastasia, one of the daughters of Nicholas II, the last Russian czar.

The story began in 1917, when the Russian Revolution overthrew the Romanovs, who had ruled Russia for 300 years. In July of the following year, the Czar and his immediate family, who were being held in Yekaterinburg in the Ural Mountains, disappeared. The Czar's family had presumably been killed by Bolshevik revolutionaries, but rumors persisted that one or more of the Nicholas' children had survived the assassination and escaped. Over the years, several women appeared claiming to be the Grand Duchess Anastasia, the fourth daughter of Nicholas and his wife, Alexandra. The best known was a woman known as Anna Anderson. Many people declared Anderson a fraud, claiming she couldn't speak fluent Russian or English like the young Anastasia. Anderson's supporters countered that they had heard her speak Russian and English, but that she rarely spoke them because of what would we now call post-traumatic stress syndrome.

For 70 years, the story of the Anastasia's reappearance inspired thousands of people who learned about her in newspapers, books and the Hollywood movie *Anastasia,* starring Ingrid Bergman as the enigmatic woman who is eventually accepted as the true Anastasia. Then in the early 1990s, Alexander Avdonin, a Russian geologist, announced that he had found the gravesite of Czar Nicholas and his family in Yekaterinburg. Actually, Avdonin had found the remains over a decade earlier, but had decided not to reveal his discovery until after the fall of the Soviet regime. Although Avdonin's forensic research seemed solid, whether the remains really belonged to members of the Romanov family was uncertain.

Figure 4.1. The Romanov family.

In 1994, an international scientific team decided to perform mitochondrial DNA analyses to test Avdonin's claims. The Czarina Alexandra was the daughter of Princess Alice, who was a daughter of England's Queen Victoria and the maternal great-grandmother of Prince Philip (husband of Elizabeth II). Therefore, the researchers knew that Prince Philip, Alexandra, and Alexandra's children should all have identical mtDNA.

By the 1990s, DNA technology was capable of extracting mtDNA samples from the human remains found in Yekaterinburg and comparing them with Prince Philip's mtDNA. The results of the experiments were definitive: Prince Philip and the individuals buried in Yekaterinburg had identical mtDNA sequences, showing that they shared the same maternal ancestry. In other words, the human remains almost certainly belonged to descendants of Princess Alice and were the Romanovs. Figure 4.2A shows the maternal lineages of Prince Philip and Anastasia. Since both are direct maternal descendants of Princess Alice, they would have identical mtDNA.

But the story wasn't quite over. The Romanov remains included only three girls, while Nicholas and Alexandra had had four daughters as well as a son, Alexei. Maybe Alexei and one of the Czar's daughters, perhaps Anastasia, had survived the assassination. It was still possible that Anna Anderson was in fact Anastasia. However, comparing Anderson's mtDNA to Prince Philip's wasn't simple. Anderson had died in 1984, and her body had been cremated. She didn't have any children, who would have inherited her mtDNA.

Eventually, the research team was able to locate a tissue sample of Anderson's in a hospital in Virginia where she had been admitted for surgery. Using this sample, they could test Anderson's mtDNA and compare it with Prince Philip's, and hence with the Romanovs'. The results of the mtDNA analysis? Sadly, the miraculous reemergence of Anastasia was not true. The genetic data clearly showed that Prince Philip and Anna Anderson didn't share the same extreme maternal ancestor.

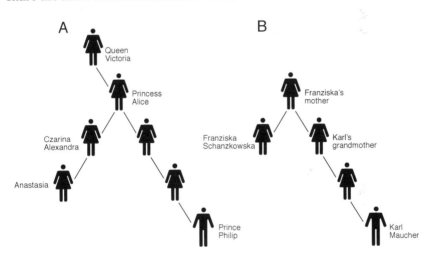

Figure 4.2. Maternal ancestry of Anastasia and "Anna Anderson." A: Maternal family tree of Prince Philip and Grand Duchess Anastasia. B: Maternal ancestry of Franziska Schanzkowska and Karl Maucher.

The researchers did find a different genetic match to Anderson's mtDNA, though. The matching mtDNA belonged to a man named Karl Maucher. Maucher was the great-nephew of one Franziska Schanzkowska. Those who had called Anna Anderson a fraud had long claimed she was most likely Schanzkowska, a Polish factory worker who had disappeared around the time of Anderson's initial appearance. Maucher, who shared the same maternal ancestry as Schanzkowska, had volunteered his mtDNA in hopes of resolving the mystery of his great-aunt's disappearance. The test results showed that the woman who called herself Anna Anderson was almost definitely Franziska Schanzkowska. Figure 4.2B shows the maternal lineages of Karl Maucher and Schanzkowska. Since they are direct maternal descendants of the same woman (Schanzkowska's mother), Karl and his great-aunt would have the same mtDNA sequences, and since Karl's mtDNA matched that of Anna Anderson, the evidence strongly pointed toward Anna Anderson and Franziska Schanzkowska being the same person.

Even after the results of the mtDNA testing of Anna Anderson's tissue were reported, a few of her most ardent followers continued to believe her story, claiming that the DNA data was incorrect or fraudulent. Finally, in 2007, a second gravesite, which had been overlooked in the original excavations, was discovered in Yekaterinburg. The second site was only seventy meters away from the original one and contained the remains of two children, a boy and girl. Mitochondrial DNA analysis again linked the remains to the family of Prince Philip and consequently to the Romanovs. The bodies of Alexei and the last Romanov daughter had been found, and with them came the sad realization that Anastasia, along with the rest of her family, had been murdered in July 1918.

BEYOND PURE PATERNAL OR MATERNAL ANCESTRY ANALYSIS

The Anastasia story shows how helpful mitochondrial DNA testing can be, if one knows the mtDNA haplotype of an extreme maternal ancestor. Often, though, mtDNA ancestry analysis is as misleading as Y chromosomal genetic testing. For example, an mtDNA genetic analysis of (extreme maternal) Jewish ancestry suggested largely European, rather than Middle Eastern, Jewish origins. The shortcoming of both Y chromosome and mtDNA ancestry analysis is that each they only trace DNA of a single ancestor, throwing away all the genetic information from the individual's other ancestors.

Clearly, the solution is to study our ancestry using the DNA of all our chromosomes. Until recently, however, such full-genome ancestry analysis was beyond science's technological capabilities. The cost of testing DNA of all our chromosomes was too high, and even if such testing could be done, determining the relative contributions to our genetic makeup from each of our ancestors was too complicated.

In the last few years the costs of whole-genome DNA testing have dropped precipitously, and new computer programs have been developed that can disentangle the genetic contributions from multiple ancestors. It is now becoming possible to perform genetic tests that study all the branches of our ancestry. The results of these studies are turning out to be more complicated and subtle than had been initially anticipated. What is being learned is that, from the perspective of a geneticist, many people don't belong to any single "racial" or "ethnic" group at all. Rather, we are complex mixtures of many different populations. For example, when Vy Higginsen had genetic ancestry analysis carried out on *all* of her DNA, she learned that her ancestry is 43% African, 33% European and 10% Asian. (Because of our still-limited ability to link DNA patterns to population groups, such estimates are approximations and don't always add up to 100%.) It's becoming increasingly clear that our genetic ancestry is more complex than what people wishing to identify with a single population or ethnic group might want to believe.

CHROMOSOME RECOMBINATION AND DNA INHERITANCE

Along with providing glimpses of the complex picture of human history, whole-genome ancestry testing is beginning to illuminate chapters from our past that have long been hidden in our DNA. One of the first fascinating insights of these tracks of human history came from a study comparing the genetic variants of natives of Puerto Rico with those from West Africans.

Scientists were able to probe the ancestry of Puerto Ricans by analyzing the pattern of chromosome recombination in their DNA. Chromosome recombination, illustrated in Figure 4.3, is the phenomenon by which the sex organs guarantee that every newborn child receives a thorough mixture of DNA from all four of its grandparents. (Chromosome recombination is a somewhat tricky concept. So if you're willing to accept that two genes that are located close to each other along a chromosome are more likely to be inherited from the same ancestor than genes that are located far apart from each other, feel free to jump ahead to the next section to see how this fundamental fact can shed light on a historical event, such as the trafficking of slaves from Africa to the Caribbean.)

To understand chromosome recombination we first need to know that each type of our chromosomes (except for the mitochondrial chromosomes, and, in men, the unpaired X and Y chromosomes) comes as part of what is called a homologous pair. For a child's DNA to be a mixture of contributions from all four grandparents, DNA from each pair of homologous chromosomes is mixed together during the formation of sperm and egg cells. Special enzymes first cut both chromosomes of each homologous pair at precisely the same locations. Next, other enzymes splice the two chromosomes back together using alternating pieces from the maternally and paternally derived chromosomes. Only after this intricate dance of recombination is completed for each pair of homologous chromosomes (and, in a female, for her two X chromosomes as well) does the cell divide into egg or sperm cells. The point of this slicing and splicing is that, without it, each chromosome in the egg or sperm cell would only be passing on genes from a single parent. For example, chromosome 1 contains over 4000 genes. Without recombination, children would get their *entire* maternal chromosome 1 from, say, their mother's father, including their maternal grandfather's variants of all 4000 chromosome 1 genes, and *none* of their chromosome 1 genes from their mother's mother (or vice versa). In contrast, recombination ensures that each chromosome the offspring receives from its mother is a mixture of genetic material from both maternal grandparents.

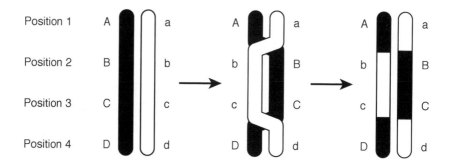

Figure 4.3. Schematic depiction of recombination and haplotypes. In this example, there are just four locations on the chromosome with genetic variants. For example, at the first location – Position 1 – there are two possible variants (or alleles): A or a. In this example, recombination only occurs in two places during the process of generating the chromosome for the egg or sperm cell. The exchanged DNA includes the region between Positions 2 and 3. Since Position 2 and Position 3 are closer to each other than Positions 1 and 3, variants B and C (or b and c) are more likely to be inherited together than A and C (or a and c). Closely linked variants that are often inherited together (like B and C or b and c) are called a haplotype.

The chromosome locations where recombination occurs appear to be random – or if they are not random, no one has figured out how they are determined. One thing that is known is that the closer two genes' locations along a chromosome, the more likely that the inherited variants for both genes will be from the same grandparent. For example, let's say that figure 4.3 is not drawn to scale and that gene B is actually 60 million base pairs from the end of chromosome 1, while gene C is 60.1 million base pairs from the end (in other words, genes B and C are quite close to each other along the 250 million base pairs of chromosome 1). In this case, a child that inherits its mother's mother's variant for gene B will likely also inherit its mother's mother's variant for gene C (and not its mother's father's variant). Such nearby genetic variants that are inherited together are examples of what we have been calling haplotypes (again see figure 4.3 for an example). In contrast, if gene A is located only 1 million base pairs from the end of chromosome 1, and consequently is far away from gene B, then, even if you know that the child inherited its gene A variant from its maternal grandmother, there is still only approximately a 50-50 chance that it will inherit the maternal grandmother's variant for gene B.

DNA ANCESTRY ANALYSIS AND THE HISTORY OF THE PUERTO RICAN SLAVE TRADE

Having learned a bit about chromosome recombination, we're ready to find out what DNA reveals about the history of the Puerto Rican slave trade. When the researchers compared the genetic variants of Puerto Rican natives with those of West Africans, they discovered that the West African variants found among the Puerto Ricans belonged to one of two distinct classes. One class consisted of large groups of variants, located along a single extended region of a chromosome. Scientists call groups of genetic variants that are located within a single large chromosomal region and have been inherited together long haplotypes. The other set of genetic variants found among the Puerto Ricans consisted of very short haplotypes: that is, they were isolated genetic variants, or small sets of variants, located within a short chromosomal region.

The intriguing observation was that the researchers found relatively few *intermediate-length* sets of jointly inherited genetic variants. This observation supported the idea that the West African DNA variants had been introduced into the Puerto Rican population during two separate historical periods. The data suggested that the variants contained within the long haplotypes had been inherited relatively recently (since the presence of long haplotypes indicated that little chromosome recombination had occurred since these variants had been inherited), while the short haplotype variants had entered the Puerto Rican population at a much earlier date.

The scientists also discovered that the genetic variants found within the short African haplotypes (inherited a relatively long time ago) all matched ones found in people from the *coast* of the West African country of Senegal. In contrast, the variants from the longer (more recently inherited) genetic regions matched haplotypes from Senegalese tribes from the interior of the country. This genetic data agreed closely with the historical record. Puerto Rican history tells us that the slave trade occurred in two separate waves. The first round of slave importation, in the 1500s, involved Africans captured in coastal Africa. The second wave of slave capture peaked in the 1700s. By that time the slave traders were having difficulty obtaining slaves near the Atlantic Coast, and had moved further inland in pursuit of people to sell into the slave trade. This history of two periods of African slave migration is the same one that can now also be read in the patterns of genetic variants in the DNA of present-day Puerto Ricans.

5

CAN WE RAISE THE DEAD?

In 2008, a Pyrenean ibex, a subspecies* of the Spanish mountain goat, was born in a laboratory near the Pyrenees Mountains that separate France and Spain. Sadly, the animal was born with a lung defect and lived only seven minutes. Nevertheless, its birth marked a dramatic scientific milestone. Before those seven minutes, the Pyrenean ibex had been extinct for eight years.

As biology's link to the past, DNA is the key to resurrecting an extinct animal. In this chapter, we'll explore the dramatic progress scientists have made toward this ambitious goal. But first, we'll look at a more modest, yet also important, research effort: resurrecting extinct *proteins*. As DNA contains the recipes for the proteins that are the cell's building blocks and the regulators of its chemical reactions, by deciphering the DNA of extinct organisms, scientists can learn what proteins they used and how they functioned. Along with providing insights into the lives of long-extinct animals, studying ancient proteins can also lead to direct tests of the theory of evolution. The reason is that evolutionary models make specific predictions about the amino acid sequences of ancient proteins. If scientists can synthesize such amino acid sequences, they can test whether those sequences have the properties of proteins found in nature.

---

* Subspecies are populations of an animal species that rarely interbreed with one another due to isolation or other factors.

To understand how such testing of evolution works, first remember from chapter 2 that a protein consists of one or more linear sequences of electrically charged amino acids. Each amino acid string spontaneously folds into a specific three-dimensional structure due to the attractive and repulsive forces among its various amino acid components, and the resulting arrangement of electrical charges determines the biological function of the protein.

What wasn't included in chapter 2 is that most linear sequences of amino acids (in contrast to proteins found in nature) *don't* spontaneously fold into 3-D structures. Random sequences of amino acids generally have electric charge distributions leading to long, floppy molecules, which don't fold into any three-dimensional structure. Amino acid sequences not found in nature –those synthesized by scientists in the laboratory – are also almost never biochemically active. And if scientists change even one of the amino acids of a naturally occurring protein, they find that 30% of the time the protein no longer folds into a 3-D structure or its ability to catalyze a chemical reaction is diminished. Only in one trial out of a thousand does randomly changing a protein increase its biological activity. In other words, proteins found in living organisms are very unusual strings of amino acids. This means that if an evolutionary model predicts the amino acid sequence of an ancestral protein, and the predicted protein is found to be biochemically active and to have a three-dimensional structure, then there's a good chance the model is correct and the predicted protein did in fact once exist in nature.

ANCIENT PROTEINS AND FUTURE CATALYSTS

Scientists predict the amino acid sequences of proteins from extinct animals by comparing equivalent proteins in multiple present-day animal species that are believed to be descendants of the extinct animal. This approach is illustrated in figure 5.1. The first step is to align amino acid sequences from a set of proteins from multiple present-day organisms. Although figure 5.1 shows only a single alignment of three present-day sequences, each seven nucleotides long, in a realistic example, scores of alignments from large numbers of related species, involving thousands or even millions of nucleotides or amino acids, might be used.

Next a model called a phylogenetic tree is created, which specifies the way in which the sequence of the ancient common ancestor is presumed to have evolved into the present-day genetic sequences. The tree includes a prediction of the most likely amino acid sequence for the protein in the extinct ancestral animal, as well as an evolutionary model describing the mutations that led to the proteins found in the modern animals. For example, the evolutionary model might be selected to minimize the total number of mutations required to have occurred between the primordial sequence and the observed present-day sequences.

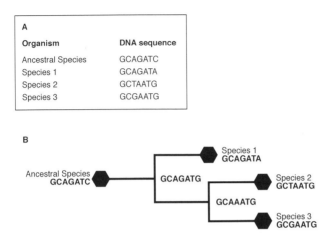

Figure 5.1. Guessing the forms of ancient proteins. Highly simplified depiction of the determination of the DNA sequence of an extinct ancestral species. A: An alignment of a seven-nucleotide sequence of a section of DNA from three related species (Species 1–3) and their hypothesized ancestor. B: Phylogenetic tree showing the hypothesized sequences of mutations by which gene in the Ancestral Species organism evolved into the present-day sequences in organisms identified as Species 1, Species 2 and Species 3. Although phylogenetic tree building can be done starting with alignments of DNA, as illustrated here, scientists more commonly build phylogenetic trees using amino acid alignments of proteins.

So far, this is all just prediction and speculation. What makes the approach interesting is that predicted ancestral proteins can be synthesized in the laboratory. Since the overwhelming majority of amino acid sequences are chemically inert and fail to form any three-dimensional structure, if an amino acid sequence derived from a phylogenetic analysis does have a 3-D structure and is chemically active, there is a good chance that the amino acid sequence corresponds to a genuine protein and that the evolutionary model that predicted it is valid.

Ancient proteins have helped researchers address questions about prehistoric creatures that previously seemed unanswerable. For example, in 2002 scientists at Rockefeller University studied the opsin family of proteins to determine whether dinosaurs had night vision. Opsins are proteins found in the eye that enable animals to detect light and see. Using phylogenetic methods similar to those described above, the researchers compared opsin amino acid sequences from four presumed evolutionary descendants of the dinosaurs (pigeons, zebra finches, chickens and alligators) for which genetic data was available. Although the predicted primordial dinosaur opsin protein had an amino acid sequence different from all four of its present-day descendants, when scientists synthesized this protein, they observed that it was biochemically active and sensitive to light just like the opsins of present-day animals. Since this ancestral protein was also observed to be sensitive to wavelengths of light found in low-light environments, most likely the dinosaurs could see at night as well. Although these experiments didn't *prove* that the predicted opsin existed in the eyes of dinosaurs, the observation that the predicted ancestral protein was light sensitive suggested that it was indeed the dinosaur ancestor of the pigeon, zebra finch, chicken and alligator opsin protein.

Ancient proteins not only teach scientists about the past, they also help researchers tackle challenging technological questions facing society today. Scientists are actively searching for chemical catalysts for many applications, ranging from more environmentally friendly energy generation to improved destruction of toxic chemicals. To find better catalysts, researchers typically begin with a natural protein and then synthesize modified versions of the protein, by changing one or a few of its amino acids. They then test whether any of the modified proteins has increased activity.

The difficulty with this approach is that there are too many possibilities to test. There are 20 different amino acids, and proteins are often a thousand or more amino acids long. So just to test the effects of a single amino acid change in a typical protein, one would need to test 20,000 or more possible changes. Since two or more simultaneous amino acid modifications are often required in order to observe an increase in a protein's activity, the number of protein variations that need to be screened soon becomes impossibly high.

Scientists have learned that to increase the likelihood of finding an amino acid sequence with novel chemical properties it is important to start with proteins that have highly stable 3-D structures and are capable of catalyzing multiple chemical reactions. Ancient proteins are believed to have had these properties because, given that the planet was hotter millions of years ago, proteins needed more stable structures to remain properly folded at those temperatures. In addition, millions of years ago, proteins had had less time to evolve and become specialized, and consequently were more versatile as catalysts. These hypotheses about the properties of ancient proteins are not mere speculations. In recent experiments, scientists synthesized predicted ancestral proteins and found they were often more stable or more chemically versatile than the present-day proteins into which they ultimately evolved.

Since ancient proteins are generally versatile and stable, scientists now often use them as a starting point in experiments in which they seek to create proteins with novel biochemical properties. One important application of this strategy has been in the search for enzymes that can destroy nerve gas toxins such as sarin. Scientists at Israel's Weizmann Institute recently synthesized the protein ancestors of two enzymes (cytosolic sulfotransferase and serum paraoxonase) as starting points in searches for proteins that more effectively degrade nerve toxins. Using this approach, the Israeli researchers discovered novel enzymes with 100,000 times greater activity at degrading toxins than they had found when they initiated their screens with present-day proteins. Hopefully, this research will soon lead to anti-toxins capable of protecting people from the terrors of nerve gas–based biological weapons.

RESURRECTING A VIRUS

Resurrecting ancient proteins and using them to discover new enzymes are major scientific accomplishments. Some scientists, though, have more ambitious goals. They want to bring entire extinct species back to life, and as the story of the Pyrenean ibex shows, that goal may soon be in reach. In fact, one type of extinct microbe, a virus called HERV-K, has already been brought back from extinction.

But wait a second. How can scientists even know that a virus is extinct? With the Pyrenean ibex, we know what it looked like, since it was still alive 30 years ago. Even if the task is to resurrect a wooly mammoth, which lived approximately 12,000 years ago, or a T. rex, which lived 65 million years ago, we at least have frozen tissue or fossilized bones to tell us what the creature once looked like. Unlike mammoths and dinosaurs, extinct viruses don't leave fossilized remains.

Or do they?

Actually, "fossils" of ancient viruses do exist, and they can be found in our own (human) DNA. You'll recall that a virus consists of a small quantity of DNA* along with a few proteins, whose "blueprints" are encoded in the virus's DNA. As part of their reproductive cycle, some viruses manipulate a host organism's cells into incorporating viral DNA into the host's genome. If a virus performs such genetic wizardry in a sperm or egg cell, then the host's offspring will also have the virus's genes embedded in its DNA.

When scientists began to intensively study the human genome in the late 1990s and early 2000s, they discovered that approximately 8% of the human genome consists of DNA sequences extremely similar to those found in viruses. For example, a hundred-nucleotide human genome segment might be identical to a viral gene except for one or more stop codons† embedded in the middle of the sequence, as a result of which the embedded viral gene could not be converted into a functional viral protein. The most likely explanation for the occurrence of such almost functional viral genes in the human genome is that a virus *did* embed its DNA into the human genome a long time ago. Over the subsequent millennia, mutations altered the viral genes in human DNA to the point that they were no longer converted into viral protein, and the virus became extinct.

One extinct virus found in human DNA is called HERV-K. In 2006, two research teams, one from Rockefeller University and the other from the Institute Gustave Roussy in France, independently resurrected HERV-K. Their approach was to compare the nonfunctional DNA sequence remnant from the presumed extinct virus with those of related active viruses found in the cells of other primate species. By means of these comparisons, the researchers identified a small number of critical nucleotides that appeared to have been altered in human DNA, as a result of which the human viral DNA remnant no longer produced active viral protein. The researchers then altered these critical nucleotides in the human DNA sequence so that the sequences would again be converted to (viral) protein in human cells. The result was that the human cells started producing viruses, which were capable of infecting human cells.‡ HERV-K had been brought back to life.

---

* Some viruses, such as HIV, encode their genes in RNA rather than DNA and have a protein with which to convert their RNA into DNA. For simplicity, we'll ignore this detail here and simply talk of viral DNA.

† Recall that a stop codon is a sequence of three DNA letters (TGA, TAG or TAA) that tells the cell's protein-making machinery to stop converting the current DNA sequence into protein.

‡ This viral resurrection was carried out very carefully to ensure that the virus could only infect a human cell one time, and hence could not lead to a viral epidemic.

ANIMAL CLONING

Having resurrected extinct proteins and microbes, scientists are now taking on the challenge of bringing complex, multicellular animals back to life. Although the task is far more challenging, here too, researchers have made dramatic progress in the last 20 years. Their strategy has been based on modifying techniques originally developed for animal "cloning": the creation of a new individual of a species that is an exact genetic copy to (that is, has the same DNA sequence) as another member of the species.

For organisms that reproduce asexually, such as plants and certain types of reptiles and sharks, cloning is no big deal. For example, whenever a gardener grows a new plant from a stem, leaf or root of another plant, they are creating a clone of the original plant. In contrast, cloning organisms, such as animals, that reproduce through the mixing of DNA from two parents is very challenging. Yet, over the last 20 years, scientists have demonstrated that animal cloning is possible.

The key to animal cloning is that virtually every cell in an animal has an identical DNA sequence. So if one could transfer the DNA from an adult animal cell into an egg cell of the same species (from which the original DNA had been removed), the egg should eventually develop into a genetically identical twin of the animal from which the DNA was originally taken. The challenge is that DNA is biochemically modified over the lifetime of an animal cell.* Unless these DNA modifications are removed, the newly created egg cell will not develop properly. Over the last few years, scientists identified the critical enzymes that can remove these biochemical modifications. Adding these enzymes to cloned fertilized eggs† has made creating viable cloned animals more feasible. Nevertheless, animal cloning is still challenging, and only 1% to 2% of cloned mouse egg cells lead to the birth of a cloned mouse.

RESURRECTING ANIMALS

Theoretically, resurrecting an extinct animal from ancient DNA is not that different from cloning a sheep or a mouse. In practice, though, it is much more challenging. First of all, a resurrected fetus must be carried by a surrogate mother of a different species. Cloned fetuses of endangered,

---

* You'll learn more about such "epigenetic" protein and DNA modifications, which occur during the lifetime of an animal, in part IV of this book.
† By adding the proper enzymes, scientists can now even use adult cells (called induced pluripotent stem cells), rather than egg cells or embryonic cells, as the recipient cells in cloning procedures.

nonextinct animals usually are implanted in surrogates of different species as well, because scientists are reluctant to experiment with rare endangered species due to the inherent risks involved. For example, when scientists at Advanced Cell Technologies in Massachusetts attempted to clone the highly endangered gaur (a wild ox found in South Asia) in 2000, the surrogate mother was a domestic cow. Similarly the surrogate mother of the Pyrenean ibex described earlier was a hybrid of a domestic goat and a Spanish ibex.

When scientists considered resurrecting the wooly mammoth, the most promising surrogate mother was an Asian elephant. But Asian elephants are themselves an endangered species, so using one as a surrogate mother raised ethical as well as technical questions. In any case, the impact on a developing fetus of having a surrogate mother of a different species is still largely unknown. Most likely mother-fetus compatibility issues exist, and this may explain why cloning of endangered and extinct species has so far been less successful than cloning sheep or mice.

A second challenge facing the resurrection of long-extinct species is that over time DNA disintegrates into its constituent nucleotides. Recent experiments have shown that even when preserved under the most favorable circumstances, such as in amber or at very low temperatures, DNA becomes completely degraded after a million years. So the dream of resurrecting 65-million-year-old dinosaurs seems destined to remain unfulfilled.

For endangered and recently extinct species, the picture is more hopeful. For example, to demonstrate the feasibility of using frozen tissue to clone animals, scientists at Japan's Riken Research Center cloned healthy mice from the frozen remains of animals that had died 16 years earlier. Encouraged by such experiments, some scientists active in the field of animal conservation have begun programs to extract and preserve DNA from a variety of animal species that are likely to become extinct in the near future.*

There may even be hope of resurrecting animal species that have been extinct for thousands of years, such as the wooly mammoth. Since the mammoths lived in very cold environments, it is possible that enough nondegraded mammoth DNA can be salvaged to resurrect a mammoth. In 2007, an entire corpse of a mammoth that lived in Siberia an estimated 10,000 to 14,000 years ago was discovered in the Siberian permafrost, and

---

* The idea of preserving DNA from endangered species in the hope of eventual resurrection is controversial within the conservation community. Many conservationists believe that such an approach is a waste of scientific and financial resources that would be better spent on preserving habitats and preventing extinctions in the first place.

using frozen DNA, in 2008, scientists succeeded in sequencing over 4 billion nucleotides of the estimated 4.7 billion nucleotides of the mammoth genome. Although it is still a long journey from 85% of a genome to the resurrection of a species, the dramatic progress that has been made gives hope that bringing the wooly mammoth, if not the brontosaurus, back to life may someday be possible.

RESURRECTING PEOPLE?

Before leaving the realm of animal resurrection, we should probably say at least a few words about the speculative world of human cryonics. The central concept of cryonics, an idea which has been embraced by over 200 hopeful people facing terminal illness, is that by freezing a person's body, or at least the brain, it may be possible to bring that person back to life sometime in the future (presumably after a cure has been found for whatever illness afflicted them).

Bringing a specific individual back to life is vastly more challenging than resurrecting an arbitrary member of a species. This fact will be obvious to anyone who has known identical twins. The cloning techniques described in this chapter essentially create an identical twin. Although identical twins appear strikingly similar to one another, they are still quite different. Identical twins often vary in fundamental traits; one may be heterosexual while the other is homosexual, or one may suffer from schizophrenia or depression, while their "identical" sibling does not.

To bring an individual back to life from a cryonic state one would, at the least, need to restore their memories. For example, if some day one could clone an individual with the DNA of former baseball legend Ted Williams (who is perhaps the best known individual to have chosen to have his body cryonically preserved) the resulting human being might well become an outstanding athlete, and perhaps even a champion baseball player. That said, unless one could also somehow implant memories into the clone's brain of having had a .406 batting average in 1941, the resulting human being would hardly be a resurrected Ted Williams.

To resurrect a person one would need to preserve and then reproduce all of the individual's neuronal connections, by which the brain stores memories. Cryonics advocates argue that such brain preservation can be accomplished by lowering the temperature of the brain. They note that methods of low-temperature organ preservation have been developed since the middle of the last century. As early as the 1950s, scientists had succeeded in preserving living hamsters at subzero temperatures and bringing them back to life by slowly and carefully warming their bodies.

Although keeping entire animals larger than rodents alive at low temperatures has proven much more challenging, preserving individual tissues and organs has become possible. Techniques developed for organ-transplant technology have enabled scientists to routinely preserve kidneys, livers and hearts at low temperatures, at least for short periods of time. In the last decade, scientists have also learned how to freeze ovarian tissue and even entire ovaries to help women facing ovarian cancer have children at a later date.

That said, current organ preservation technology is nowhere near being able to preserve brains until that future day when restoring cryonically preserved brain tissue may be possible. Supporters of cryonics counter that even if brain transplantation won't be possible for another thousand years, cryonic brain preservation is still worthwhile. They argue that by freezing the brain, we can preserve it for however long is necessary until memory-restoration technology has been developed. Maybe. Perhaps someday scientists will discover techniques for preserving brains and for restoring human memories from them. Unfortunately for those whose brains are currently being preserved, those brain-preservation methods are not likely to be those being used today.

6

WHO OWNS OUR PAST?
WHO CONTROLS OUR FUTURE?

The village of Supai is one of the most isolated towns in the United States. Located deep within the Grand Canyon of Arizona, Supai can only be reached by helicopter or by eight miles of hiking or mule or horseback riding. Supai is the cultural and economic center of the Havasupai people, an Indian tribe of approximately 650 people which has lived in the Grand Canyon for at least 800 years.

Life in the Grand Canyon has never been easy, and in recent years, the Havasupai have been faced with a new threat, diabetes. As of 1991, 38% of the adult men and 55% of the women were afflicted with the disease. Since the origins of diabetes were unknown, no one knew whether the high incidence of diabetes among the Havasupai was the result of genetics, diet or other factors. What was clear was that diabetes was threatening to destroy the tribe.

In 1989, a member of the tribe asked John Martin, an anthropologist from Arizona State University (ASU) who had been studying the tribe for decades, for help. Martin knew that a genetic variation predisposing people to diabetes had been discovered among the Pima, another Arizona Indian tribe. Martin proposed that his colleague, ASU geneticist Therese Markow test the DNA of the Havasupai. If the Havasupai also had a genetic variant linked to diabetes risk, perhaps that information could be used to develop a treatment for the tribe.

The following year, Markow began collecting blood samples from over 200 Havasupai. DNA analysis revealed that the Havasupai did not carry the diabetes-linked genetic variant found among the Pima Indians. Markow was also unable to find any other genetic variant in the Havasupai DNA associated with diabetes risk. Although these results were disappointing, if this had been the end of the story, it simply would have been another of the many dead ends that occur in research. But it wasn't the end of the story.

In 2003, Carletta Tilousi, a member of the tribe, learned that the Havasupai DNA had been used in other genetic studies that were unrelated to diabetes research. Havasupai DNA was being analyzed for genetic links to mental illness as well as to study the extent of inbreeding among tribe members. Havasupai DNA was even being used in genetic anthropology research showing that the Havasupai's ancestors had migrated from Asia, contradicting the tribe's beliefs that they had always lived in the Grand Canyon. As a result, many of the Havasupai felt betrayed and were furious at ASU and its researchers.

DNA CONNECTS US WITH OUR PAST AND FUTURE

DNA is our biological link to the past. DNA can determine someone's paternity. DNA can inform us about our family's medical history. DNA can teach us about our racial and ethnic origins, helping scientists discover the Jewish roots of the African Lemba tribe or pinpoint the African origins of the victims of the transatlantic slave trade.

DNA also gives us a glimpse into our future and even that of our children. For some diseases, such as Huntington's disease, DNA can reveal whether a person will succumb to the disease years in the future. And by looking at the DNA of prospective parents, doctors can determine the chances that their child will get diseases such as cystic fibrosis or sickle cell anemia.

Despite the usefulness of all this genetic information, sometimes people would prefer to keep their DNA private. Perhaps they are afraid that knowledge of their racial or ethnic ancestry may lead to discrimination or persecution. Or they may believe that their employment or insurance prospects will be affected if the medical secrets of their DNA are revealed. As DNA data becomes more routinely incorporated into our medical records, concerns that people or institutions will learn about us through our

DNA without our permission are increasing. To address such fears, many countries are introducing laws to help people keep the information in their DNA private. In 2007, the United States enacted a law prohibiting health insurance companies and employers from using genetic data in making enrollment or employment decisions. Although this law was an important step in protecting privacy in this new world of genetic information, challenging ethical and legal issues remain. Not only can it be difficult to enforce DNA privacy laws, but the current laws are still limited; for example, providers of disability, life and long-term-care insurance are all exempt from the law.

GENETIC RESEARCH AND INFORMED CONSENT

As scientists become more adept at deciphering DNA, increasing numbers of people are being asked to contribute their DNA for genetic study. Although research participants are often willing to let scientists probe their DNA to help answer medical or scientific questions, they may not want their DNA used in other research, such as tracing historical origins.

To balance the needs of researchers with the rights of research participants, researchers use what's called the principle of "informed consent." Under informed consent, each research subject must be told how their data will be used and must give explicit permission for that use. Although in principle informed consent is a simple idea, implementing it fairly can be challenging. Often, research participants are not true volunteers but patients who have come for medical treatment. Patients may be under considerable stress and may feel they should give consent to please their doctors. They may have limited formal education or language skills and may not understand the complex language often found in informed consent forms. They may not understand whether they are allowing their DNA to be used for a variety of research purposes or just to study the condition for which they are being treated.

Even when properly implemented, informed consent agreements have drawbacks. Many devastating diseases, including cancer, diabetes and heart disease, are genetically complex. Investigating their genetic roots involves testing thousands of people. Requiring a separate informed consent form to collect the same genetic data for multiple studies can dramatically slow the pace, and increase the cost, of disease research. Although study participants may be willing to allow their DNA to be used in multiple studies, scientists may not be able to predict their future research needs. Consequently, they aren't able to explicitly list all their future studies in a single informed consent form. Meanwhile, the alternative of contacting participants years later to obtain additional consent is also often prohibitively expensive and complicated.

PRIVACY THROUGH ANONYMIZED DATA

Because of the potentially crippling effect of narrowly defined informed consent on research, many scientists argue for less restrictive language on consent forms. They point out that research participant concerns are usually that genetic data could be used to discriminate against them, for example in terms of insurance coverage. A simpler solution than participants consenting to every use of their DNA is to just "anonymize" the research data, so that DNA data can't be linked to the participant's name.

Unfortunately complete data anonymizing also limits the utility of the data, especially in studies of diseases that evolve over many years. For example, consider a study designed to identify genetic variants that predict whether an individual will suffer from Alzheimer's disease. Such a study needs to store research subjects' identities in order to determine whether they in fact eventually contracted the disease. To deal with this problem, rather than remove the link between each DNA sample and the subject's identity, scientists label samples with an identifying code (rather than a name) that is kept secure and confidential. As long as the codes are kept secret, the privacy of subjects' genetic information is maintained, but if the codes were to be stolen or misused, the participants' anonymity would be lost.

Even in cases when research codes are not stolen or misused, participant anonymity may be compromised. Law enforcement agencies and courts have attempted to force researchers to reveal the results of genetic tests in police investigations and legal proceedings. To address such concerns, the United States passed a law granting scientific researchers "certificates of confidentiality" allowing them to refuse to disclose participant data, even under subpoena. Although certificates of confidentiality help maintain research subject anonymity, their effectiveness has limits. First of all, they don't *compel* researchers to maintain participant privacy but just *allow* them to refuse to release such data.

Even if a researcher wants to protect a subject's anonymity, their ability to do so under a certificate of confidentiality is uncertain. The law establishing certificates of confidentiality has been challenged in court, most recently in 2005. In this case, three women accused a family member, one

John Trosper Bradley, of sexually abusing them when they were children. One of the women had, coincidently, participated in a Duke University research study of mental health in adolescents, many years earlier. In the trial, Bradley's attorneys sought to force Duke University to release the woman's study records to determine whether she had reported her abuse in the study. Duke refused, citing its certificate of confidentiality, and a North Carolina appeals court agreed that Duke was not obligated to release its research data. The court's decision was a narrow one, though. It ruled only that Bradley had not met the "threshold requirement of materiality," meaning that he had not proved that the data was relevant to the case. As a result, it remains unclear whether certificates of confidentiality will protect researchers from being forced to release their research data in future trials.

DNA AND OUR CHILDREN'S FUTURE

Privacy and informed consent issues are particularly acute when children are involved, and nowhere are these issues more sensitive than in the genetic screening of newborns. Newborn screening began in the United States with a test for a medical condition called phenylketonuria (PKU). A person with PKU lacks an enzyme required to degrade the amino acid phenylalanine. Approximately 1 in 15,000 children in the United States is born with PKU, and if not treated, PKU can lead to seizures, intellectual disability and other serious medical symptoms.

In the early 1950s, Horst Bickel, a German pediatrician, discovered that PKU can be treated by eliminating phenylalanine from a child's diet. To be effective, phenylalanine dietary restriction must be started at a very early age, typically before the onset of any symptoms. As a result, programs for newborn testing for PKU were instituted in the 1960s, and by 1975, mandatory PKU testing of newborns was being performed in 43 of the 50 states. The rationale for mandatory testing was that infants can't make such decisions themselves and that the dire medical consequences of untreated PKU overrode the parents' rights to control their children's medical treatment. Over the years, the PKU testing program has helped thousands of children avoid devastating illness, and although some parents protested the mandatory nature of the testing, by and large the legitimacy of the program has not been challenged.

With the success of the PKU testing program, newborn screening for other diseases was soon started. Some of these programs were less medically justified and more controversial. In the early 1970s, several states in the U.S. instituted newborn screening for sickle cell anemia (SCA), even though no treatment was available. Making matters worse, the genetic screens not only identified infants with SCA but also those who were simply carriers* of the disease. These children were disease-free and would receive no benefit from screening even if treatment became available. In fact, they might well be stigmatized against as a result of being carriers. Further contributing to the sickle cell screening controversy was the fact that the sickle cell genetic defect is primarily found in Africans and African Americans, a population particularly sensitive about discrimination. Because of these issues, SCA testing programs for newborns had largely been abandoned by the late 1970s.

The question of whether newborns should be screened for SCA reemerged in the late 1980s. Doctors had learned that children with SCA were susceptible to severe bacterial infections. Identifying such children at an early age and giving them preventative antibiotic treatment protected them from the devastating effects of such infections. As a result, newborn screening for SCA was reintroduced, and by 1994 newborns were being screened in 42 states. Although DNA testing continued to also identify healthy children who were merely carriers, screening was considered worthwhile because of the increased protection that it offered to infants who did have the disease.

With the dramatic decreases in the cost of DNA testing, the number of genetic conditions being screened in newborns has increased. In 1995, 5 conditions were generally included in genetic screens for newborns; by 2005, many states were testing for 24 or more genetic conditions, including some for which early diagnosis didn't provide a clear medical benefit. As a result, newborn genetic screening has again become controversial. These controversies are likely only to become more contentious in the future. Soon whole-genome DNA screening will be less expensive than performing individual genetic tests. Society will need to decide what to do with all this genetic data. How much genetic data on a newborn should be given to the child's parents? Should parents be only given data relevant to early medical intervention, or should they be informed of genetic variants that will only affect their child as an adult? Should parents be told of genetic variants that only increase the *probability* that their child will be affected – genetic

---

* A "carrier" is someone who has one normal and one disabling variant of a gene, underlying a recessive genetic disease. Although a carrier will not have the disease, if two carriers have a child, there is a 1 in 4 chance that the child will inherit the disabling variant from each parent and hence be afflicted.

information that requires extensive education to properly interpret? And how should society balance the demands of parents to be given all information about their child with the wishes of children who may later prefer that their parents didn't know their genetic secrets, and who may themselves not want to know everything about their genetic makeup?

PLAYING GOD WITH UNBORN CHILDREN

Even more daunting ethical questions have come with our ability to read DNA prenatally and to use that information to decide who should be born. Abortion has long been a challenging and controversial issue, but with our increased ability to use fetal DNA to glimpse into the potential child's future the moral issues have multiplied.

In some situations, using prenatal DNA data to decide whether to abort an infant has become relatively noncontroversial. Doctors often perform prenatal DNA tests for serious childhood diseases, such as Tay-Sachs disease or Down syndrome, and when they are found, many prospective parents choose abortion without moral misgivings. But what about prenatal testing for Huntington's disease? Huntington's disease is a terrible and lethal disease that affects approximately 1 in 10,000 people and is perhaps best known for causing the death of the folk singer Woody Guthrie. Although Huntington's is a devastating, incurable disease, it only affects people in middle age. A treatment for Huntington's disease might exist by the time a child born today reaches middle age. Should a fetus with the Huntington's genetic variant be aborted?

Even more troubling are cases where genetic variants only increase the *probability* that someone will get a disease. Should such fetuses be aborted if the probability the person will get a terrible disease is 90%? 50%? 10%? These ethical issues are likely to only get more difficult to resolve as our understanding of DNA improves. Prospective parents may soon not only be able to learn whether their future child will be affected by a devastating disease but also by less severe disabilities, such as hearing impairment or moderate intellectual disability. How should we as individuals or as a society respond to prospective parents who want to have an abortion in such situations?

SAVIOR SIBLINGS

Along with the moral issue to aborting of "undesirable" fetuses comes the related question of whether parents should be allowed to genetically select "desirable" ones. Although the idea of picking a child on the basis of genetic characteristics desired by the parents may seem morally reprehensible, in a few extreme situations it has already become a part of medical practice. Certain severe diseases can currently be treated only through a bone marrow transplant from a healthy individual with a compatible blood type, and sometimes the only possible donor is a sibling. But what if the patient doesn't have a healthy compatible sibling? Is it morally acceptable for parents to have another child simply for the purpose of creating a bone marrow donor for their sick child?

Such questions are not merely theoretical. In 1984, a four-year-old girl in an Italian family we'll call the Agostinos* was suffering from a severe form of leukemia, for which the best treatment was a bone marrow transplant from a sibling. Unfortunately, the Agostinos didn't have a child who could be a donor, so they wanted to have another child, hoping that the baby would have a blood type compatible with its sister's. The Agostinos' odds of having a blood type–compatible healthy sibling were only 25%, but they were lucky. Mrs. Agostino gave birth to a healthy boy with the right blood type, and 19 months later, the older sister received a bone marrow transplant from her brother. The procedure was a success: the girl was cured, and the brother suffered no adverse consequences. In fact, a recent medical report stated that 25 years later both siblings are healthy and doing well. Nevertheless, one wonders what the Agostinos would have done if prenatal testing had shown that the fetus was not compatible with its sister (which had a 3 out of 4 chance of occurring). Would they have aborted this healthy fetus and tried again? Would they have continued having children until one of them finally was a compatible donor?

Seventeen years after the birth of the Agostino baby, scientists in the United States introduced an arguably less controversial way of giving birth to a "savior sibling." The method combined in vitro fertilization (IVF) and preimplantation genetic diagnosis (PGD). IVF was originally developed to help women with fertility disorders have children. In IVF, a woman is given drugs that stimulate her ovaries to produce multiple eggs. These eggs are removed from the ovaries and fertilized with the father's sperm in glass dishes or tubes, leading to the popular term "test tube babies." Each of the eggs is allowed to become an embryo of four to eight cells and then one or more of the embryos is implanted in the mother's uterus. Typically 10 to 12

---

* For reasons of patient privacy, the research article describing this family did not include the family's name.

embryos are produced, and 2 or 3 are implanted in the mother (to increase the likelihood that at least one will develop successfully).

Preimplantation genetic diagnosis adds another step to the IVF procedure and is generally used to prevent the birth of children with serious genetic diseases, such as Tay-Sachs disease, cystic fibrosis or sickle-cell. With PGD, embryos created by IVF are genetically tested, and only healthy embryos are implanted. This way, prospective parents who are carriers of severe genetic diseases can be confident that their child will not suffer from the disease. PGD with IVF has the advantage over other methods of prenatal screening for genetic diseases in that abortion is not required; an embryo with disease-causing mutations simply isn't implanted in the mother.

Preventing genetic diseases is not the only reason IVF with PGD is performed. In 2001, PGD was used for a very different reason. The family that requested PGD had a young daughter, called Molly N. in the medical literature, suffering from Fanconi anemia. Bone marrow transplantation from a healthy donor is the best treatment for Fanconi anemia. The donated bone marrow must be compatible with the patient's blood type, which often limits potential donors to the patient's siblings. Since Molly didn't have any compatible siblings, her parents wanted their doctors to perform IVF with PGD on egg cells from Molly's mother. Molly's mother went through four cycles of IVF, producing 33 embryos. Five of the embryos were healthy and compatible. Unfortunately, the initial embryo transplantations were unsuccessful, and only on the fourth attempt was the medical team able to implant a healthy, compatible embryo. Nine months later Molly's mother gave birth to a healthy child, and as part of the normal procedure of giving birth, the umbilical cord connecting the mother and the child was extracted. Some of the baby's blood from the umbilical cord was saved and was subsequently used to successfully treat Molly. In this case, the medical end result for both siblings was positive. Just as had been the case with the Agostinos 17 years earlier, the baby was truly a savior sibling.

Although selecting a child to save a sick sibling might be morally acceptable, IVF with PGD could someday also enable parents to choose a child on the basis of more questionable criteria, such as its sex or its likelihood to excel in sports or mathematics. If such a scenario seems unbelievable, one need only consider that in many parts of the world parents often choose to abort a fetus just because it is female. This practice is most common if the parents already have a daughter. In India, for example, in families where the first-born was a girl, the sex ratio at birth of the second child changed from 110 males per 100 females* in 1990 to 120 males per 100 females in 2005. Meanwhile, the sex ratio at birth of firstborns and of second children in families in which the firstborn was a boy didn't change. In other words, despite the illegality of sex-selective abortion, the simultaneous change in second-child sex ratios and the increased availability of prenatal ultrasound is not likely to be coincidental.

DNA-based sex determination will soon make it possible to know a fetus's sex by the seventh week of pregnancy rather than needing to wait until week 12 or 13 for an ultrasound examination. As a result, the moral issues raised by sex-selective abortions are only likely to become more acute. By combining prenatal genetic testing with in vitro fertilization, it will soon be possible to select fetuses without requiring abortions. The temptation to play God and choose a child on the basis of a variety of nonmedical considerations may become too strong for some parents. The potential consequences are disturbing.

DNA OWNERSHIP AND THE HAVASUPAI

The legal and ethical conflicts over the control of our DNA often have painful consequences for everyone involved. This was certainly true in the case of the Havasupai Indians. After the Havasupai learned that their DNA was being used for research unrelated to diabetes, they tried to reclaim their DNA samples and prevent their use in any genetic studies. In 2003, they issued a "banishment order" stating that Arizona State University and "its Professors and employees are, from this date forward banished from the Havasupai Reservation." The following year, they initiated a lawsuit against ASU, and the individual researchers involved, demanding that they stop using their DNA and asking for $50 million in damages for the hurt and injury that the genetic studies had caused. ASU's representatives responded that the Havasupai participants had signed a consent agreement stating that their DNA could be used to "study the causes of behavioral/medical disorders." The Havasupai countered that their tribal members had only limited formal education, and for many of them English was their second language. Ultimately, the courts required ASU to pay $700,000 in damages to the Havasupai and to return the blood samples. ASU had spent $1.7 million in its legal defense.

---

* In the absence of any sex selective abortions, approximately 105 boys are born for every 100 girls. Why this ratio isn't closer to 1-1 is unknown and is an area of scientific debate and research.

The legal battles between the Havasupai and ASU show how complex and emotional the issues around ownership of one's DNA can be. The story also illustrates the limitations of the methods that we, as a society, have developed for addressing these issues. For the Havasupai, increased data anonymization would not have made any difference. Clearer communication regarding the consent agreement might have avoided the conflict, but as the gap widens between the capabilities of modern genetic technology and most people's ability to understand those capabilities, similar conflicts are likely to occur in the future.

PART III

DNA AND RNA:
HOW GENES ARE
REGULATED

7

CAN A GENE KEEP YOU FROM THINKING CLEARLY?

The first time the boy we'll call Billy was examined by Drs. James Purdon Martin and Julia Bell at the National Hospital in London, in the 1920s, he was only 18 months old. Billy had appeared normal at birth, and for the first year of his life, he seemed to be a healthy little boy, crawling and taking his first steps at approximately the same age as most children. But now he was a year-and-a-half old and he still hadn't started talking. Although sometimes healthy children don't talk until they are two to three years of age, in Billy's case, there was additional reason to be concerned.

As Billy's doctors began to study his family history, they learned that Billy had five first cousins who all suffered from severe intellectual disability. They were described as having a "mental age" of three to five years. None of them spoke more than single words. They all eventually needed to be institutionalized, being unable to live on their own. As Martin and Bell learned more about Billy's family, they discovered that three of Billy's second cousins as well as two first cousins of his mother were also severely intellectually disabled (or, as they would have been described in the medical language of the time, were "imbeciles"). Remarkably, all 11 of Billy's severely afflicted family members were male. Only a single female family member, out of 40, was diagnosed with mild intellectual disability.

Martin and Bell spent the next 16 years studying Billy and his family, eventually publishing their observations in a landmark paper in the *Journal of Neurological Psychiatry,* in 1943. Unfortunately Martin and Bell weren't able to do more than describe their observations; the cause of the disease remained mysterious, and no treatment was available. Subsequently, other families similar to Billy's were found. In these families as well, multiple family members, most of them male, suffered from severe intellectual disability, often compounded with the inability to speak or with social impairments that we now label as autism. Later epidemiological studies revealed that approximately 1 in every 3,000 to 4,000 boys was afflicted with what was initially called Martin-Bell syndrome; among girls, Martin-Bell was found approximately half as often.

For nearly a quarter-century, very little was understood about this terrible disease. Not until the late 1960s did scientists discover that the chromosomes of individuals with Martin-Bell syndrome looked quite unusual when viewed under a microscope,. Their X chromosomes appeared constricted and had a tendency to break. For this reason, the name of the disease was changed from Martin-Bell to fragile X syndrome. By the 1980s, scientists had tools with which to search for the specific genetic variation – the mutation – that was causing fragile X syndrome. Finally in 1991, a team of researchers from Erasmus University in Rotterdam pinpointed the precise genomic location where the afflicted boys' DNA differed from that of their healthy relatives, and from the DNA of healthy unrelated boys, as well. The mutation was indeed on the X chromosome, near a previously unknown gene that the researchers called *FMR1,* for fragile X mental retardation gene 1.

THE *FMR1* GENE AND THE FMR1 PROTEIN

In our exploration of DNA in the previous chapters, we mainly focused on DNA's role as a link to the past and the future. We've largely ignoring DNA's primary function, which is to encode the amino acid sequences that make up a cell's proteins. In this chapter and the following ones, we'll return to DNA's role of storing genes, the blueprints for our proteins. In chapter 2 you learned about the *CCR5* gene and saw how a mutation in that gene could change the shape of the CCR5 protein and thereby make someone immune from HIV. The mutation that the Rotterdam researchers found in boys with fragile X syndrome was different from the CCR5 mutations in a fundamental way: the mutation was located *near* the *FMR1* gene, but not actually within the gene. As a result, boys with fragile X syndrome and healthy boys produce identical FMR1 *protein.*

Since proteins are the molecules that perform the cell's primary biological functions, it may seem surprising that a DNA modification that doesn't change a protein, can cause a disease. The root of this apparent paradox is that the previous description of how DNA specifies the functioning of our cells was incomplete. Chapter 2 focused on genes, that is, the segments of DNA that specify a protein's amino acid sequence, and only briefly mentioned two other players that play key roles in the production of proteins: RNA and the DNA that is not contained in genes, called regulatory DNA. In this chapter, we'll explore the roles of these two important molecules and see how a genetic mutation that disrupts the proper interactions among DNA, RNA and protein can lead to the devastating effects of fragile X syndrome.

MESSENGER RNA

Chapter 2 stated that a cell reads the gene "blueprint" in its DNA to synthesize the specified protein. Although this description is partly correct, it leaves out an important part of the process: a critical intermediate step between the DNA blueprint and the final synthesized protein. This intermediate step involves DNA's sister molecule RNA, or ribonucleic acid.

Using RNA enables the cell to overcome two serious logistical challenges. First of all, while the cell's DNA is stored in chromosomes, which are packed inside a cell structure called the nucleus, cells often need to produce proteins in locations other than in the nucleus. Although most cells are tiny, measured in fractions of a millimeter, some cells, such as nerve cells, can be quite large. As an extreme example, certain nerve cells in a giraffe can be up to 15 feet in length. This is a long distance for proteins to travel when they are needed quickly. The second challenge facing cells is that they may need to produce proteins simultaneously in several different parts of the cell. Since each cell has only two copies of each gene (one inherited from the father and the other from the mother) and since both copies are in the cell's nucleus, a cell couldn't synthesize proteins in multiple cellular locations if it had to do so directly from its two DNA blueprints.

The cell solves both of these problems by dividing the process of making a protein into two steps. In the first step – called gene transcription, or simply transcription – specialized proteins within the cell's nucleus copy the sequence from its DNA master blueprint onto intermediate short-term blueprints consisting of one or more molecules of RNA. In the second step, called translation, a second class of specialized proteins – the ribosomes introduced in chapter 2 – synthesize the protein from the intermediate

RNA blueprint. The RNA molecules, which the cell uses for the transfer of genetic information, are called messenger RNAs, or mRNAs, to distinguish these molecules from other types of RNA with different functions. Figure 7.1 shows this two-step pathway proceeding from DNA to RNA to protein. It also illustrates the important relationships among the structures of DNA, RNA and protein. DNA's structure is a linear sequence composed of (base-paired) nucleotides. This leads to RNA's structure as a linear sequence of (unpaired) nucleotides, which in turn leads to a protein's structure, expressed as a linear sequence of amino acids.

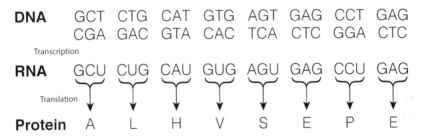

Figure 7.1. Transcription and translation. Schematic illustration of the decoding of DNA by the cell. The top section shows a segment of double-stranded DNA. The middle section is the corresponding messenger RNA molecule, produced through the process of transcription. Note that the original DNA sequence of the top strand has been retained except that the thymine nucleotides (T) have been converted to uracil (U). The bottom row depicts the corresponding part of the translated protein, produced by a ribosome. Each triplet of consecutive nucleotides specifies one amino acid in the protein. A, L, H, V, S, E and P are abbreviations of 7 of the 20 amino acids used by the ribosome to build proteins. (See figure 2.2 for the complete list of amino acids and their abbreviations.)

RNA is chemically similar to DNA, also consisting of a string of nucleotides, using essentially the same four building blocks as DNA. Three of the RNA nucleotides, adenine (A), cytosine (C) and guanine (G), are identical to the bases in DNA, while the fourth, uracil (abbreviated as U), is very similar to the fourth DNA nucleotide, thymine (T). Beyond this substitution of U for T, RNA also has a somewhat different chemical backbone, and unlike DNA, is rarely bound together in complementary pairs. Lastly, RNA molecules are much shorter than DNA molecules because they just contain the blueprint for a single gene.*

---

* In bacteria, and in rare cases in multicellular organisms, cells may copy two or more gene sequences onto a single RNA molecule.

How do messenger RNAs enable a cell to overcome the logistical challenges described above? A cell can synthesize multiple mRNAs from a single gene. These mRNAs can be transported to the millions of ribosomes that are distributed throughout the cell. Whenever some protein is needed, ribosomes can bind to nearby mRNAs that code for the required protein and begin producing it. As a result, large quantities of protein can be rapidly synthesized throughout the cell from mRNA copies, even though there are only two DNA "master copies" of the protein's blueprint in the cell's nucleus.

REGULATORY DNA

In order to understand how Billy's DNA led to his intellectual disability, I also need to correct a second oversimplification from the earlier description of how DNA is converted into protein. Previously, it was suggested that DNA solely consists of blueprints of *how* to make proteins. But if DNA is to incorporate all the information needed to build an organism, it also needs to encode *how much* of each protein to make, as well as *under what circumstances* the cell should make the protein.

For organisms that consist of many different types of cells, each cell type also needs a way to specify which proteins it should synthesize and which ones it shouldn't make, even though every cell type has the identical DNA sequence in its chromosomes. For example, pancreatic beta cells need to produce insulin, while your muscle cells and nerve cells, which also have the insulin gene encoded in their DNA, must *not* synthesize insulin.

In other words, DNA is much more than a set of blueprints for proteins. DNA is more similar to a coordinated collection of *recipes*: genes specify the correct "ingredients" to make proteins, while the regulatory DNA determines how much protein should be made. Depending on the specific sequence of nucleotides within any given section of regulatory DNA, special proteins, called transcription factors, may bind to it and begin the transcription of a gene. This part of the genetic code is just as important to the proper functioning of a cell as the part that specifies how to build a protein from amino acids. Yet in spite of decades of research, science's understanding of how the regulatory DNA works is still limited.

We can get a sense of how important, and how complex, this regulatory component of the genetic code is by looking at how much of the cell's DNA is devoted to this task. Less than 2% of our DNA is used to encode proteins. We don't yet know how much of the remaining 98% is used for gene regulation, because we still have only a partial understanding of how the regulatory genetic code works. But we do now know that a substantial part of this DNA is used for gene regulation. In part, we know this from comparing our DNA with that of other animals. When scientists compare the number of nucleotides used for specifying protein sequences between human DNA and DNA of, say, the roundworm, they find that humans and worms aren't all that different. Worm DNA uses approximately 25 million nucleotides to encode protein sequences, about half as many as the roughly 50 million bases used in human DNA. In contrast, the remaining (non–protein encoding) worm DNA contains only about 75 million nucleotides, while in humans that number is over 3 *billion* nucleotides. So even though we still don't entirely understand all the functions of this remaining DNA, if we are going to search for the biological reasons why humans are more complex than worms, it seems likely that some important answers will be found in DNA that does not code for proteins.

HOW GENETIC VARIATION CAN AFFECT A PROTEIN

Knowing that a gene codes for a protein, messenger RNA transports the coded message to the ribosomes, and regulatory DNA controls when this happens gives us a more complete picture of how cells make proteins. We'll soon see how these processes can go awry and lead to disastrous results, but first, to get a better understanding of how alterations in DNA sequence can affect the cell's production of proteins, let's take a look at the example in figure 7.2. (If you just want to follow the story of how impaired DNA regulation leads to fragile X syndrome, you can skip ahead to the next section.)

Figure 7.2A shows a single strand of a region of DNA from an (arbitrarily chosen) reference genome sequence that includes one hypothetical gene we'll call gene A. The DNA sequence is located approximately 1,000,000 bases from the end of a chromosome. The actual DNA molecule would consist of this strand of DNA bound to its complementary pair (which is left out to make the figure easier to read). We see that at locations 1,000,000–1,000,002 are the bases A, T and G, indicating the beginning of the code for some protein. At positions 1,000,042–1,000,044 we find TAA (a stop codon) indicating the end of the prescription for gene A. The start and stop codons are underlined. So gene A is 42 base pairs long and the resulting protein will be $42/3 = 14$ amino acids long.

Figure 7.2B shows the same genetic region for the DNA of a person we'll call Mary. The top row is the sequence of a single strand of the DNA Mary inherited from her father; the bottom row shows a single strand of the DNA she inherited from her mother. In both cases, the complementary strands are omitted.

Notice there is only one place (indicated by an arrow), where Mary's paternal DNA sequence differs from the reference sequence in figure 7A. At position 1,000,006, Mary's paternal sequence substitutes a T for an A. As a result, the sequence now has a stop codon (TGA) at positions 1,000,006–1,000,008, so protein A will be truncated after just two amino acids. Since protein A will now be missing 10 of its 12 amino acids, its structure will be totally different from its normal one, and as a result it will almost definitely be nonfunctional.

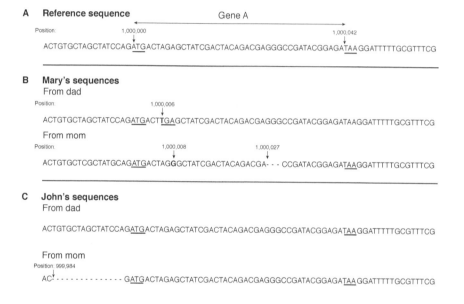

Figure 7.2. Genetic variants near a hypothetical gene and their consequences. A: Reference sequence for the gene. B: DNA sequence of the two copies of this gene inherited by Mary. C: The DNA for the copies of the gene inherited by Mary's brother John. In each example, only one strand of the double-stranded DNA molecule is shown. (See the text for a more detailed description of the schematic.)

The lower sequence in figure 7.2B shows one strand of the DNA sequence Mary inherited from her mother. It differs from the figure 7.2A reference sequence at two places. At location 1,000,008 a G has been substituted for an A. Consequently, the third codon of gene A, at positions 1,000,006–1,000,008, changes from AGA to AGG. Since both AGA and AGG happen to code for the same amino acid (arginine), this DNA change won't alter the resulting protein. The other change is the deletion of the three G's starting at 1,000,027. This will make the resulting protein one amino acid shorter than usual. Whether this amino acid change significantly changes the shape, and consequently the function, of protein A is unclear. If the deleted amino acid does affect protein A's function, it could well have an impact on Mary's cells, since we've seen that the copy of gene A, which Mary inherited from her father, is also nonfunctional.

Figure 7.2C shows the DNA sequences, at the same location, inherited by Mary's brother John. The figure shows John has not inherited the same gene A copy from their dad as Mary. The copy that John has inherited (shown at the top of figure 7.2C) is identical to the reference sequence. Most likely it is functional, though on rare occasions the reference sequence has been taken from an individual who has a disabling mutation at some gene. So one can't assume that every gene in a reference sequence is necessarily functional.

The bottom schematic of Figure 7.2C shows the copy of gene A that John has inherited from his mother. In this case, too, John did not inherit the same copy of gene A that his sister Mary did. John's copy has only a single difference from the reference sequence, namely the deletion of 15 base pairs just before the start of gene A. Since there are no differences from the reference sequence within gene A itself, John's maternal gene specifies exactly the same amino acid sequence for protein A that the reference sequence does. Nevertheless, the relatively large DNA deletion immediately in front of gene A (where regulatory DNA is often located), suggests that the production of protein A from gene A will be misregulated. This misregulation may cause this copy of gene A not to produce any protein A at all. In this case, John's cells would make approximately half the usual amount of protein A, since he has only one functional gene A. Depending on how protein A is used, this change may affect the functioning of John's cells.

From figure 7.2, we can also deduce that gene A is *not* on the X or Y chromosome. How do we know this? Figure 7.2C shows that John has inherited *two* copies of gene A, one from his mother and one from his father. While we inherit two copies of most genes, men inherit only one X chromosome (from their mother), while inheriting a Y chromosome from their father. Since figure 7.2c shows John with a copy of gene A from both parents, gene A is not on the X chromosome.

Let's now imagine instead that gene A *is* on the X chromosome. In that case, John would have only one copy of gene A, which he inherited from his mother. And if for some reason, the single copy of gene *A* that John inherited did not produce any protein A (perhaps because of a DNA deletion immediately adjacent to gene *A*), he would not have a second copy to make up the deficiency. His cells would not produce any protein A whatsoever. It is exactly this kind of X chromosome–related protein deficiency that is at the root of fragile X syndrome.

DNA REGULATION: THE X CHROMOSOME AND FRAGILE X SYNDROME

Males like Billy with fragile X syndrome have an insertion of approximately 400 extra base pairs near the beginning of the *FMR1* gene on their sole X chromosome. As a result, their cells produce little or no *FMR1* messenger RNA and don't produce FMR1 protein. Billy's mother also had an X chromosome with a nonfunctional *FMR1* gene; that's how Billy inherited his nonfunctioning *FMR1* gene. Being female, though, Billy's mother had a second X chromosome. Since she didn't have symptoms of fragile X syndrome, her second X chromosome almost certainly had a *FMR1* gene with a normal regulatory region, and therefore her brain cells produced FMR1 protein. Even limited production of FMR1 protein is sufficient to prevent the most devastating consequences of fragile X syndrome, which is why fragile X occurs primarily, and with more severity, in males: males don't have a "backup" X chromosome. As you might expect, any condition caused by the loss of a protein encoded by an X chromosome gene will be much more common among men than among women. Fragile X is one such disease. Duchenne muscular dystrophy and hemophilia are other serious diseases that mainly afflict males because they are caused by disabling X chromosome mutations. Even the higher prevalence of red-green color blindness in men can be traced to a genetic variation on the X chromosome.

We are still left with the question of how the loss of a single protein can cause the devastating effects of fragile X syndrome. Recent experiments on mice have shown that FMR1 regulates protein production by binding to the messenger RNAs. As much as 4% of all the messenger RNA produced in our brain cells is bound by FMR1 protein. In particular, FMR1 regulates the production of proteins used to process signals from a neurotransmitter called glutamate. We'll talk more about glutamate and its role in brain signaling in later chapters. For now it suffices to know that when a healthy brain cell detects a glutamate signal, it triggers a biochemical pathway involving FMR1, which leads to the production of a small amount of protein (figure 7.3). If a brain cell lacks FMR1, a glutamate signal causes the cell to produce too much protein (figure 7.4). Somehow, this excess protein production overloads the brain's signal processing capability, which leads to the symptoms of intellectual disability, inability to speak and autistic-like social behavior that are associated with fragile X syndrome.

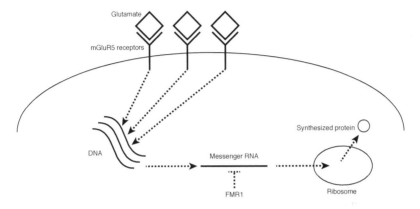

Figure 7.3. Conceptual illustration of the mGluR5 glutamate signaling pathway in brain cells. Glutamate is bound by mGluR5 receptors on brain cells, triggering a series of biochemical reactions leading to the production of messenger RNAs for many proteins. The quantity of messenger RNAs reaching the ribosomes is regulated by FMR1 protein.

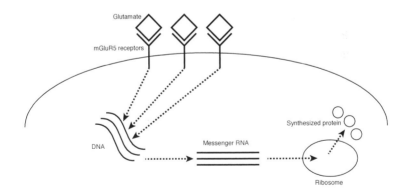

Figure 7.4. A brain cell in an individual lacking FMR1 protein. In this case, too much messenger RNA reaches the ribosomes and too much protein is produced. The end result is fragile X syndrome.

The positive aspect of these discoveries is that they suggest a way of treating fragile X. Since the disease is caused by uncontrolled glutamate signaling, then perhaps fragile X could be treated with a drug that blocks glutamate signaling. This line of reasoning has recently led the drug companies Roche and Novartis to develop experimental drugs that bind strongly to a brain glutamate receptor called mGluR5, thereby decreasing the sensitivity of the brain to glutamate signaling (figure 7.5).

In mice that lack functional FMR1 protein, these new drugs not only slow the onset of fragile X symptoms when administered to young mice, they even appear to reverse the symptoms of "mouse fragile X disease" in older mice. Because of these encouraging results, human trials of glutamate-pathway inhibitors for the treatment of fragile X syndrome have been started recently. While it is still too early to tell whether these drugs will be successful in treating people with fragile X syndrome, there is finally hope that effective treatments for this devastating disease may become available in the not-too-distant future.

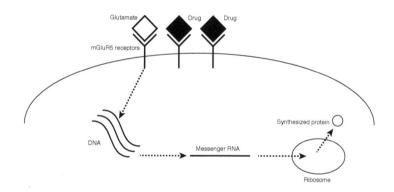

Figure 7.5. The strategy underlying new drugs that have successfully treated fragile X disease in mice and are currently in human clinical trials. The drugs block mGluR5 receptors, so fewer receptors are available for glutamate signaling. Consequently there are lower concentrations of messenger RNA, even though there is no FMR1 protein available for RNA regulation.

8

HOW MUCH SLEEP DO YOU NEED?

What would you do with two more hours every day? Spend more time with your family? Take care of errands? Just relax? For some people this is not just a hypothetical possibility; they gain two extra hours per day simply by sleeping less than six hours at night. Just to be clear, I don't mean getting by on six hours of sleep, as do many overworked people who don't sleep enough, are tired all week and finally collapse into bed to sleep half the weekend. No, I'm talking about living life fully on six hours of sleep each night and being no more tired than people who get a full eight hours.

Such people do exist. Just how many isn't known; most estimates range between 1% and 3% of the population. While large-scale surveys indicate that almost 10% of the population averages less than six hours of sleep per night, most suffer physically or psychologically from their lack of sleep. In contrast, true "short sleepers" rarely get more than six hours of sleep per night yet don't feel any more tired or sleepy than other people, nor do they suffer from health problems related to their lack of sleep.

Scientists have long suspected that genetic variations may contribute to being a healthy short sleeper, but identifying the key genes has been difficult. Consequently, finding two short sleepers from the same family was a real opportunity for sleep researchers. In the early 2000s, two women we'll call Lois Williams and her daughter Clara came to the sleep laboratory at the University of California, San Francisco (UCSF). Lois was 69 years old at the time and had been sleeping less than six hours per night since her early twenties. Clara was 44 and had been sleeping approximately six and a half hours since childhood. No one else in the Williams family was a short sleeper. If the UCSF scientists could find a genetic variant shared by Lois and Clara, but not by any other family member, they might have an important clue as to their unusual sleep patterns. When the UCSF scientists tested Lois and Clara's DNA sequences for a gene called *DEC2*, they discovered that both women had a rare mutation that changed a single proline amino acid to an arginine amino acid in the DEC2 protein. None of the other members of the Williams family had this mutation, nor could the UCSF researchers find the mutation in the DNA of 250 other people they tested.

HOW ANIMALS TEACH US ABOUT PROTEIN FUNCTION

Although finding the same rare variant in the two short sleepers certainly raises suspicions that the variant might be responsible for their sleep patterns, it isn't proof. We all have thousands of rare variants in our genomes, any one of which might be linked to how much sleep we need. The UCSF scientists needed to determine how the DEC2 protein acts in the brain and whether it really affects how much sleep we need.

Discovering how a protein acts in our cells is a critical step in learning how gene variations affect us. It's a question we've encountered before, though we haven't focused our attention on it. For example, the last chapter noted that people with certain mutations in the DNA that regulates the *FMR1* gene have intellectual disabilities, but we didn't look at how scientists discovered what the FMR1 protein does or how it affects human intellectual ability.

To determine the function of a protein, scientists usually study how a biological system changes when the protein is added or removed while everything else is left unchanged. A system in which all but one thing is kept constant is called a controlled system. In biology, the easiest way to create a controlled system is by using a collection of identical cells called a cell culture. With cell cultures, scientists can study millions of cells and then see what happens if they modify a single gene or a single aspect of the cells' environment while keeping everything else unchanged.

Useful as they are, cell culture experiments often don't explain how a protein functions in a person. After all, we're a lot more complicated than individual cells. So scientists also perform experiments on animals. By studying animals in a laboratory setting, scientists are able to maintain many of the advantages of a controlled system. They can raise the animals in identical environments, for example, by giving each animal the same kind of food in exactly the same quantities. The animals can all be given the same amount of time outside their cages, the same learning stimuli and equal amounts of time with their mothers while they are young. In this way, if scientists modify one aspect of the animals' environment, say doubling their food ration, and then observe some consistent change in, say, the animals' behavior or in how long they live, they can be confident that the change was the result of the increased food consumption.

Along with controlling for differences in a laboratory animal's environment, scientists can also control for variations in an animal's genetic makeup. In fact, humans have been manipulating the genetic composition of animals for thousands of years by using selective breeding. By only breeding animals with some desired trait, the gene variants that underlie that trait become more common in the population. By repeatedly breeding sibling animals with one another, one can eventually create groups of animals, called inbred strains (or simply strains) that are genetically the same. In other words, all the animals in a single strain are essentially identical twins.*

Of course, people are different from mice, worms and other laboratory animals. On the other hand, animals are much more similar to people than cell cultures are. In fact, as we'll see throughout this book, the biology of animals, even of mice and rats, is remarkably similar to human biology. Mice develop gradually from infancy to adulthood. Their brains learn new things in ways that are similar to ours. They exhibit behavior that can be eerily reminiscent of human emotions, such as fear or anger or despair. They get diseases similar to ours, and (unless they're killed first) they grow old and die.

LEARNING FROM TRANSGENIC MICE

The real power of using inbred strains of laboratory mice was realized when scientists learned how to modify individual mouse genes. The original method for genetic modification of mice, which is still widely used, involves disabling a single gene so that it doesn't produce any functional protein. This

---

* Animals belonging to a single strain aren't exactly like human identical twins. For one thing, males and females aren't identical, since the males have a Y chromosome, while the females have two X's. Unlike human twins, some animals within a strain are older than others. Indeed, parents and their offspring belong to the same strain, but any new mutations will cause an offspring's DNA to differ from that of its parents. That said, animals from a single strain are genetically very similar.

is called making a gene knockout mouse. Depending on what they are trying to learn, researchers may disable both copies of a gene or just disrupt one copy while leaving the other one intact. For example, the DNA of a mouse could be altered so that it had one normal and one disabled *DEC2* gene* while otherwise being genetically identical to other mice of its strain. In this way, scientists can study the biological effects of a disabled *DEC2* gene, yielding clues to the consequences for humans of *DEC2* genetic variants.

Creating a knockout mouse is very challenging. When scientists first proposed producing knockout animals, their ideas were met with skepticism and disbelief. One of those scientists, Mario Capecchi, a researcher at the University of Utah, tells the story of how his original grant application to develop a knockout mouse was rejected by the U.S. National Institutes of Health, with the comment that the methods had a "vanishingly small probability of success." Nevertheless, Capecchi and a few other scientific visionaries persevered, and by 1989 they had succeeded in creating a strain of mice in which a single gene had been disabled. In 2007, Capecchi, along with Martin Evans and Oliver Smithies, received the Nobel Prize in Medicine for their development of genetically engineered mice.

Since 1989, the technology for genetically manipulating mice has exploded. Scientists can now open catalogs from specialized laboratories and order mice with any one of hundreds of genes disabled. Mice with thousands of other knocked-out genes can be produced in laboratories by using modified mouse embryo cells. In fact, just about the only kinds of genes that can't be targeted for gene disruption are the "essential" mouse genes, which if disabled, result in a mouse that is no longer able to live.

More complex and subtle genetic manipulations than just disabling a single gene have also become possible. By knocking out two genes in a single mouse, scientists can test whether two proteins (let's call them protein A and protein B) can compensate for one another. We can imagine an experiment in which a mouse with the gene for protein A *or* the gene for protein B knocked out appears completely normal, while a mouse with the genes for *both* proteins A and B knocked out does not. In this case one could conclude that proteins A and B are redundant and having functioning versions of either one is sufficient for normal development.

Researchers also study proteins with multiple functions by engineering "conditional" knockout mice. In a conditional knockout mouse, a gene is

---

* More precisely, the scientists would disable the gene that is the mouse equivalent of the human *DEC2* gene – that is, the mouse gene that produces a protein with an amino acid sequence very similar to that of the human DEC2 protein.

modified so that it is disabled only in certain types of cells. For example, the gene might be disabled in brain cells while still acting normally in heart or muscle cells. Conditional knockouts are especially important when studying essential genes, since mice with essential genes completely knocked out don't survive. With a conditional knockout, the essential gene can be disabled only in certain tissues or during a specific time during the animal's development. As long as the gene is intact in cells in which it's essential, the animal will survive and the impact of gene's absence in other cells can be studied.

Last, but certainly not least, scientists can now also modify a mouse's DNA so that rather than a protein being disabled, the protein has modified – or even enhanced – functionality. In addition, entirely new genes can be added to a mouse's DNA. Such genetically enhanced mice are typically referred to as transgenic mice or, if they have had a gene added to their DNA, as "knockin" mice.

In chapter 19 we'll look at how transgenic mice have facilitated the understanding of an important protein called FOXP2. Scientists had discovered that people with a rare variant of the *FOXP2* gene suffered from a severe language disorder, but no one had any idea why. It turned out that humans and mice have almost identical *FOXP2* genes. So scientists were able to create a knockin strain of mice whose *FOXP2* gene had been replaced by the same gene variant that people with the speech disorder had. By doing so, they were able to study the properties of the FOXP2 protein.

TRANSGENIC RATS AND MONKEYS

Scientists have learned how to genetically manipulate animals other than mice. For example, a number of universities and companies recently established the Knockout Rat Consortium, with the goal of enabling researchers to order rats with a specific gene knocked out, just as they can with mice. Going from mice to rats may not seem like a big deal, but in terms of physiology and behavior, rats are often more similar to people than mice are.

Sometimes rodents are simply too different from humans to be useful in biological experiments. Consequently, there have also been attempts to genetically manipulate monkeys. Although creating transgenic primates is still extremely challenging, the technology hurdles are gradually being overcome.

One motivation for developing transgenic primates has been to understand Huntington's disease. Although scientists had discovered in 1993 that Huntington's disease was caused by variations in a single gene, which they called huntingtin, no one has figured out what the protein coded by the huntingtin gene does. Nor have scientists been able to understand how the dysfunctional form of the huntingtin protein leads to the disease. As a result, there is still no effective treatment for people with Huntington's disease.

Mice also have a huntingtin gene, and scientists have been able to engineer mice with the same huntingtin gene variant as that found in people with Huntington's disease. In the language of medical research, the scientists had developed a mouse model of Huntington's disease, an important step in understanding the biology underlying any disease. Unfortunately, the mouse model for Huntington's disease had a distinct limitation; the neurological symptoms of the human disease were difficult to observe in rodents. In this case, our differences from rodents were significant. For scientists to properly study Huntington's disease they were going to have to choose animals, such as monkeys, that are more similar to humans. Consequently the engineering of the first transgenic monkey, in 2001, was a major milestone. By 2008, transgenic monkeys had been designed with the human-disease-causing variant in their huntingtin genes, and these monkeys did exhibit many typical symptoms of Huntington's disease. The science of creating transgenic monkeys is still in its infancy, and the moral implications of changing monkey DNA make many people (including me) uncomfortable. That said, if research using transgenic monkeys might lead to successful treatment of this horrific and otherwise untreatable disease, it's hard to say that it shouldn't be done.

CAN HAMSTERS TEACH SCIENTISTS ABOUT SLEEP?

Animal experiments have also taught scientists about sleep, and they eventually led to *DEC2*. The story began in the late 1980s with the lucky discovery of the first gene linked to sleep patterns in mammals. Martin Ralph was a graduate student in the laboratory of Michael Menaker at the University of Oregon, studying sleep patterns in hamsters. Hamsters are excellent animals for studying sleep because when they're awake they're continuously running on their exercise wheels. So designing instruments to automatically monitor when they're awake isn't difficult.

Ralph and Menaker wanted to study the sleep patterns of animals that have neither sunlight nor artificial light to tell them when it's day or night. Their hamsters were kept in total darkness 24 hours a day. Remarkably, the Oregon researchers found that even in continuous darkness all the hamsters would go to sleep at essentially the same time every day and would also wake up at the same time. Well, almost all the hamsters. One odd hamster, we'll call him Fred, woke up two hours earlier every day. Rather than being on a 24-hour daily cycle, without sunlight to synchronize his internal biological clock, this hamster was living a 22-hour day.

Since the hamsters were all kept in the same environment and were from the same genetic strain, the scientists suspected that Fred had acquired some rare new mutation. To identify the mutation, the researchers began to inbreed Fred's offspring, systematically selecting hamsters with short daily cycles, and eventually obtaining hamsters that went to sleep once every 20 hours.

By performing cross-breeding experiments and counting how many offspring had daily cycles of 24, 22 and 20 hours, the scientists discovered that a single gene was causing Fred and his offspring to have their unusual daily sleep and wake cycles. The researchers determined that nearly all hamsters have two copies of a genetic variant that causes them to have a 24-hour internal clock. Fred, though, had one copy of a rare variant that caused his daily cycle to be 22 hours; his inbred descendants with two copies of the variant had a 20-hour daily cycle.

Although Ralph and Menaker demonstrated the existence of a body-clock setting gene, at the time, with neither the genome sequence of the hamster nor the technology to engineer transgenic hamsters, they were unable to determine the gene's DNA sequence. Not until 2000 did a team of scientists, including Ralph and Menaker, finally identify the genetic variant in a regulatory protein called CK1ε that caused Fred to have a 22-hour internal clock.

In the meantime, other scientists were searching for additional proteins that affect an animal's cycle of sleeping and waking, and by the early 2000s, they had identified approximately a dozen of them. Some were found by looking for proteins whose concentration in the brain increased and decreased synchronously every 24 hours. Others were found because they interacted biochemically with the synchronously varying proteins. So when the UCSF researchers decided to look for mutations that might explain why Mrs. Williams and her daughter were short sleepers, they tested all the genes that had been linked to sleep-wake cycling in animals. One of those genes was *DEC2*.

DOES *DEC2* REALLY REGULATE SLEEP?

Finding a mutation in a gene linked to daily-rhythm fluctuations definitely suggested that the *DEC2* mutation was responsible for Mrs. Williams' short sleeping. To be certain, the UCSF scientists needed a controlled experiment and the help of knockout and transgenic animals. Since techniques for engineering transgenic hamsters don't yet exist, the scientists used transgenic mice. Essentially, the strategy was to see if they could engineer mice with modified *DEC2* genes that were short sleepers.

Because mice don't run continuously on exercise wheels when awake, measuring how much time a mouse sleeps per day is more difficult in mice than in hamsters. Mice also have unusual sleeping pattern. Rather than sleeping for several hours at a time, mice sleep in multiple short stretches that might be described as catnaps. Although this makes measuring mouse sleep intervals more difficult, the scientists could monitor how much a mouse slept by making videos of the mouse's activity as well as by attaching electrodes to the mouse's brain and measuring changes in its brain waves. (The patterns of mouse brain waves, just like those of human brain waves, change when the animal is asleep.)

With a reliable way of measuring how much the mice were sleeping, the UCSF scientists were able to compare the sleep requirements of normal mice with ones with modified *DEC2* genes. They engineered *DEC2* knockout mice, as well as transgenic mice with an extra *DEC2* gene having the same proline-to-arginine substitution found in Mrs. Williams' DNA. The researchers found that the transgenic mice were active an hour more per day than either normal or *DEC2* knockout mice. The brain-wave recordings confirmed that the transgenic mice were indeed sleeping less. The scientists also engineered transgenic mice in which they added the proline-to-arginine *DEC2* gene but knocked out the normal *DEC2* genes. These mice were really short sleepers; they "catnapped" two and a half hours less per day than normal mice.

Not only do the *DEC2* transgenic mice sleep less than other mice, they also are more resilient to sleep disturbances. In a second set of experiments, the mice were prevented from sleeping during the first six hours of light (when mice really like to sleep). After the six hours of sleep deprivation, the mice were allowed to sleep and, unsurprisingly, both normal and transgenic mice slept more than they normally do – not unlike humans who sleep more after they've been up all night. But the *DEC2* transgenic mice needed less than half as much extra sleep to return to their normal activity levels than the other mice did.

Finally, the researchers studied sleep behavior in transgenic fruit flies (yes, flies do sleep). These flies also had their DNA modified to include either the normal mammalian *DEC2* gene or the proline-to-arginine version found in Mrs. Williams. And sure enough, the transgenic flies with the mutated *DEC2* gene slept approximately two hours less per day than the flies with the normal *DEC2* gene.

Precisely *how* DEC2 protein regulates our need for sleep still isn't clear. Other genetic variants may well contribute to short sleepers' ability to live well with little sleep, and new candidates for human sleep-regulating genes have recently been detected. The functions of these genes still need to be confirmed in mice, or by other means, before scientists are certain that these genes do regulate how much sleep we need. At that point, we may know why 1% to 3% of the population is able to do just fine with less sleep than the rest of us.

Of course, identifying genetic variants that enable some people to be short sleepers doesn't help the 97% of us who don't have these genetic variants and need eight hours of sleep. Nevertheless, just as discovering the *CCR5* genetic variant that gives a few people HIV immunity led to novel AIDS drugs that may ultimately help people without a *CCR5* mutation, learning how rare genetic variants enable a few individuals to live healthy lives on six hours of sleep may someday lead to drugs enabling the rest of us to enjoy more waking hours in our lives.

9

WHAT IS LOVE?

How much would you pay to discover your genetic predisposition to kindness or to marital fidelity? How much would you pay to find out your *partner's* predisposition to these traits (possibly useful information before deciding to tie the knot)? Back in 2008, Genesis Biolabs, a small company in Arizona thought an appropriate price was $99. For $99 and a sample of blood or saliva from which to extract DNA, Genesis would send you the results of a genetic test that it claimed would assess one's propensity for kindness and fidelity.

Deciphering the biological origins of human traits such as love or marital fidelity is not easy; our behavior is influenced by our individual experiences, something that can't be measured by a genetic test. Yet beneath our varied experiences, perhaps some interactions between DNA and protein, analogous to those which make certain people less susceptible to AIDS or more vulnerable to intellectual disability, might predispose some individuals to be more loving or kind. Learning how scientists are unraveling how genes affect our feelings of love and kindness is the central theme of this chapter and the next. Although ultimately our interest will concern human love and kindness, we'll also focus on love in animals. Hopefully by looking at the simpler worlds of animals we'll get clues about how genes affect the experience of love in humans as well.

LOVE AND COMMITMENT IN THE ANIMAL WORLD

To study "loving feelings" in animals requires knowing how to measure such feelings in an animal. How can one do that? We can't ask an animal to describe its emotions. We have to infer an animal's feelings from its behavior. For example, how long do mating animals stay together? How much time does a parent spend grooming or otherwise nurturing its young? Of course, there is always the danger of jumping from observations of behavior to projecting human emotions onto animals. Nevertheless, by observing the way an animal treats its mate and its young, we do learn about *some* characteristics of the animal that one might well argue resemble what we call love.

By observing animals, scientists have learned a great deal about loving behavior, and no animals have taught us more than small rodents called voles. Voles look a lot like mice, and in some places are even referred to as field mice or meadow mice. A variety of vole species exist; over 150 species are described in the scientific literature. In general, voles look like one another, and vole behavior among the various species is similar as well. Genetic tests indicate that, by and large, vole species have similar DNA sequences as well.

But vole species also have striking differences, especially regarding what we might describe as feelings of love. In most vole species, the males are not exactly model lovers. Male voles are generally promiscuous, and after mating, they lose interest in their partners and move on in search of their next romantic conquest. They are solitary creatures that rarely make long-term associations even with other male voles. And certainly, they aren't interested in wasting their time taking care of little vole pups.*

Not all voles behave this way, however. Some vole species, such as prairie voles, are very loving. Male and female prairie voles form long-term pair-bonds, and once pair-bonded, show little interest in new romantic partners. After pair bonding, a male prairie vole is aggressive toward other males that approach its partner, a form of behavior referred to in the scientific literature as "mate guarding" but that could also be described as jealousy. Both males and females take active roles in raising their pups. Altogether, a male prairie vole displays the sort of commitment to marriage and family that most human parents hope for in their daughters' mates. Figure 9.1 shows a montane vole (one of the promiscuous vole species) and a pair of prairie voles.

One might think that the differences in prairie and montane vole behavior are all "cultural," with each vole species raising its pups with the

---

* These are typical behaviors of males of many species, which may not surprise observers of human male behavior.

"values" of the parents. This hypothesis is unlikely to be correct, though, since it can't explain why prairie and montane voles behave differently immediately after birth. A prairie vole pup placed in isolation a few days after birth is clearly distressed. The pup will call out using the prairie vole's ultrasonic vocalization, and will have elevated concentrations of stress hormones in its blood. In contrast, a newborn montane vole placed in isolation does not call out, nor do its stress hormone levels go up. Newborn montane voles seem quite content when left alone.

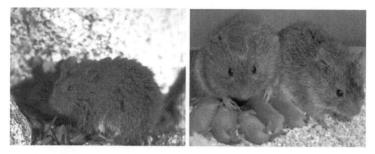

Figure 9.1. Montane vole (left) and prairie voles (right).

RODENT LOVE, OXYTOCIN, AND VASOPRESSIN

As montane voles and prairie voles already differ in social behavior at birth, their differences probably reflect innate biological differences. Since their behavioral differences involve pair-bonding and pup rearing, hormones related to sex and reproduction might contribute to their different behavior. These observations led scientists to focus on two such hormones, oxytocin and vasopressin, in their efforts to understand the differences in montane voles and prairie voles.

Oxytocin and vasopressin are closely related proteins, each consisting of just nine amino acids, of which seven are identical. Most mammals use them as signaling molecules, and they play essential roles in giving birth and maternal feeding. In experimental animals, oxytocin and vasopressin have also been shown to influence parenting behavior, and presumably even parental feelings. Consider, for example, the (female) laboratory rat. If the rat has pups, she will exhibit all the maternal nurturing behavior toward her pups that one expects from a mother animal. In contrast, an adult female virgin rat isn't affectionate to baby rats at all; she'll avoid or even attack them. If a virgin female rat is first given estrogen, though, and is then injected with oxytocin, her behavior undergoes a striking change. Suddenly the virgin rat becomes nurturing and takes care of unrelated pups as if they were her own. Similarly, the *lack* of oxytocin can prevent "mother love" in rats. For example, if mother rats are given a drug that blocks the brain's oxytocin receptors, they stop displaying maternal behavior.

Oxytocin induces similar behavior in mice as in rats. To test oxytocin's influence on mice, scientists engineered mice in which either the gene for the oxytocin receptor or the gene for oxytocin itself was knocked out. Female oxytocin knockout mice were able to give birth, and they exhibited normal maternal care of their young. Presumably, their brains were able to recruit vasopressin or another related hormone to replace oxytocin as a signaling molecule. In contrast, the female mice lacking oxytocin receptors didn't nurture or care for their pups. Without oxytocin receptors, the mice were unable to express motherly love to their pups.

Like oxytocin, vasopressin influences social behavior in rodents. In this case, the evidence comes from voles. As we've learned, after mating, male prairie voles are devoted partners. Despite this fact (or perhaps because of it), male prairie voles take a while before selecting a female partner. But inject a male prairie vole with vasopressin, and it's as if he's been shot by Cupid's arrow. The next time he comes in contact with a female, the vole will act as if he has already bonded with her, even if they have never mated. Apparently, for male prairie voles, a shot of vasopressin has as big an impact on the psyche as sex. Not so for the montane vole. When male montane voles are injected with vasopressin, nothing happens. Even after a hit of vasopressin, the male montane vole won't "commit" to a female and continues to play the field. It seems that, at least for montane voles, once a Romeo, always a Romeo...

VOLE LOVE AND VOLE GENETICS

Hopefully by now you're convinced that, in the laboratory, oxytocin or vasopressin can change animal pair-bonding, parental nurturing and other loving behavior. But what about animals in the wild? Why are male prairie voles such devoted mates and caring parents, while male montane voles are the opposite? One might guess that the amino acid compositions of oxytocin or vasopressin differ between the species, but they are identical. One might also suspect that the oxytocin or vasopressin receptor proteins differ between the species. In fact, the montane vole and prairie vole receptor proteins do differ, but the differences are minimal and don't change the functioning of the receptors. In other words, montane vole and prairie vole receptor proteins are essentially interchangeable.

So what's the difference? The answer is in the DNA regulatory region of one of the vasopressin-receptor genes, *AVPR1A*. The regulatory region in the prairie vole *AVPR1A* gene has an extra piece of DNA that is missing in montane vole DNA. This extra DNA, which varies in length by as much as 140 base pairs among individual prairie voles and is approximately 400 base pairs long, changes the amount of AVPR1A protein a cell produces. The *AVPR1A* gene regulatory regions are illustrated schematically in figure 9.2. Because of the difference in regulatory regions, prairie vole cells produce almost twice as much *AVPR1A* messenger RNA and AVPR1A protein as the montane vole cells. This additional AVPR1A receptor protein makes all the difference in social behavior.

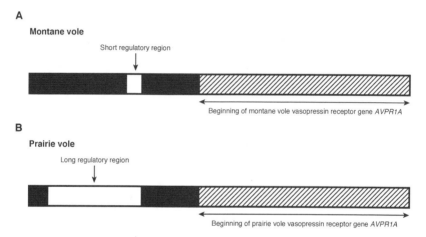

A

Montane vole

Short regulatory region

Beginning of montane vole vasopressin receptor gene *AVPR1A*

B

Prairie vole

Long regulatory region

Beginning of prairie vole vasopressin receptor gene *AVPR1A*

Figure 9.2. Vasopressin-receptor genes in voles. Simple schematic illustrating the section of the montane and prairie vole chromosomes near the beginning of the vasopressin-receptor gene. The only difference is the insertion of a large section of DNA (approximately 400 base pairs long), in the prairie vole regulatory region compared to the montane vole regulatory region.

To prove this, scientists engineered transgenic mice in which the mouse *AVPR1A* gene was replaced with the prairie vole *AVPR1A* gene. (They used mice because mice are easier to modify genetically than voles and display non-monogamous behavior similar to that of montane voles.) The impact of the *AVPR1A* gene transfer on the behavior of the mice was clear. The male transgenic mice with prairie vole vasopressin-receptor genes were "perfect gentlemen" – no more playing the field for these mice. Instead they displayed committed social behavior. In other words, their social behavior was that of male prairie voles.

Repeating this gene-swapping test between prairie voles and montane or meadow voles would be interesting. Unfortunately, scientists are not yet able to genetically manipulate vole DNA as they do mouse DNA. What is feasible is to simply add DNA into vole brain cells. So in 1999, scientists inserted DNA from the prairie vole *AVPR1A* gene into the brain cells of meadow voles (another promiscuous vole species). As with the transgenic mice, the meadow voles exhibited strikingly different behavior after the addition of the prairie vole *AVPR1A* gene. The normally promiscuous meadow voles changed their ways and displayed more committed, prairie-vole-like behavior.

Even more dramatic vole experiments didn't require any genetic manipulation at all. These experiments simply exploited natural genetic variations among prairie voles. As mentioned earlier, the extra piece of DNA that appears in the regulatory region of the prairie vole *AVPR1A* gene varies by as much as 140 base pairs among prairie voles. Since the promiscuous meadow voles and montane voles don't have this inserted region at all, scientists hypothesized that prairie voles with shorter insertions might behave more like meadow or montane voles than prairie voles with the longer *AVPR1A* insertions.

To test this idea, scientists compared two groups of prairie voles; one with two copies of the *AVPR1A* gene with long insertions, the other with two *AVPR1A* copies with short insertions. The two vole groups were created by separately breeding voles with either the longest or the shortest insertions.* The researchers found that prairie voles with shorter *AVPR1A* insertions had lower concentrations of vasopressin receptors in their brains. Remarkably, the social behavior of these short-insertion prairie voles, in terms of monogamy and pair-bonding, was also similar to that of the montane vole "playboys," which lack the insertion altogether. In other words, genetic variation in the vasopressin-receptor gene not only leads to behavioral differences between different vole species but also among different members of the same species.

Proteins that interact with oxytocin or vasopressin can also influence nurturing and loving behavior in animals. One example is CD38, a cell-membrane protein that functions both as a neurotransmitter receptor and as an enzyme. Although CD38 has been primarily studied for its role in the immune system, recent experiments show that CD38 also influences the

---

* This didn't take long since the gestation period for voles is only three weeks.

brain. In particular, CD38 plays a role in regulating oxytocin. In a series of experiments in 2007, scientists engineered knockout mice lacking functional CD38 protein and observed that the mice lacking the *CD38* gene had only half as much oxytocin in their blood plasma as normal mice. The *CD38* knockout mice also displayed behavior similar to that of oxytocin-receptor knockout mice. Female *CD38* knockout mice did not display maternal nurturing toward their pups. Male *CD38* knockout mice were less sociable than normal male mice. In one experiment, *CD38* knockout mice displayed the mouse equivalent of the all-too-common human experience of not recognizing someone they have met before, what psychologists call "social amnesia." While most mice respond differently to mice they have encountered previously (in terms of sniffing and other social interactions), the *CD38* knockout mice did not. Even though their sense of smell was just fine, the *CD38* knockout mice reacted toward all other mice as if they were strangers.

OXYTOCIN AND AUTISM

The observation that oxytocin and vasopressin signaling affects pair-bonding and nurturing behavior in rodents raises the question whether they also influence human social behavior. Although it's worth reiterating that discoveries in mice or voles don't guarantee that similar biological mechanisms exist in people, in the case of the oxytocin/vasopressin network other evidence suggested that the animal experiments might be relevant to humans. For one thing, oxytocin and vasopressin are involved in the physical changes and processes that accompany childbirth and motherhood, such as dilation of the cervix during childbirth and the production of milk for breast-feeding. So perhaps the feelings, such as bonding, nurturing and even love, that accompany becoming a parent might also have roots in these biological pathways.

How can scientists study feelings of love in people? One approach is to examine situations in which those feelings are impaired, for example, in autism. Autism is a complex disease with a wide range of symptoms, but one striking characteristic is difficulty expressing and receiving affection. Science's understanding of autism is limited, but evidence exists that impaired oxytocin signaling may contribute to autism. Autistic individuals often have lower blood oxytocin levels than unaffected people, and administering oxytocin can diminish autistic symptoms. For example, in one experiment, 13 young adults diagnosed with high-functioning autism were shown images of human faces on a computer monitor. This test was performed twice, once after they had been given oxytocin in the form of a nasal spray and once after they had received a placebo. The researchers found that the autism patients spent more time gazing* at the facial region around the eyes (as people without autism do) after they had been given oxytocin than after they had been given a placebo. In a second experiment, the same patients were asked to distinguish between "good" (generous or kind) and "bad" cartoon characters in a computer game. The patients were better able to distinguish between the good characters and the bad ones after intranasal oxytocin administration – or, at least their answers became more similar to those of people who don't have autism.

On the basis of such tests, a few physicians have begun to treat autistic children with intranasal oxytocin. So far, only a couple of such case studies have been reported in the medical literature, but the initial results have been positive. In 2012, doctors at the University of Fukui, in Japan, gave a 16-year-old autistic girl oxytocin over a period of two months. The girl was described as having relatively high functioning autism but showing "marked impairment in social interactions." She had not responded positively to drugs or other medical interventions. After two months of regular oxytocin administration, the girl was able to "express gratitude" for the first time. She also "listened to conversations" and was able to calm down more quickly after a fit of anger. These results were encouraging enough that the Japanese team is now carrying out a clinical trial of long-term oxytocin treatment on a larger group of autistic children.

Since oxytocin signaling has been linked to autism, perhaps DNA variants in oxytocin-pathway genes might be at the root of some cases of autism. Variants in 25 genes have already been provisionally linked to autism susceptibility.† A few people with autism have a deletion on chromosome 3, which includes the oxytocin- receptor gene. Other autistic individuals have oxytocin-receptor genes chemically modified in a way that prevents them from producing oxytocin-receptor protein. Because of these findings, scientists have been searching for more common genetic variants that might be linked to autism. Families with autistic members have been tested for length variants in the genetic region close to the *AVPR1A* gene (that is, for

---

* Eye gazing is measured by specialized electronic and optical instruments that nonintrusively record a person's eye movements.

† That autism can be caused by a genetic mutation may seem surprising, since autism rarely runs in families and people with autism often have difficulty finding marriage partners. In fact, in those cases in which mutations have been linked to autism, the mutations have typically not been inherited but rather are de novo mutations occurring during the production of the egg or sperm cells.

genetic variants similar to those found in the DNA of voles). Though the results haven't been completely consistent, the evidence has suggested that, in some cases, autism may result from vasopressin-receptor gene variations similar to those that distinguish the monogamous prairie voles from the promiscuous montane voles.

OXYTOCIN AND LOVING FEELINGS IN HEALTHY PEOPLE

The possible link between impaired oxytocin signaling and autism suggests that oxytocin might also affect the ability of normal people to establish and maintain loving relationships. This hypothesis has been supported by experiments showing that, when given intranasal oxytocin, healthy people also spend more time looking at the eyes in images of human faces, suggesting that oxytocin signaling may facilitate making eye contact in nonautistic people as well.

There is even evidence suggesting that oxytocin levels reflect the intensity of everyday romantic relationships. Experiments carried out in 2012 by researchers from Bar-Ilan University, in Israel, compared blood oxytocin levels between single adults and individuals who had recently started a romantic relationship. The researchers found that, on average, the lovers had oxytocin levels almost twice as high as the singles.

When the Bar-Ilan team checked the relationship status of the young lovers six months later, they found that the average oxytocin levels of the lovers who had remained together were almost 20% higher than the levels of couples that had separated. Admittedly there was considerable variability within the data; some couples with lower oxytocin levels were still together, while others with higher oxytocin blood concentrations had split up. But, at least in this experiment, blood oxytocin levels could predict the stability of a romantic relationship better than would be expected by chance. Since the Bar-Ilan researchers hadn't measured the lovers' oxytocin levels before they fell in love, they couldn't determine whether romantic love increases oxytocin levels or whether some people naturally have higher oxytocin levels and that such people (for some reason) are more likely to have enduring relationships. That said, the experiment again suggested that oxytocin is involved in human romantic relationships.

OXYTOCIN AND COMMUNICATION

Further clues pointing to oxytocin's role in human romantic love came from experiments in which oxytocin was given to people in long-term relationships. One experiment tested 47 couples who were having difficulties communicating. Each couple was asked to name a few topics that they consistently disagreed about, such as finances or how to spend their leisure time. The researchers then asked the couple to talk about one of the topics* during a 10-minute videotaped session. After the session, the subjects' level of cortisol (a stress hormone) was measured from saliva samples, and the participants were asked whether they thought their conversation had been more or less stressful than usual. A psychologist also reviewed the videotaped session and independently rated whether the participants' communication appeared supportive or hostile on the basis of, for example, the number of sarcastic or kind remarks.

The interesting part of the experiment was that half of the couples had been given doses of intranasal oxytocin before the session, while the other couples received an inert nasal spray. Neither the participants nor the evaluating psychologists knew which couples had been given oxytocin and which had not. Yet both the reviewing psychologists and the subjects themselves reported that the couples that had received oxytocin communicated "more positively" than those who had been given the placebo. The participants who had received oxytocin had lower cortisol levels, as well, also suggesting that the sessions had been less stressful for them. Although the differences in communication ratings and stress hormone levels were modest, the consistent trend of the results was intriguing.

OXYTOCIN AND HUMAN PAIR-BONDING

Another experiment on oxytocin's effects on romantic relationships was even more provocative. In this study, the research subjects were 57 men, some of whom were single while others were in stable monogamous heterosexual relationships. Each subject was brought into a room with an attractive young woman who was a member of the experimental team. The experiment began with the subject and the experimenter facing each other, approximately 6 feet apart. The experimenter would then ask the subject to walk toward her until he felt that the distance between them was "ideal." At this point the distance between them was measured.

Half of the single men and half of the men in couples had been given a dose of oxytocin before the experiment started, while the others received a placebo. Neither the subjects nor the experimenter knew who had been

* The researchers specifically avoided selecting highly charged topics to avoid having the sessions turn into explosive confrontations.

given oxytocin. The single men on average chose a distance of a little less than 2 feet as ideal, whether or not they had been given oxytocin. The men in long-term relationships who had not been given oxytocin also chose an average ideal distance that was a just under 2 feet. In contrast, the men in long-term relationships who had also been given oxytocin selected ideal distances from the attractive experimenter that were approximately 5 inches farther away. In other words, oxytocin appeared to restrain the men in long-term relationships from wanting to get too close to the pretty young woman.

Taken at face value, the results seemed to suggest that oxytocin reinforces pair-bonding behavior in men in long-term relationships, just as it does in prairie voles. These experimental results do need to be replicated. But if future experiments confirm these findings, then perhaps male government leaders (like former U.S. President Bill Clinton and former Italian Prime Minister Silvio Berlusconi) might be well advised to take oxytocin if they want to avoid having romantic temptations disrupt their personal and professional lives.

AVPR1A GENE VARIANTS AND HUMAN PAIR-BONDING

If oxytocin influences a person's romantic life, then perhaps variations in oxytocin-pathway genes might explain why some people are more "lucky at love" than others. After all, genetic variation in the *AVPR1A* gene affects pair-bonding in voles. Although people don't have the same variations in length of the *AVPR1A* regulatory region as voles, we do have other *AVPR1A* genetic variants: 16 different length variants exist in the human *AVPR1A* gene, all of which are relatively common. A postmortem study of human brains also showed that the concentration of *AVPR1A* mRNA (and most likely, therefore, of AVPR1A protein as well) depends on which *AVPR1A* gene variants one has, just as in voles. And since different rates of the AVPR1A protein synthesis lead to different behavior in prairie voles, perhaps a similar connection existed in people.

These observations led a team of Swedish scientists to test whether variations in the *AVPR1A* regulatory region might be linked to differences in human pair-bonding.[*] In their widely cited study, published in 2008, approximately 1000 human volunteers filled out questionnaires assessing their "pair-bonding status." Participants were asked their marital status, whether they had ever been unfaithful to their partners, and how many

---

[*] Other studies have attempted to link vasopressin-receptor gene variants to such traits as trust, compassion and altruism.

sexual partners they'd had. Saliva samples were taken from each of the volunteers, and their DNA was screened for variants in the chromosomal region that includes the *AVPR1A* gene. Although the Swedish study reported "statistically significant" correlations between self-reported pair-bonding history and *AVPR1A* gene variants in men (but not in women), its findings were modest. Only 1 of the 16 *AVPR1A* variants was significantly associated with pair-bonding scores. Even in this case, the effect was small, and the variant associated with poorer pair-bonding was not the shortest one (as was the case with prairie voles), but an intermediate length variant.

LOVE AND NURTURING BEYOND THE OXYTOCIN PATHWAY: PHEROMONES AND *MHC* GENES

Oxytocin signaling is not the only brain pathway that plays an important role in animal love. Signaling triggered by the sense of smell also contributes to animal nurturing, mating and sexual behavior. This signaling pathway is mediated by small molecules called pheromones, and one of the key pheromone receptor proteins is TRPC2. As a result, TRPC2 is a key regulator of animal behavior. For example, male *TRPC2* knockout mice are as likely to court and mount males as females. Virgin male *TRPC2* knockout mice display caring behavior to pups, while normal virgin adult male mice are more likely to kill any pups they find. Meanwhile, female *TRPC2* knockout mice spend less time nursing their pups and protecting them from outsiders than normal mother mice.

Whether humans use pheromones and smell-based signaling is much more controversial. The idea that people use pheromones was first proposed by Martha McClintock in 1971. McClintock, at the time an undergraduate student at Wellesley College, observed that the women in her dormitory were having their menstrual periods at the same time of the month. McClintock hypothesized that menstruating women secrete pheromones in their sweat, triggering the onset of menstruation in other women. Several research teams attempted to reproduce McClintock's findings. Some claimed success; others could not replicate her results and criticized her methodology. Forty years later, whether humans use pheromones and whether they regulate menstrual cycles remain debated scientific questions. At the least, we do know that if humans use pheromones, our mechanisms for pheromone detection must be different from those used by other animals. The human *TRPC2* gene has mutations that make it nonfunctional; it's a relic of a once-functional gene, just as the human appendix is a relic of a once-functional organ.

Although the relevance of pheromones to menstrual timing is still debated, there is other evidence, albeit also controversial, suggesting that pheromones affect human sexual and romantic behavior. The data comes from the laboratory of Claus Wedekind at University of Bern, in Switzerland. Wedekind's work was based on mouse experiments that had shown that mice choose mating partners by scent. Based on smell, they select partners with variants of certain genes (called *MHC* genes) that were *different* from their own. Since humans also have highly variable *MHC* genes, Wedekind speculated that human mate selection might also be influenced by *MHC* gene variants. To test this hypothesis, Wedekind's team recruited 44 men, each of whom wore the same T-shirt for two nights. On the following day, female participants smelled the shirts and indicated which of the sweaty T-shirts smelled the most "pleasant" to them. (Honest – I couldn't make this up even if I wanted to.) The researchers reported that women* gave the "most pleasant smell rating" to T-shirts worn by men with the most *dissimilar MHC* variants. Two years later, Wedekind's team repeated their experiments with essentially the same results, showing that at least for this research team, the results were reproducible.

Do *MHC* variants also affect the way people choose partners in real life? To find out, scientists have measured whether couples in long-term relationships have more dissimilar *MHC* variants than randomly selected pairs of opposite-sex individuals, and a 1997 study by researchers from the University of Chicago claimed to have found just such an association. However, attempts to replicate these findings have been less successful. Although two additional studies reported correlations between "smell-test" ratings and *MHC* gene dissimilarity, two others didn't find any. Only one out of nine investigations searching for elevated *MHC* gene dissimilarity within long-term couples reported a link that couldn't be explained by chance. Nor could anyone explain just how *MHC* variants influenced human body odor. It's true that MHC proteins affect the smell of urine. And since mice are known to deliberately smell each another's urine, that *MHC* variation might influence mate-selection in mice is not surprising. On the other hand, people don't generally communicate by smelling each other's urine (at least not consciously), so just how MHC proteins might affect human partner choice remains unclear.

Despite the mixed results, the search for links between human *MHC* variants and romance continued. In 2006, a research team from the

---

* Actually, this was true only of the women who were not taking oral contraceptives. The researchers hypothesized that birth control pills somehow interfere with women's pheromone-sensing pathways.

University of New Mexico reported that a couple's *MHC* variants might influence their romantic *success*, if not their romantic choices. The New Mexico team recruited 48 young heterosexual couples who were married, living together, or dating exclusively. Both members of each couple filled out questionnaires intended to assess their relationship satisfaction. The team also determined three major *MHC* variants from each participant's DNA and tested whether the similarity of a couple's *MHC* variants was correlated with their relationship satisfaction. Remarkably, the team found that "as the proportion of *MHC* alleles couples shared increased, women's* sexual responsiveness to their partners decreased, their number of [other] sexual partners increased, and their attraction to men other than their primary partners increased." In other words, the results suggested that even if people don't pick their partners on the basis of *MHC* variant dissimilarity, perhaps they *should*.

GENETIC VARIATION AND THE DATING GAME

To date, no one has replicated the University of New Mexico *MHC*–sexual-satisfaction experiment, and without independent replication, proving that its dramatic results weren't caused by chance is difficult. Nevertheless, the results were so striking that they have already influenced the world of 21st-century matchmaking. With so many people willing to pay for information to lead them to their perfect mate, several companies have proposed to help people find their "match" with genetic testing. We've already met one of these companies, Genesis Biolabs, with its "marriage worthiness" test. The Genesis Biolabs test was inspired by the studies of *AVPR1A* gene variants in voles and the Swedish study linking *AVPR1A* variants and human pair-bonding behavior. Apparently the results of the human studies didn't succeed in convincing the public, though, and as of 2012, Genesis Biolabs was no longer in business.

Other companies extolling better romance through genetic testing have had more commercial success. Two of them, ScientificMatch.com and GenePartner.com, base their services on the human *MHC* experiments described above. They will test your *MHC* variants from a saliva sample and use the results to help select potential partners for you. Another company, Switzerland-based BasisNote.com, skips the *MHC* testing step and simply sends you a set of odors to sniff. After you identify which odors

---

* No correlation between male sexual satisfaction or responsiveness and MHC variation was found in study.

you liked best, the company searches its database for "smell-compatible" partners. Meanwhile, Chemistry.com, a division of Match.com, the large Internet dating service, takes a more comprehensive approach. Chemistry. com doesn't do any DNA testing or smell analysis but instead sends you a questionnaire. Your answers supposedly reflect which receptor variants you have for various hormones and neurotransmitters, including testosterone, dopamine, serotonin and noradrenaline. Chemistry.com then uses that information to recommend compatible matches from its database.

The existence of all these companies shouldn't be construed as evidence that DNA-based matchmaking is actually useful. In the words of Larry Young of Emory University: "I do not believe that any service that claims to use genetic information, or any estimation of neurochemistry (based on personality or genotype) has any basis in reality." Professor Young is as likely to know as anyone; it was his laboratory that carried out the key experiments on *AVPR1A* variants and pair-bonding in voles.

Nonetheless, as our understanding of the biological basis of love and pair-bonding grows, we may yet find that oxytocin, vasopressin, pheromones and their receptors do play central roles in human feelings of love. Science may soon also discover other proteins that contribute to the biological origins of loving behavior in people. Finding these proteins is likely to again come from studying animals, though this time the animals will probably be not just rodents but also dogs, wolves and foxes. And it's to these studies that we'll turn in the next chapter.

10

WHAT IS KINDNESS?

Kindness. Compassion. Gentleness. Devotion. Such "prosocial traits," as the psychologists call them, represent much of what we most value in a mate, a friend, and in people in general. They are characteristics that make someone truly special. Indeed some religious persuasions believe that these traits ensure a path to heavenly bliss, or even to becoming a saint.

The notion that prosocial traits may be influenced by our genes is a difficult or even heretical one for many people. Yet, as we've seen in the last chapter, traits such as fidelity and parental nurturing are influenced by DNA and proteins in animals, and appear to affect human loving behavior as well. In this chapter we will consider what science has discovered about the impact of genes on kindness and compassion. The focus will be primarily on animals, since animals are much easier to study than people, but since similar genes are often found in animals and people, what we learn from animals may well be relevant to understanding human behavior as well.

PROSOCIAL TRAITS AND DOGS

The idea that one could learn about kindness or compassion by studying animals might seem strange. Indeed, the fact that we often call them "personality" traits reflects the extent that we associate kindness and compassion with people. Yet any dog lover will tell you that you can't find a better teacher of gentleness or kindness than a dog. Or, in the words of one wag, "If you really want to know who loves you more, your spouse or your dog, just lock them both in the trunk of your car for a few hours. Then open the trunk and see which one's happier to see you."

Dogs, or at least the study of dogs, can help us learn about kindness, devotion and other prosocial traits because they have been genetically bred to display these traits. In other words, dogs are the result of an extended genetic "experiment" carried out by humans to artificially select the very personality traits that we value in them. Geneticists also study dogs because dogs are a relatively young species. Most scientists estimate that humans began breeding wolves for gentleness and tameness 15,000 to 30,000 years ago. In contrast, humans are believed to have diverged from chimpanzees (our closest living evolutionary relatives) four to nine million years ago.* This recent emergence of dogs is important because fewer DNA changes between dogs and wolves have had time to accumulate. Since five million years is 200 times as long as 25,000 years, one can estimate that there will be approximately 200 times as many random differences between the human and chimp genomes as between the dog and wolf genomes. Consequently, important genetic differences between dogs and wolves, including those that led dogs, but not wolves, to be gentle around people, should be easier to find than differences between humans and chimps.†

Identifying the critical genetic changes between dogs and wolves requires knowing the DNA sequences of both species. In the case of the dog genome, this task was completed in 2005 when the first genome of the dog (or more precisely of one dog, Tasha, a female boxer) was sequenced. Although the genetic sequence of the wolf is not yet available (complete genome sequencing currently still costs many thousands of dollars), comparing wolf and dog DNA at multiple genetic locations is already possible.

---

* How do scientists estimate how long ago two species, such as humans and chimps, diverged from a common ancestor? One method is based on counting differences in the genomes of the two species and estimating how many genetic changes occur every generation. Species divergence is also estimated by nongenetic techniques, such as by carbon dating of bone remnants and by geological analysis.
† This comparison between primate and dog genetic differences is an overestimate because mutations typically occur in the production of the egg or sperm cells that become the next generation, and the approximately 20 to 25 year human generation is roughly 10 times the 2-3 year dog/wolf generation time. Even taking this into account, finding important variations between wolves and dogs should be at least 20 times easier than finding them between humans and chimps.

POSITIVE SELECTION AND DOG BEHAVIOR

Shortly after Tasha's sequence was obtained in 2005, researchers began to compare differences between wolf and dog at multiple genetic locations. The primary method they used was to look for signs of positive selection. Positive selection is the biological phenomenon that occurs when either a mutation or a change in the environment suddenly enables someone with an uncommon genetic variant to have more offspring than individuals with more common variants. When this happens, the previously rare variant becomes widespread within the population over a comparatively short time; in scientific parlance, the variant is positively selected.

Figure 10.1 illustrates how scientists search for genetic variants that have been positively selected. Figure 10.1A shows a schematic representation of the DNA from a single region of one chromosome from eight individuals in a population. Applying this illustration to our example of the breeding of wolves into dogs, we'll assume that this chromosome sampling was taken from wolves before the people began breeding them to become dogs. We see that along this chromosome (as elsewhere in the genome) all the individuals have mostly the same DNA bases, though occasionally there are locations where the wolves' DNA varies. These differences are called neutral variants, or neutral mutations, because they neither make the wolf more or less likely to survive and produce offspring; these neutral variants are indicated in the figure by small rectangles. Let's also assume that one genetic variant in this region (indicated schematically by the circle) makes a wolf gentler. Since this original chromosome sampling was made before human breeding of wolves began, this variant has occurred on one chromosome just by chance.

Figure 10.1B depicts the results of performing a similar DNA test of the same chromosomal region 2000 years later. The wolf population has become a dog population, and those initially rare individuals with the critical circle variant that makes them gentler have produced more offspring than those that lacking mutation. As a result, after this time period the entire dog population consists of descendants of wolves with the previously rare variant.

In Figure 10.1C, another 3000 years have passed. Although after this much time recombination events and mutations have randomly created some neutral variations near the critical circle variant, by and large most dogs' DNA in the region will still look similar to one another. In contrast, in the rest of the genome, because of mutations and recombination events that have accumulated over *millions* of years, the distribution of genetic variants will still resemble Figure 10.1A. In other words, scientists can detect positive selection by finding genomic locations where dogs all have the same variants whereas wolf populations have a diverse set of variants.

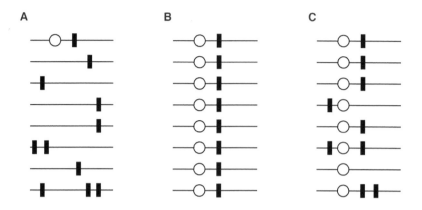

Figure 10.1. How positive selection of a genetic variant causing an advantageous trait leaves a DNA trace. A: DNA from a single chromosomal region from eight individuals in a population. The circle in the topmost chromosome indicates an advantageous new mutation. B: The same region 2000 years later. The advantageous variant, along with its nearby variants, has now spread to the entire population. C: After another 3000 years, a few new random mutations can be found in the region, but most individuals still have not only the circle mutation but also the neighboring variants that were located next to the mutation when it occurred 5000 years earlier.

Even with the great similarity between the wolf and dog genomes, detecting positive selection is challenging: thousands of differences exist within the approximately two and a half billion base pair genomes of the two species. These differences need to be sorted out from all the genetic variations among different dogs or different wolves. Despite these challenges, some intriguing hints are emerging from the DNA comparisons of the wolf and dog genomes. Although the vast majority of the DNA sequences of the two species are extremely similar, a few genetic regions have been detected where the wolf and dog DNA are consistently different in a manner suggesting positive selection. Two of these regions are near genes that, on the basis of studies in humans and mice, appear to be involved in memory or behavioral development. Another genetic region showing strong signs of positive selection is especially intriguing, because it is close to the dog equivalent of what is called the Williams Beuren syndrome region, or WBS region, in the human genome.

FRIENDLINESS AND WILLIAMS BEUREN SYNDROME

In humans, the WBS region is located on chromosome 7 and includes approximately 27 genes. The region gets its name because it is deleted* in the DNA of approximately one in 10,000 people, which leads to a medical condition known as Williams Beuren syndrome, or Williams syndrome for short. Individuals with Williams syndrome tend to be short compared to other members of their families, they often have cardiac problems, and they generally have an unusual facial appearance with a small upturned nose and wide lips. People with Williams syndrome also have unusual mental traits: although intellectually disabled in some tasks, they are often adept at verbal skills and occasionally are talented musically as well.

Perhaps the most striking trait of people with Williams syndrome is their unusual friendliness, even toward strangers. People with WBS can be so friendly and trusting around strangers that one is tempted to say to them, as to a young child, "Now be careful and don't talk to strangers!" Indeed, the behavior of individuals with Williams syndrome around strangers is so reminiscent of the gentle, indiscriminately friendly behavior of a dog that some scientists wondered whether this prosocial behavior might have a common biological origin in people and dogs.

Identifying the biological origins of the extreme friendliness in Williams syndrome is difficult, in part because so many genes are deleted. Unraveling which genes are involved in social behavior and which cause the other WBS traits, such as short stature or heart problems, is challenging. Recently, though, scientists have begun to make progress by using our friend, the knockout mouse.

That we can learn so much about Williams syndrome from the mouse is a consequence of evolution. The theory of evolution predicts that mouse DNA should not only contain similar genes to the human WBS genes but that those mouse genes should be located near one another on a single chromosome, just as they are in humans. This prediction of the locations of the mouse equivalents of the WBS genes (like most predictions of the theory of evolution) turns out to be correct. Consequently, scientists can engineer mice that lack a single chromosomal region (which happens to be on mouse chromosome 5) containing the mouse equivalents of the missing human genes in Williams syndrome. The striking result is that these mice exhibit symptoms similar to those of people with WBS. For example, the

---

* More precisely, people with WBS have this chromosomal region deleted in one of their two chromosome 7s.

mice have facial characteristics reminiscent of people with WBS and often have cardiovascular problems. The WBS-region knockout mice are also unusually friendly.*

WBS-region knockout mice have given scientists an animal model that mimics the symptoms of Williams syndrome, helping them understand and develop treatments for WBS in humans. Equally valuable has been the ability to engineer mouse strains in which only one, or a few, of the 27 WBS genes are disabled, thereby letting researchers study which WBS region genes influence specific traits. One of the 27 WBS-region genes is called *Gtf2ird1*. Although the precise functions of the Gtf2ird1 protein still aren't completely understood, Gtf2ird1 is known to influence cell-signaling pathways in the brain. Consequently, Gtf2ird1 might well play a role in our emotions and behavior. In fact, a mouse with just its *Gtf2ird1* gene knocked out displays the same "friendly" behavior as mice missing the entire WBS chromosomal region. At least in mice, the friendliness associated with Williams syndrome appears to be linked to a single gene, *Gtf2ird1*.

Gtf2ird1 may play a similar role in affecting human emotions. In 2006, researchers from New York's Mount Sinai Hospital and Costa Rica's National Children's Hospital reported on a young girl we'll call Gloria, who was brought to an autism clinic in San José, Costa Rica. Gloria was unusual because, along with autistic symptoms, she also had the indiscriminate friendliness typical of Williams syndrome. When Gloria's DNA was analyzed, the doctors found that she had a large deletion in one copy of her chromosome 7 DNA. Gloria's chromosome 7 deletion wasn't the one generally found in Williams Syndrome. As shown in Figure 10.2A, Gloria's deletion overlapped only a small portion of the usual WBS region, and included just two WBS-region genes, *Gtf2ird1* and *GTF2I*. Gloria's deletion also included 14 other genes outside the WBS region. Two of these missing genes, *HIP1* and *YWHAG*, are associated with brain signaling and are believed to be the source of Gloria's autism. More relevant to our discussion is that Gloria displayed Williams syndrome friendliness even though only two of her WBS genes were missing.

---

* How do scientists determine that a mouse is "unusually friendly"? One method is to isolate a mouse in a cage and later introduce a second mouse into the enclosure. Typically the first mouse won't interact with the newly introduced mouse, and if it does, the first mouse is likely to be aggressive. In contrast, when a second mouse is added to the cage of a mouse with a deleted WBS region, the WBS-deleted mouse is less aggressive and interacts in a "mouse-friendly" manner (for example, by following or sniffing the second mouse).

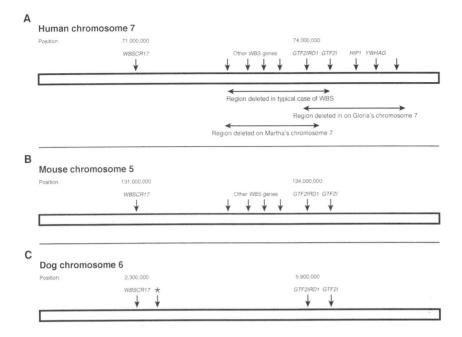

Figure 10.2. Williams Beuren syndrome region of chromosome 7. A: The locations of genes of the WBS region. A few important WBS genes are annotated (for simplicity most are not). The dotted lines with double-headed arrows indicate the regions on chromosome 7 typically deleted in individuals with Williams syndrome as well as in Gloria's and Martha's genomes. B: Region on mouse chromosome 5 corresponding to the WBS region on human chromosome 7. C: The corresponding region on dog chromosome 6. The chromosomal region showing signs of positive selection in the dog genome is located adjacent to the *WBSCR17* gene. Although this genetic location is three and a half million base pairs away from *Gtf2ird1* and *GTF2I*, it may still affect the regulation of the *Gtf2ird1* and *GTF2I* genes.

Before jumping too strongly to a conclusion, it's worth remembering that generalizing from a single patient is notoriously risky. In fact, in 2008, a research team led by doctors from the University of Utah identified another young girl – referred to in their paper as Subject 5889 but whom we'll call Martha – with a different rare DNA deletion in the WBS genetic region. As illustrated in Figure 10.2A, Martha's chromosome 7 deletion included nearly the entire WBS region, including *Gtf2ird1,* but did not include *GTF2I*. The Utah researchers examined Martha using a wide array of physiological and psychological tests. They found that although Martha exhibited nearly all of the symptoms of WBS, she did not display the overly friendly behavior to strangers usually found in WBS children. Perhaps loss of *GTF2I* rather than *Gtf2ird1* was the cause of the outgoing social behavior in WBS, and maybe the "friendly" behavior observed in *Gtf2ird1* knockout mice is not quite the same phenomenon as the friendliness in WBS? Nevertheless, even if the precise mechanisms by which variations in *GTF2I* or *Gtf2ird1* affect friendliness are not yet known, the fact that unusual friendliness in dogs,* mice and people seems to point to the same genetic region is intriguing.

FRIENDLINESS AND THE SILVER FOX

Whether more detailed genetic analysis linking tameness in dogs (versus wolves) eventually focuses on the *Gtf2ird1* and *GTF2I* genes remains to be seen, and it probably won't be known for several more years. Although dogs were domesticated from wolves "only" 15,000 to 30,000 years ago, that's still enough time for thousands of genetic changes to have arisen between the two species. Sifting through those genetic differences to find the ones that make dogs such gentle companions may take a while.

If dogs and wolves had separated as species more recently, there would have been less time for "irrelevant" genetic differences to have arisen, and hunting for gene variant(s) endowing dogs with their special personalities would be easier. Remarkably, just such a recent gateway into the genetics of gentleness has been opened by another cousin of the dog, the Siberian silver fox. The story began in the 1950s at a research center near the Siberian city of Novosibirsk, in what was then the Soviet Union. For generations, local farmers had been raising Siberian silver foxes for their fur. Raising silver foxes has its drawbacks, though. The animals can be quite aggressive and difficult to handle. So in the 1950s a Soviet geneticist, Dmitry Belayaev, decided to try to breed silver foxes that were tamer and easier for farmers to manage.

Belayaev's approach was taken straight from the days before fancy genetic engineering. He simply bred wild foxes for tameness, in the way that people have bred domestic animals for millennia. He carried out his breeding program in as systematic and objective a manner as possible. For each new generation of foxes (a silver fox generation is approximately three to four years), Belayaev and his associates allowed only the tamest 5% of the male foxes and the tamest 20% of the females to mate. (Belayaev developed elaborate behavioral tests for the foxes so that he could objectively select the

---

* Although the genetic location where dog and wolf DNA differ is actually at the opposite end of the WBS region from Gtf2ird1, this region may still affect the regulation of the Gtf2ird1 gene.

tamest ones for breeding.) To make better comparisons, they also set up a parallel experiment in which the most aggressive foxes were allowed to mate with one another.

The results of the experiment were astonishing. By the mid-1960s, after only four generations, the foxes that were selected for tameness were already showing signs of domestication, meaning that the animal's gentleness and willingness to be around people was becoming "hardwired" into its genetic makeup. A domesticated animal will react tamely to humans without any special training. In contrast, a lion cub, raised in captivity by humans and fed by hand, may grow up to be tame, but when that lion offspring are born they will be just as wild as their ancestors.

Belayaev's team demonstrated that the selected foxes were not merely tame but in fact domesticated. They showed that the foxes were gentle even when raised without any people. They even transplanted embryos of domesticated animals into the wombs of wild, surrogate mothers, and vice versa, to show that the tameness was the result of genetics and not the fox's prenatal environment.

Soon the tame foxes had become as friendly and approachable as dogs. Remarkably, these foxes also became more doglike in their physical appearance, with floppy ears and fur color patterns reminiscent of dogs. Since even some of the foxes that were not among the top 5% in tameness were becoming very doglike, allowing these foxes to be killed for their fur became increasingly painful for the scientists. So some researchers began taking a fox or two home, as pets. Eventually, the domesticated foxes became so tame, and so adorable, that the research group began selling them as pets, not only to raise money for the project, but also to find loving homes for the foxes.*

For many scientists, the most exciting potential of domesticated fox research will be identifying the genetic variants underlying the behavioral differences in wild and domestic foxes. Since only a few generations separate wild and domestic foxes, in principle, finding these variants shouldn't be difficult. The technical hurdle is that, in contrast to the dog and mouse genomes, the fox genome has not yet been sequenced, making the detection of genetic variants occurring solely in the gentle foxes difficult. But as the cost of DNA sequencing continues to fall, this barrier will soon disappear.

---

* In case you'd like a pet fox, you could contact the project through the Laboratory of Evolutionary Genetics of Animals, at the Russian Institute of Cytology and Genetics (http://www.bionet.nsc.ru/booklet/Engl/EnglLabaratories/LabEvolutionaryGeneticsAnimalsEngl.html). But be forewarned: the cost of a pet fox, including transportation from Siberia, can reach $4000 or more.

In the meantime, scientists have been analyzing fox DNA using less expensive methods. One approach involves simply locating chromosomal regions where tame and wild fox DNA consistently differ. Another strategy is to look for genes that are converted to protein at different rates in the brains of tame or wild foxes. Though less effective than comparing entire DNA sequences, these approaches are far cheaper and can provide hints of the genetic origins of fox tameness. At the time of writing, the first tantalizing results for such experiments are emerging. For example, one genetic location that differs between tame and wild foxes also differs significantly between wolf and dog DNA.

"EMPATHY" IN ANIMALS?

With the dramatic success of the fox experiment, domestication projects for other animal species were started, both in Novosibirsk and elsewhere. Within a few years, domesticated, human-friendly minks and rats had been bred as well. Another project attempted to create a domesticated river otter, but so far hasn't been successful. Possibly the critical genetic variants that exist in the fox, rat and mink genomes simply aren't present in otter DNA.

Considering the negative image of rats in traditional stories and fables, it may seem especially surprising that scientists have been able to breed domesticated rats. (At least this may be surprising to those who haven't had a pet rat.) In fact, even undomesticated laboratory rats can show behavior that looks remarkably like kindness or even empathy, as was demonstrated in an extraordinary recent experiment carried out by researchers at the University of Chicago.

Here's how the experiment worked. During 12 one-hour testing sessions, two rats were placed within a relatively spacious cage. One of the rats (always the same one) was free to wander throughout the cage. The second rat was trapped inside a small cage within the larger enclosure. Judging from the trapped rat's alarm calls, being inside the small cage was quite distressing.

The cages were designed so that the "free" rat, and only the free rat, could open the smaller cage and release the trapped rat. Opening this cage was not easy (at least not for a rat), and considerable effort was required to figure out how to do it. In fact, the free rats typically spent most of the first several sessions unsuccessfully trying to open the other rat's cage. But after approximately a half-dozen sessions, most of the free rats learned how to open the cage.

Although no reward (or punishment) was given to a free rat for releasing a trapped rat, three-quarters of the free rats chose to open the cage and let the other rat escape. The researchers also found that a greater proportion of female rats than male rats became "door openers." Perhaps the gender gap in compassion that sometimes seems to exist between men and women exists in rats as well.

In a variation of the experiment, in addition to the cage with the trapped rat, the experimenters placed some chocolate chips into the enclosure.* This time, the free rats again released the trapped rats. Remarkably, more than 50% of the time, the free rats also saved some chocolate chips for the trapped rats. When the same experiment was performed repeatedly using the same pair of rats, the scientists observed that the free rat saved chocolate chips for the trapped rat more frequently in the later trials, after the rats had presumably "gotten to know each other," than it did in the earlier trials.

The results of these experiments were disturbing to people who believe that only we humans (and certainly not rats) are capable of empathy and compassion. Even within the research community, some scientists challenged the interpretation of the experiment as showing "empathy." The critics, including Alex Kacelnik and his collaborators at Oxford University, didn't dispute the remarkable behavior reported by the University of Chicago researchers. Rather they said that the rats' behavior, though prosocial, didn't demonstrate empathy because the Chicago researchers hadn't demonstrated that the rescuing rat was "driven by the psychological goal of improving [the other rat's] well-being."

The critics pointed out that if all such prosocial behavior were to be interpreted as indicating empathy, then one might need to conclude that even *ants* feel empathy. In 2009, scientists from the University of Paris had shown that ants would release a fellow ant that experimenters had trapped in the sand by tying its leg with a nylon thread. The ants showed remarkable sophistication in their rescue efforts. They would dig sand from around the victim's leg, move the sand away and then repeatedly bite at the nylon thread until the trapped ant was free. The researchers also discovered that ants are particular about what kind of "victim ant" they'll help. The ants wouldn't free a trapped ant if it belonged to a different species. They wouldn't even free a trapped ant of their own species, unless it belonged to the same *colony*

---

* By offering non–food-deprived rats both chocolate chips and rat chow, and observing that the rats ate the chips and not the chow, the scientists had discovered that rats are quite partial to chocolate chips.

as the rescuers. So perhaps rather than feeling empathy, the ants, or even the rats, were exhibiting prosocial behavior simply to recruit a future ally for more self-serving goals such as finding food or procreating offspring. It's hard to say. But then one could make a similar case against humans. For example, when someone is comforting a crying infant, can we be sure whether they are motivated by compassion rather than simply a desire for the infant to be quiet?

Whatever interpretation one chooses, observing such empathy-like behavior in a rat is remarkable. Besides being intrinsically fascinating, the rat experiments point the way to powerful new tools for discovering the molecular mechanisms underlying empathy. One can envision experiments in which researchers would disable one of the rat's brain-pathway proteins and then measure changes in the level of rat empathy. Or different rat strains might be tested for empathy. Any consistent differences between strains could lead to tracking down DNA variants linked to differences in empathy (or at least in rat behavior that looks a lot like empathy).

By combining behavioral experiments on laboratory rats with comparative genome studies of dogs, wolves and foxes, scientists may soon identify genetic variants in animals' underlying prosocial traits, including kindness, gentleness or compassion. And as the critical genetic variants are discovered, understanding the biological mechanisms by which these genetic variants influence animal behavior should become possible.

At that point, scientists will be able to determine whether the gene variants leading to gentleness or compassion in animals also exist in people. Perhaps the results will point to *Gtf2ird1*, the hypothesized "friendliness gene" of Williams syndrome. The research may also lead to other genes of the brain's neuronal pathways. Scientists will then be able to investigate whether variants in these human genes are associated with differences in what we perceive as human friendliness or gentleness. At the moment this is pure speculation, but with the genetic analysis tools already available, we may be able to address such profound questions soon.

11

WHAT IS SEX?

The patient, a man in his mid-40s who had been referred to the San Raffaele Clinical Genetics Laboratory, in Milan, Italy, appeared to be completely normal. We'll call him Mario. Mario was healthy, had a normal male sex drive and appeared to be happily married. What had brought him to the clinic was that he was childless, despite having attempted to father a child throughout his 10-year marriage. A standard medical examination revealed that Mario's testosterone levels were low and that his testes were abnormally small. Further laboratory testing showed that his sperm count was zero, so Mario's inability to father a child was not surprising. But Mario's DNA tests did reveal one surprise. Mario had two X chromosomes and no Y chromosome. According to traditional genetics, Mario was a woman. In the terminology of the medical profession, Mario was completely sex-reversed.*

The X and Y chromosomes, the sex chromosomes, usually determine whether we are male or female. Women generally have two X chromosomes and men have one X and one Y chromosome. Yet people like Mario, whose chromosomes appear to be incompatible with their physical appearance, show us that the relationship between our chromosomes and our sex is not

---

* I'll generally describe physical differences between males and females using the term *sex*, while referring to behavioral differences with the word *gender*. The distinction will become important in chapter 20.

so simple. As we'll see in this chapter, our sex is determined less by our chromosomes than by individual genes and the proteins that are encoded by those genes. Healthy (albeit infertile) men may not have a Y chromosome, while normal females can have a Y chromosome. Complete sex reversal is rare, though estimating just how rare is difficult. Sex-reversed individuals may happily live their lives ignorant that they don't have the "correct" set of sex chromosomes. They may only become aware of their condition if they seek help for infertility or if they happen to have a medical condition that doctors diagnose by analyzing a patient's chromosomes.

It's also possible to inherit a set of genetic variants that lead to partly male and partly female sexual characteristics. Such individuals are described in the scientific literature as being partially sex-reversed, or intersex. They may have primarily female physical characteristics, with some male features, or they may be males with some female features. Even their internal sexual organs may be ambiguous. Rather than having clearly identifiable ovaries or testes, they may have internal sexual organs that have characteristics of both ovaries and testes. In other words, physical examination may be unable to determine an intersex individual's sex, or whether it is even meaningful to label them as male or female.

DEFINING SEX

To understand how genetic variation can lead to sex reversal, we need to know how to label a person's sex in the first place. The answer might seem simple – put crudely: "Just look in their pants." In some cases, though, it's hard to tell whether a sexual organ is a large clitoris or a small penis. In fact, in approximately 1 out of every 2000 births, even the doctor performing the delivery is unsure of the sex of the newborn. Even after they reach adulthood, partially sex-reversed individuals may have physical characteristics that appear to be a mixture of male and female. Their internal sexual organs (or gonads) may also be a combination of male (testes) and female (ovaries).

Because of these difficulties, scientists sometimes try to specify the sex of an individual by how many germ cells* they produce. Since female animals usually produce germ cells singly or in relatively small numbers, while male animals produce sperm in large numbers, germ cell quantity might be a way to specify animal sex. However, sex-reversed individuals, such as Mario,

---

* Germ cells (the unfertilized egg and the sperm) are the cells that combine during sex to produce a fertilized egg that eventually becomes a newborn animal.

often don't produce either sperm or egg cells, so this method of classifying sex also can fail.

Consequently, scientists have sought other ways of defining sex, and in the early 20th century they discovered that a person's chromosomes seemed to specify their sex. Theophilus Painter, a scientist at the University of Texas, noticed that, under a microscope, the cells of women and other female mammals had a pair of small structures, what we now call X chromosomes. The cells of males had only one X chromosome, along with an even tinier structure that came to be known as the Y chromosome. At that time, the functions of the chromosomes were completely unknown. Nevertheless, by simply looking at them, scientists thought they could determine a person's "real" sex, even if "looking in their pants" or examining their sex organs didn't yield a clear-cut answer.

Unfortunately for those who want to classify all people as male or female, the sex of individuals like Mario is difficult to determine unambiguously, no matter what definition one uses. Anatomically they may belong to one sex, while their chromosomes indicate the other sex. They may not have clearly delineated ovaries or testes, and their gonads may not produce either egg or sperm cells. As a result, the definitions of sex based on types of gonads or germ cells fail as well. So even though, for simplicity, we'll be talking in this chapter about "males" and "females," we need to remember that nature is diverse, and simple labels don't always fit the real world.

THE THREE STAGES OF SEXUAL DEVELOPMENT

To understand how people like Mario can become sexually reversed, we need to look at the process of human sexual development. This process, by which a fetus becomes a male or a female, can be conveniently separated into three phases:

- Sex determination: the initial genetic trigger to becoming male or female
- Sex differentiation: the development of testes or ovaries
- Hormone-directed development: the growth of external genitals and other sexual characteristics due to hormones released by the testes or ovaries

Sex determination occurs immediately after conception and starts the embryo down either the male or the female development pathway. In humans, sex determination is usually triggered by the presence or absence of a Y chromosome in the fertilized egg cell. During this initial stage the embryo produces an undifferentiated, unisex sex organ. During sex differentiation, the second phase of sexual development, the undifferentiated gonads become either testes or ovaries. Finally during hormone-directed development, the testes or the ovaries secrete hormones that instruct the body to create the external genitalia and the other physical characteristics we associate with males or females. (This model is oversimplified, but it does provide a useful framework for describing how an animal becomes male or female. In particular, this model will help us see how sex reversal is a natural consequence of how our bodies develop.)

SEX DETERMINATION IN MAMMALS AND BIRDS

Just as in humans, in nearly all other mammals, chromosomes determine an animal's sex. Females have two identical sex chromosomes, called X chromosomes. Male mammals have one X and one Y chromosome, just like male humans. Birds also have two sex chromosomes. With birds, it's the males that have two identical sex chromosomes (called Z chromosomes since they are different from the mammalian X and Y chromosomes), while females have differing sex chromosomes: Z and W.

Although generally mammals and birds use two sex chromosomes to determine sex, there are exceptions. Consider the platypus. The platypus is an unusual animal by any measure. It's a mammal (platypus mothers suckle their young), but the females lay eggs, just like birds. Sex determination in the platypus is also remarkable. Instead of having 2 sex chromosomes, the platypus has 10! A female platypus has 10 X chromosomes while a male platypus has 5 X chromosomes and 5 Y's. So when a male platypus is manufacturing sperm, his body has to ensure that each sperm cell has either exactly 5 X chromosomes and no Y's, or 5 Y chromosomes and no X's (otherwise the baby platypus won't have exactly 10 X's or 5 X's and 5 Y's).

Some mammals have even stranger sex-determining mechanisms than the platypus. Spiny rats and mole voles have lost the Y chromosome altogether; both males and females have identical sex chromosomes. Just what determines whether a spiny rat becomes a male or a female is still unknown. The leading hypothesis is that there is a genetic variation on some other (non-sex) chromosome, and which variant a spiny rat has at that genetic location determines whether the rat becomes a male or a female.

Sex reversal and intersex exist in animals as well as in humans. Perhaps the most striking example of ambiguous sexual characteristics is a gynandromorph – an animal that develops as a male on one side of its body and as a female on the other side. Gynandromorphy has been observed in birds, insects and other types of animals, though never in mammals. Recent chromosomal analysis of gynandromorph birds revealed that they have cells with two Z chromosomes on one side of their bodies and cells with one Z and one W chromosome on the other side. This strange mixture among the gynandromorph's cells apparently comes about because of the fusion of two eggs within the bird's mother at the time of fertilization. Figure 11.1 shows just how bizarre a gynandromorph bird can appear, and illustrates the pitfalls involved in trying to label some animals as being male or female.

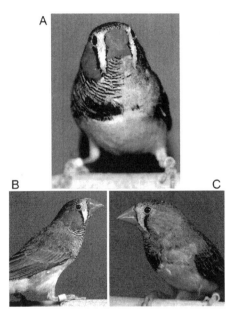

Figure 11.1. Gynandromorph zebra finch. A: Both sides are shown. B: Right side of the bird, which has the appearance of a male bird, with thin black striations on its chest and a bright orange patch on its cheek. C: The left side of the bird, which has the female zebra finch's uniform gray color.

SEX DETERMINATION IN HUMANS

Except for the platypus, spiny rat and a few other oddities, mammalian sex is determined by the X and Y chromosomes. But just how do these chromosomes specify sex? For many years after the discovery of the X and Y chromosomes, no one knew. Nor was it clear whether the critical factor in becoming a male was having a Y chromosome or *not* having two X chromosomes.

Answers began to emerge in the 1930s and 1940s with the discovery of people who had neither one X and one Y chromosome nor two X chromosomes. In 1942, Dr. Harry Klinefelter, a physician at Massachusetts General Hospital, in Boston, wrote his first paper describing an individual with two X chromosomes *and* one Y chromosome (a condition now known as Klinefelter syndrome). Shortly thereafter Klinefelter found other people with a single Y and two X chromosomes.

In fact, Klinefelter syndrome isn't that rare; an estimated 1 in 1000 men (for these individuals appear male) has it. These men are generally healthy, though they are unable to father children. They do have certain distinguishing characteristics including tall stature, long arms and legs, little chest hair and a somewhat "feminine" physique. They tend to have small testicles and enlarged breasts. They also seem less aggressive on average than "normal" men, and their blood testosterone levels are usually lower than other men's.

Klinefelter syndrome was not the only unusual sex-related chromosome condition discovered around this time. A few years earlier, in 1938, Dr. Henry Turner, an endocrinologist at the University of Oklahoma, published a report describing seven women whose cells had one X chromosome and no Y chromosome. These women were short, vulnerable to osteoporosis and unable to bear children. An estimated 1 in 2000 to 1 in 5000 girls are born with this condition, which became known as Turner syndrome.

Since people with two X chromosomes and one Y appeared male, while individuals with one X and no Y chromosome mostly appeared female, scientists concluded that the presence or absence of a Y chromosome, rather than the presence of two X chromosomes, determined whether a fetus would become a boy or a girl. Yet no one understood just how the Y chromosome determined sex, nor whether the entire Y chromosome was necessary to specify "maleness," or if just a part of the Y chromosome was sufficient.

Only after 30 years of further research was a team from the National Institute for Medical Research, in London, able to prove that the critical region of the Y chromosome consisted of just one gene, which became known as the *SRY* gene (for Sex determining Region on the Y chromosome). The experiments were performed on fertilized mouse egg cells with two X chromosomes. (In other words, left to their own devices, the fertilized eggs would have developed into female mice.) The London researchers then added a single short piece of DNA from the mouse Y chromosome, which included the *SRY* gene, into each of the fertilized eggs. The resulting embryos developed as normal (though sterile) male mice with testes and male sex characteristics, even though the embryos had two X chromosomes and no Y chromosome.

SRY AND SEX DETERMINATION IN HUMANS

By the 1990s, the *SRY* gene had been established as the determinant of sex, at least in mice. Did *SRY* also determine sex in humans? The answer became apparent as people with sex reversal began to have their *SRY* gene tested. The test results showed that approximately 10% of the cases of sex reversal are caused by some unusual variant in the *SRY* gene. Scientists also discovered that different *SRY* mutations had varying impacts on the "maleness" of the individual. People with certain *SRY* mutations had a male appearance, though with underdeveloped testes. Individuals with other *SRY* mutations looked like women.

Scientists also learned that on rare occasions the DNA containing the *SRY* gene can move from the Y chromosome to the X chromosome during the manufacture of sperm. The result is one sperm with a Y chromosome missing its *SRY* gene and a second sperm with a *SRY* gene on its X chromosome. If the first sperm winds up fertilizing an egg, a relatively normal girl will be born, though with a Y chromosome as well as an X. If the second sperm fertilizes an egg, the result will be a relatively normal baby boy with two X chromosomes and no Y chromosome. Such transfers of the *SRY* gene from the Y to the X chromosome are very rare; they are estimated to occur in only 1 in 80,000 male births. Nevertheless, this means that there are thousands men without Y chromosomes who are unaware that their chromosomes claim they are female.

SEX DIFFERENTIATION

SRY initiates the male sexual-development pathway; without SRY protein, the developing embryo progresses as a female. This is just the first piece of the puzzle, though. SRY is simply a regulator protein that controls what *other* proteins in the sexual-development pathway are produced. These other proteins regulate the formation of the ovaries or the testes during the sex-differentiation stage. Building these sex organs requires multiple proteins, all of which are encoded by genes. Variations in any of these genes can lead to abnormal sexual development. As a result, only 10% of people with sex reversal are found to have mutations in their *SRY* gene.

What other proteins are involved in human sexual differentiation? At the time of the discovery of SRY in 1990, the identities of only a few were known. Twenty years later, dozens of sex-differentiation proteins had been identified, and the list continues to grow. When sex-reversed individuals seek medical assistance today, their doctors can often find a causal mutation in one of their sex-determining genes. Nevertheless, in over 50% of the cases of intersex or sex reversal doctors are still unable to find the causal variant. Figure 11.2 gives an overview of how the presence or absence of SRY usually triggers the three stages of sexual development in males and females. Figures 11.3 and 11.4 outline how variations in SRY or other sex-related proteins (which we'll learn about shortly) can cause an initially female embryo to become male, or vice versa.

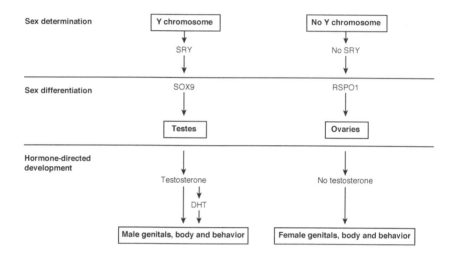

Figure 11.2. Sexual development pathways in humans and other mammals. The male sex development pathway is outlined on the left; the female pathway on the right. The top third of the schematic illustrates the sex-determination stage in which the Y chromosome with the *SRY* gene leads to activation of the SOX9 protein in males (or the lack of SRY leads to producing RSPO1 protein in females). The second stage, sex differentiation, includes the activation of SOX9 or RSPO1 leading to the development of testes or ovaries, respectively. Finally in the third stage, hormone-directed development, the presence, or absence, of testosterone leads to the actual development of the male, or female, genitals, body and behavior. DHT = dihydrotestosterone (a hormone).

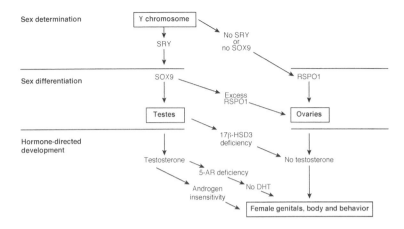

Figure 11.3. Male-to-female sex reversal. The vertical arrows on the left outline the normal path of male sex development. The five sets of diagonal arrows pointing downward and to the right illustrate ways this path can be altered toward female development, for example, by a nonfunctional *SRY* or *SOX9* gene, by an overactive *RSPO1* gene, or by an androgen or androgen-receptor deficiency.

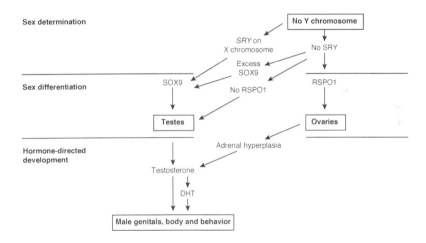

Figure 11.4. Female-to-male sex reversal. The figure depicts how sex development without a Y chromosome, which typically would lead to a female, can be altered. As with male-to-female sex reversal, the change can occur during the sex-determination, sex-differentiation, or hormone-directed development stage and can lead to either complete or partial sex reversal depending on how the usual development pathway is modified. Various sex-reversal pathways are illustrated by the diagonally downward pointing arrows and are described in the text.

SOX9

One of the most important genes in male sex differentiation is a gene on chromosome 17 called *SOX9*. *SOX9* is directly regulated by SRY. The SOX9 protein, in turn, triggers the production of hormones that cause the undifferentiated fetal sex organs to become testes, rather than ovaries. The importance of SOX9 became apparent in the 1990s, when scientists identified several girls who had a Y chromosome with a functional *SRY* gene. The reason they appeared female was that each had a disabling *SOX9* mutation.

The discovery of young women with functional *SRY* genes was soon followed by mouse experiments using *SOX9* knockout mice. The *SOX9* knockout mice appeared female even though they had Y chromosomes with normal *SRY* genes, just like the women with the *SOX9* mutations. In a later experiment, transgenic *SOX9* mice were engineered that had a mutation near their *SOX9* gene, causing their cells to produce *SOX9* messenger RNA whether or nor SRY protein was present. As a result, these transgenic *SOX9* mice all appeared male, even the ones that "should" have been female because they lacked a Y chromosome. Apparently, SOX9 is sufficient to make a mouse male even if it doesn't have SRY.

SOX9 also was the key to why Mario is a man and yet has no Y chromosome. Mario had inherited a rare mutation that affected his *SOX9* gene. When Mario's doctors took his family history they learned that Mario's brother and his uncle (his father's brother) were also unable to father children. Genetic testing showed that the brother and the uncle also didn't have a Y chromosome. They too were "genetic females" like Mario.*

All three men had inherited a rare genetic variant located close to the *SOX9* gene on chromosome 17. Since this mutation was not within the *SOX9* gene itself, their SOX9 proteins were normal. Instead their genetic variant altered the regulation of *SOX9* so that Mario's cells produced SOX9 continuously, whether or not SRY protein was present. Consequently Mario, his brother and his uncle became men, but since they didn't have the other Y chromosome genes needed to produce sperm, they were unable to father children.

RSPO1 AND THE FEMALE SEXUAL PATHWAY

Both SRY and SOX9 are critical for male sexual differentiation. Without functioning SRY and SOX9 proteins, a fetus, even if it has a Y chromosome,

---

* We'll use the term "genetic female" to describe individuals with cells having two X chromosomes and "genetic male" to indicate someone with one X and one Y.

will become female. Other genes are essential for female sexual differentiation, and someone with nonfunctional variants of these genes will develop, at least partly, into a male. One such gene, *RSPO1*, encodes a protein critical for developing ovaries; individuals with nonfunctional RSPO1 develop testes instead. Depending on what *RSPO1* mutation they have, the person may appear primarily male with some female characteristics or primarily female with some male traits.

RSPO1 also plays important biological roles unrelated to sexual development, and people with nonfunctioning RSPO1 often have medical problems. In fact it was a medical issue unrelated to sex or reproduction that led the 45-year-old Sicilian fisherman we'll call Gino to consult his physician. Gino suffered from severe skin problems and had a history of throat cancer as well. When Gino's doctors performed genetic tests, they found that he had two X chromosomes and no Y chromosome. Gino was genetically female. Through additional genetic testing, his doctors discovered that three of Gino's brothers, all of whom suffered from similar skin problems, also were genetically female. DNA testing revealed that all four brothers had inherited two copies of a mutation that completely disables *RSPO1*. Since disabling variants of *RSPO1* are uncommon, inheriting two nonfunctional variants is very rare. Gino's parents, though, had been first cousins, so his chance of inheriting the same disabling mutation from both parents was much higher.

Subsequently other proteins have been discovered that also play critical roles in determining whether the gonads become testes or ovaries. In each case the protein's function was demonstrated using transgenic mice (in which the corresponding gene has been either knocked out or enhanced) resulting in either female mice with a Y chromosome or males without a Y chromosome. For several of these proteins doctors have also found individuals, like Mario and Gino, with a genetic mutation causing sex reversal.* Undoubtedly more such sex-reversal causing mutations will be discovered in the coming years as genetic testing in cases of sex reversal becomes routine.

HORMONE-DIRECTED DEVELOPMENT

Typically, by the third month in the womb, sex differentiation is complete and the gonads have developed into either ovaries or testes. Sex development

---

* Although the two case histories we've described involved men without Y chromosomes, in clinical practice, sex-reversed women with a Y chromosome are more common than men without Y chromosomes..

enters its third stage, regulated by the sex hormones. If the fetus is male, the testes secrete testosterone into the blood stream. Testosterone and its metabolites signal the body to make the male external genitalia and develop other male characteristics. If the fetus has ovaries, estrogen is produced instead, and female sex development occurs.

The importance of the testes in male sexual development has been known for thousands of years. Farmers have long been aware that removing the testes from a male animal made the animal sterile, less aggressive and altogether less "male." Aristotle wrote about the effects of castration, the removal of testes from a male, as early as 350 BCE in his *History of Animals*. Yet the critical role of the testes in determining whether an animal becomes male or female wasn't demonstrated until the 20th century.

In the 1940s, Alfred Jost, a scientist at the College de France, in Paris, observed a small, undifferentiated sex organ in the embryos of mammals that later developed into the testes in males and the ovaries in females. In his critical experiments, Jost surgically removed the undifferentiated sex organ from a developing embryo. The resulting animal (Jost used rabbits but the results would be the same with other animals) developed a uterus, vagina, and other female genitalia. Jost had demonstrated that without internal sex organs a developing animal would become female, whether or not the animal's cells had a Y chromosome.

THE SEX HORMONES IN HUMANS

Although scientists obviously can't perform experiments on the effects of testosterone secretion on sexual development in human fetuses, they can study the role of sex hormones by observing medical conditions in which abnormal amounts of these hormones are secreted. Excess or deficient levels of estrogen or testosterone may be caused by rare genetic variants, by a tumor in a hormone-producing organ or as side effects of drugs. People with such atypical hormone concentrations may develop ambiguous genital organs or unusual secondary sexual characteristics, such as facial hair in women or enlarged breasts in men.

Hormonal imbalances that lead to unusual sexual development often have complicated names like 17β-hydroxy-steroid dehydrogenase-3 deficiency or congenital adrenal hyperplasia. Don't worry about the names – the underlying conditions are not difficult to understand, nor is it hard to see how they can lead to sex reversal. Take 17β-hydroxy-steroid dehydrogenase-3 (17β-HSD3) deficiency, for example. 17β-HSD3 is a protein that the fetus needs to synthesize testosterone. Individuals with a disabling mutation in both copies of their *17β-HSD3* gene can't produce testosterone, even if they have a Y chromosome and normal testes. When such an infant is born, she (for it will almost definitely be considered a girl) will have female external genitalia. Typically, the fact that these girls are not genetically female will not be discovered until they reach puberty and don't menstruate.

5α-reductase deficiency is a genetic condition with symptoms similar to those of 17β-HSD3 insufficiency. 5α-reductase (5-AR) is an enzyme required to convert testosterone into dihydrotestosterone, which is also essential for male sexual development. Depending on the severity of their *5-AR* mutations, individuals with 5-AR insufficiency may produce lower amounts of dihydrotestosterone, or none at all. As a result, they may have relatively normal male external genitalia, female genitals or a mixture of both.

To be affected by 17β-HSD3 deficiency one must have a mutation in both copies of the *17β-HSD3* gene. Similarly 5-AR deficiency only occurs in individuals with disabling variants in both of their *5-AR* genes. Since these genetic variants are not common, inheriting a mutation in both genes is very unlikely and these conditions are rare (less than 1 per 100,000 births). The exception is among populations where first-cousin marriages are common; for example, in certain communities in the Dominican Republic, the Middle East and Papua New Guinea, as many as 1 in every 100 newborn children are sex-reversed as a result of either 17β-HSD3 or 5-AR deficiency.

Male development requires not only testosterone production but also its successful detection by testosterone receptors. Consequently, defects in the androgen receptor, which detects testosterone, can also lead to sex reversal. Over 400 different mutations in the androgen-receptor gene are known. Some mutations only reduce the receptor's sensitivity; others completely disable it. Depending on what androgen receptor mutations a genetically male person has, he may appear male, he may have ambiguous genitalia, or he may even look female, a condition called complete androgen insensitivity syndrome. Children with this syndrome typically grow up as girls. They even develop breasts when they reach puberty, though they are unable to menstruate, since they have no female internal organs.

Our last example of sex-reversal occurs when a genetically female fetus produces testosterone. This can occur because the testes are not the only organs that can produce testosterone. The adrenal glands, which normally synthesize the chemicals adrenaline and cortisol, can also produce testosterone. To make cortisol, the adrenal glands need an enzyme called 21-hydroxylase (21-OH). People with mutations in both copies of their *21-OH* genes have adrenal glands that can't produce cortisol. Their adrenal glands synthesize testosterone instead. This condition is called congenital adrenal hyperplasia, or CAH, and occurs in approximately 1 out 10,000 to 15,000 births. In genetic males this extra testosterone is usually not noticed, since the male fetus produces large amounts of testosterone anyway. In genetic females, though, congenital adrenal hyperplasia may have physiological or behavioral effects. Women with CAH often have irregular menstrual periods or don't menstruate at all. They also tend to behave more like typical males, both as children and as adults. Some genetic females with CAH also are partially sex-reversed and have genitals that appear both male and female.

ENVIRONMENTAL SEX DETERMINATION IN ANIMALS

In people and other mammals, genes determine who will become male and who will be female. In most cases, SRY and the Y chromosome are the determining factors. Even among the rare exceptions, people like Mario, Gino or women with complete androgen insensitivity syndrome, sex is determined by a genetic variant. Yet genetic sex determination isn't universal in the animal kingdom. In some animal species, males and females are genetically identical. Certain reptiles, amphibians and fish do away with the sex chromosomes altogether and have the same genes on their non-sex chromosomes as well. Instead the *environment* determines an animal's sex.

Whether a marine turtle becomes male or female depends on the temperature at which the turtle egg is incubated. If the incubation temperature is cool, below 80 degrees Fahrenheit, 100% of the turtle hatchlings become females. If the incubation temperature is above 86 degrees, they become males. Only at incubation temperatures between 80 and 86 degrees will a mix of males and females be born.

In alligators, sex determination is also determined by temperature. As with marine turtles, low temperatures lead to females, while at higher temperatures the hatchlings are male. The alligators throw in another twist, though; if the incubation temperature is very high, then the hatchlings are all female again! Needless to say, species that employ environmental sex determination, such as marine turtles and alligators, are very sensitive to climate change, as they can't continue reproducing unless both males and females are born.

If environmental sex determination seems strange, then environmental sex reversal, which occurs in some reptiles, amphibians and fish, is truly bizarre. Animals that use environmental sex reversal do have sex chromosomes and sex genes. Those sex chromosomes determine whether the animal is born male or female. But these animals can spontaneously change sex *after* birth in response to changes in the environment. In some cases female animals become males; sometimes males turn into females. For example, hatchlings of some fish species can switch between becoming males or becoming females as the temperature or the acidity of the water changes.

My favorite example of environmental sex reversal occurs in fish called cleaner wrasse, which live in Indian and Pacific Ocean coral reefs. (They are called "cleaner" fish because they remove parasites from the scales of other fish.) Cleaner wrasse are born as females with ovaries, though they also have small, undeveloped testes. They live in social groups consisting of a single dominant aggressive male and a group of three to six females (called a harem for obvious reasons). If the male dies or is removed from the harem, a neighboring male fish may abduct the females into his harem. But if no other male is around, something much stranger happens. The biggest, most aggressive (dare I say the nastiest?) female spontaneously changes sex and becomes a male who then takes over the harem! The testes of this newly male fish are soon spawning fish sperm, whereas just a few weeks earlier, its ovaries had been spawning eggs. Just how this remarkable transformation happens in wrasse and other species that use environmental sex reversal is not completely understood, though it is known that hormones play an important role.*

Sex specification in marine turtles, alligators and cleaner wrasse is our first example of a key trait in animals that is determined by the environment rather than by genes or DNA. Genes are not the only important influence in shaping an animal's life. We've focused on genes until now, primarily because science's understanding of the impact of genetic variants on cells is much more advanced than our understanding of environmental influences. Yet to address many scientific questions, such as whether a marine turtle will become male or female, understanding the effects of the environment on our proteins and our cells is critical. In the last few years, scientists have made dramatic progress in learning how the environment affects us at the cellular level. An entire new scientific discipline called epigenetics has developed to describe these discoveries. And it is to epigenetics and the methods by which scientists are learning to include the impact of environmental factors on our development that we'll turn to next.

---

* For example, if one gives female bluehead wrasse (a species related to cleaner wrasse) a hormone similar to testosterone called 11-KT, the fish adopt the coloration and behavior of male wrasse.

PART IV

EPIGENETICS: HOW OUR ENVIRONMENT CHANGES OUR CHROMOSOMES

12

CAN YOU GET CANCER FROM A GENE?

Angelina Jolie is an award-winning actress who's been called the world's most beautiful woman. An international celebrity, she's known equally for her work as a humanitarian activist and her high-profile marriage to screen idol Brad Pitt. Yet for all the publicity that surrounds her, Jolie succeeded in keeping one corner of her life private until she revealed it in a *New York Times* op-ed article on May 14, 2013. In the article, Jolie described the toll that cancer has taken on her family and the impact it's had on her own life. Jolie's mother died of ovarian cancer at age 56. Jolie's grandmother also died of cancer. Only two weeks after Jolie's article appeared, her aunt succumbed to breast cancer. In her article, Jolie revealed that she possesses an unusual DNA variant of a gene called *BRCA1*. Most likely, her mother, aunt and grandmother all had the same genetic variant. Although the mechanisms by which the BRCA1 protein functions are still poorly understood, researchers now know that this genetic variant makes a woman far more susceptible to breast and ovarian cancer.

In previous chapters, we've seen how a genetic variation can determine whether someone becomes male or female or is afflicted with fragile X syndrome or Huntington's disease. We've seen how DNA variations can affect how much sleep one needs or (at least among animals) whether one is a promiscuous or devoted mate. In each of these examples, protein changes caused by genetic variation directly led to differences in one's health or behavior. Often, though, life is not so simple. As we've seen, alligator and marine turtle sex is determined by the environment, not by genetics. And although human sex is determined by our genes, other aspects of our lives are greatly influenced by events in our environment. If you get run over by a truck, your chances of survival will be low no matter what genetic variants you inherited. If your body is infected by pathogenic bacteria, you're likely to get sick. If you eat nutritious foods and exercise regularly, you're more likely to be fit and healthy than if you eat poorly and never exercise.

In Part IV we turn our attention to the impact our environment has on our cells and on our lives. As a start, we'll need to see how scientists disentangle the genetic and environmental influences on our lives. In this chapter we'll look at how scientists unravel whether a disease or human trait is caused by genetics, environmental factors or some combination of the two. In particular, we'll explore the progress that has been made in discovering the origins of breast cancer and the impact these discoveries are having on women, like Angelina Jolie, whose lives are threatened by this terrible disease.

GENOTYPES AND PHENOTYPES

The first step scientists must take to unravel the effects of our genes and our environment is to determine whether a trait is influenced by our DNA at all. If it is, the next task is to identify which genes are responsible. These difficult tasks link two basic aspects of a person:* their genotype and phenotype. An individual's genotype is simply the sequence of nucleotides that he has at one (or more) locations in their genome. Essentially, a genotype describes how one person's DNA sequence *differs* from some reference DNA sequence. For example, John's genotype might show that at nucleotide number 150 in the *CCR5* gene he has a C, while in a reference genome there is a G at that location.†

In contrast to a genotype, a phenotype is just about any characteristic or trait that the person has. A phenotype might be someone's eye color, how

---

* We're referring here to genotypes or phenotypes of *people,* but the terms are used more generally to describe the DNA and traits of animals, plants or any other living creature.

† The reference DNA sequence against which a genotype is compared is often arbitrary; it may simply be the genome of the first individual who's DNA happened to get sequenced. Nevertheless, having *some* reference against which to compare an individual's DNA sequence is very useful.

tall they are or whether they are afflicted with Alzheimer's disease. Usually the term *phenotype* is applied to traits that are influenced by genetics, that is, traits that are influenced by one's genotype. So someone's eye color is considered a phenotype, while whether they are Boston Red Sox fans is not. That said, for many of the most interesting human phenotypes, such as intelligence or sexual orientation or whether a person tends to be happy or depressed, the extent to which that the trait is influenced by genetic variation is still unknown.

How do scientists determine if a phenotype is caused, if only in part, by an underlying genotype? Occasionally, simply careful study of the phenotype provides clues. For the phenotype of HIV immunity, scientists knew that HIV infects its victims via the T cells of the immune system. Since they knew that HIV infects a T cell by binding to a receptor protein, it was plausible that a variant in a T cell receptor gene might be linked to HIV immunity. This wasn't *necessarily* true, but it did make T cell receptor genes promising candidates. In this case, the scientists were lucky. When the AIDS-protecting variant was eventually found, it was in *CCR5*, a gene for a T cell receptor protein.

Finding a phenotype in an experimental animal can also help identify its molecular roots. One example we've seen is the identification of the genetic variant causing short sleep periods in hamsters, described in the chapter 8. Another example is the discovery of the cause of narcolepsy, a sleep disorder in which people sleep too much and fall asleep at inappropriate times. Since the 1970s, scientists had known that narcolepsy also occurs in Doberman pinscher dogs, and by performing experiments on affected dogs, they were able to determine the genetic causes of canine narcolepsy. By 1999, researchers knew that a gene variant for the receptor of a neurotransmitter called hypocretin caused narcolepsy, at least in Dobermans. Since humans also use hypocretin, it was a reasonable guess that impaired hypocretin signaling might also cause human narcolepsy, and scientists subsequently did discover that hypocretin deficiency leads to narcolepsy in people.

Another clue that a trait may be genetic is when it occurs in infancy or early childhood. If an infant is born deaf or blind, a genetic cause is likely. Finding an unusual trait in a young child, such as the physical strength possessed by the astonishing four-year-old we'll meet in chapter 16, also suggests that the phenotype wasn't acquired through years of practice. Sometimes, though, traits can appear at an early age for reasons unrelated to genetics. For example, some infants are diagnosed with intellectual or physical disabilities that have nothing to do with genes and everything to do with their prenatal environment (say, if the mother had AIDS or drank excessively during the pregnancy).

TRACING GENES THROUGH FAMILIES

In contrast to the above examples, often there are no obvious clues whether a phenotype is inherited, comes from the environment, or is a combination of the two. In such cases, scientists often look for families in which more than one family member has the trait, since traits resulting from inherited genetic variants will typically cluster in families. Of course, nongenetic factors can also run in families; for example, entire families could be exposed to the same toxic chemical or infected by the same virus. That said, if multiple individuals in each of several families share the same phenotype but the families don't share a common environment, then some relevant genetic variant is probably being inherited.

Studying twins can be an especially effective way to determine whether a phenotype has a genetic component. The reason is that if a trait is genetic, it should more frequently be shared by identical twins than by nonidentical twins. Researchers also study identical twins who have been separated and raised in different environments. If separately raised twins share an uncommon phenotype, such as becoming champion athletes or being homosexual or suffering from cystic fibrosis, chances are it's because of shared genetic variants.

Scientists have long known that families with two or more women who have had breast cancer occur more often than would be expected by chance. A young woman with no family history of breast cancer has about an 8% chance of developing breast cancer at some point in her life.* If her mother or sister had breast cancer, her chances of developing breast cancer increase to approximately 13%. If she has two sisters with breast cancer, her odds go up to over 20%. Since families with recurrent breast cancer have been found in widely varying locations, the data strongly suggests that genetic variations contribute to some forms of breast cancer.

Learning that breast cancer has a genetic component was just a first step in deciphering the genetic factors underlying the disease. What scientists really wanted to know was which genetic variant(s) are responsible. But without better understanding of cancer biology or a good

---

* The precise risk calculated in breast cancer studies has varied somewhat between populations and over time. However the *relative* increased risk for women with a family history of breast cancer has been observed consistently.

animal model, pinpointing the responsible genetic variants, the so-called causal variants, was a monumental needle-in-a-haystack challenge. Each person's DNA has six billion base pairs, and millions of genetic differences exist between any two people (except identical twins). Since any one of those millions of variations could lead to increased disease vulnerability, how could scientists hope to track down causal variants in such an ocean of variation?

FINDING THE BRCA1 GENE

Scientists discovered that the link between *BRCA1* and breast cancer by using methods similar to those described in chapter 4 for tracing the African ancestry of Puerto Ricans. The key observation was that DNA recombination during sperm and egg production causes genetic variants located close to one another to usually be inherited together, while distant genetic variants, or ones located on different chromosomes, are generally inherited independently. To use this fact to search for genetic variants underlying a human trait, researchers begin by identifying a large number of chromosomal locations where people often have different nucleotides. For example, at a location precisely 100 bases in front of some gene, 50% of people may have an A nucleotide, 30% have a G, 15% have a T and 5% have a C. Such a location is called a marker location, or simply a marker. Nowadays genetic research studies often use a million such marker locations.

Next, the researchers find families with the phenotype of interest – for example, breast cancer – and determine the DNA base at each marker location for all the family members. They then ask whether there are any marker locations where all the family members with breast cancer have, say, an A, while all the family members without breast cancer have a G. If so, it does *not* necessarily mean that these variants are causal variants for breast cancer susceptibility. However, because nearby genetic variants tend to be inherited together, while distant ones are generally inherited independently, consistently finding certain DNA variants at a marker location means that the location is probably *close* to a causal variant. If the same marker location turns up in multiple families, there is an even better chance that the marker really is close to a causal variant.

By the 1980s the costs of performing DNA testing at multiple marker locations (a process called genotyping) had fallen to the point where it was possible to genotype large numbers of people. Scientists from a number of university research centers, led by Mary-Claire King, a genetics professor then at University of California, Berkeley, took on the task of genotyping individuals from families with recurrent breast cancer. In 1990, they announced that certain marker locations within an eight million base pair region on chromosome 17 were consistently different in family members who had been afflicted with breast cancer and in (female) family members who had never had breast cancer.

This was a significant achievement. Going from the three billion potential locations in the human genome to a single region of eight million bases was a huge step forward. Nevertheless, this region still contained 20 genes. Until someone found the specific gene that was linked to the increased cancer vulnerability, this information was unlikely to be clinically useful. A race began to identify the causal variant(s). Several academic groups, along with various companies, began testing increasing numbers of marker locations on chromosome 17 in families with recurrent breast cancer. Finally, in 1994, scientists at a company called Myriad Genetics announced that they had identified the gene in which the causal variants were located. It was a gene about which very little had been previously known. The researchers named the gene *BRCA1*, for breast cancer gene 1.

In some ways, the name *BRCA1* was an unfortunate choice. It soon became apparent that the BRCA1 protein plays an important role in ovarian cancer as well as in breast cancer. More importantly, even though no one knew what function the BRCA1 protein served, BRCA1 clearly did *not* cause breast cancer. On the contrary, all the causal *BRCA1* variants (and nearly every afflicted family turned out to have a different causal variant in the *BRCA1* gene) either led to nonfunctional BRCA1 protein or insufficient quantities of BRCA1 protein. Rather than causing cancer, BRCA1 somehow *protected* women from breast cancer. The *BRCA1* gene variants that led to increased breast cancer susceptibility were doing their damage by disrupting the action of this protective protein.

Identifying the *BRCA1* gene and its disabling variants was a major scientific milestone. The discovery enabled women from families with recurrent breast cancer to find out whether they had inherited the disabling genetic variant that ran in their family. However, the discovery didn't explain what function the BRCA1 protein normally carries out, nor how it helps to protect women from breast cancer.

To study the functioning of the BRCA1 protein, scientists relied on knockout and transgenic mice. In early experiments scientists engineered knockout mice in which one of their two copies of the *BRCA1* gene had been disabled. The hypothesis was that these mice would have phenotypes similar to those of people who had inherited one nonfunctional copy of the *BRCA1* gene. Unfortunately, the hypothesis turned out to be wrong. The researchers discovered that these knockout mice didn't have any increased incidence of tumors; in the case of *BRCA1*, a simple mouse model didn't accurately mirror what happens in humans. Scientists next tried to produce mice with *both* copies of their *BRCA1* gene disabled, but BRCA1 turns out to be an essential protein in mice, and the mice were stillborn. Eventually scientists engineered conditional knockout mice in which the *BRCA1* genes were disabled only in the mammary glands. They also developed transgenic mice with *human BRCA1* genes in their mammary glands. By performing experiments on these animals, scientists are finally starting to understand how the BRCA1 protein acts in the development of normal mammary gland tissue, as well as in the repair of the damaged DNA that occurs in cancer.

IT'S MORE THAN JUST OUR GENES

Despite all the insights gained from studying transgenic mice, the mechanisms by which BRCA1 protein protects cells from breast and ovarian cancer are still only partly understood. What is clear is that to unravel the origins of breast cancer scientists will need to understand how BRCA1 interacts with all the other proteins in its biochemical pathways. Each of these proteins is also encoded by a gene, which may have a diversity of variants affecting a woman's vulnerability to breast cancer. Scientists are also learning that nongenetic factors contribute to breast cancer. In fact, the epidemiological studies that helped identify the link between BRCA1 variants and cancer susceptibility demonstrated that the majority of breast cancers *don't* run in families. The importance of nongenetic factors in breast cancer is also evident from the rapid increase in the incidence of breast cancer over the last century. In the 1940s, approximately 1 in 1000 women was stricken by breast cancer in her 50s. By the year 2000, that number had risen to nearly 3 women out of 1000.* Changes in human DNA occur over thousands of years; so whatever has caused this sharp increase in breast cancer incidence must be related to something nongenetic. It has to be linked to some change in the environment.

---

* Although it is difficult to estimate the extent to which more accurate diagnosis accounts for some of the increased number of cases, it seems likely that at least part of the observed threefold change is due to a real increase in breast cancer incidence.

Scientists have known for decades that cancer may just as easily be triggered by something in our environment as by an inherited genetic variant. For example, the existence of chemicals in tobacco smoke that can trigger biological pathways leading to lung cancer is well known. Identifying environmental factors that affect breast cancer has been more challenging. Increased use of oral contraceptives and hormonal replacement therapy for post-menopausal women appear to play a role. Increased alcohol consumption and modern fatty diets have also been linked to increased susceptibility to breast cancer.

Animal studies have also identified environmental pollutants and toxins, of which the best known are the dioxins, which in large doses can cause breast cancer in animals. Demonstrating that dioxins or other environmental pollutants contribute to human breast cancer has been more challenging. Obviously scientists can't perform experiments in which women are exposed to dioxins, but long-term studies of breast cancer rates in women who were exposed to high levels of dioxins during an industrial accident suggest that dioxin exposure also contributes to human breast cancer. That said, there is still insufficient evidence to prove that these environmental factors are the cause of the dramatic rise in nonhereditary breast cancer.

Breast cancer appears to arise from a complex interplay of genetic and environmental factors. Although science's understanding of the genetic factors underlying the disease has improved significantly in the last two decades, our knowledge of the environmental components is still quite limited. As a result, there are still no easy answers for women like Angelina Jolie who have a *BRCA1* mutation. Such women know that the *BRCA1* variant they've inherited gives them a greater chance of getting breast or ovarian cancer; epidemiological studies have shown that a 20-year-old woman with a disabling *BRCA1* mutation has a 65% chance of getting breast cancer by the age of 70 (and a 39% chance of getting ovarian cancer), while 20-year-old women without the *BRCA1* mutation have only an 8% chance of getting breast cancer before turning 70.

The problem facing Jolie and other women with inherited BRCA1 mutations is that using that information to decrease their cancer risk isn't simple. A vulnerable woman can be very diligent about having regular mammograms and otherwise monitoring her health. That won't stop her from getting cancer, though it may help her doctors detect the disease earlier, when it is more treatable. She may decide not to have estrogen replacement treatment when she reaches menopause, since estrogen replacement may increase her risk of breast cancer. She could even choose the extreme measures taken by Angelina Jolie of preventive mastectomy and/or oophorectomy (surgical removal of the ovaries), which might prevent her from getting cancer but could well have a major negative impact on her quality of life. Such choices would be difficult for anyone.

13

WHAT IS LEARNING? WHAT IS MEMORY?

As she later wrote in her *New York Times* bestseller *Brain on Fire*, in 2009, Susannah Cahalan was a 24-year-old with a promising future. A healthy, bright and attractive woman, Cahalan was enjoying her first year as a reporter for the *New York Post,* as well as life with her boyfriend in New York City. Then Cahalan's life abruptly changed. She began to suffer from a strange neurological disorder in which she experienced seizures, paranoid delusions and amnesia. Yet unlike the individuals with the brain disorders we've considered previously – fragile X syndrome and Huntington's disease – Cahalan was not suffering from a genetic disease. Cahalan's condition, called anti-NMDA receptor encephalitis, is an autoimmune disease, and though its fundamental causes are still poorly understood, what is known suggests that the disease is caused less by one's genetic variants than by one's environment.

As we'll see in this chapter, our environment can change our lives by altering the amounts and types of proteins produced by our cells. Scientists refer to changes in our cells that are *not* the result of sequence variations in our DNA (not genetic) as epigenetic changes.* Epigenetic changes may

---

* *Epigenetic change* has more than one definition within the scientific literature. Some scientists reserve the term for cellular alterations that can be inherited or that modify chromosomes. In contrast, we'll refer to any cellular modification that doesn't alter the cell's DNA sequence as an epigenetic change.

come from events in our external environment, such as a viral infection or smoking a cigarette. They can also result from changes in our body's *internal* environment, the local environment surrounding a cell. Epigenetic factors can alter a protein's biochemical activity or the rates at which proteins are synthesized or degraded. Often, all of these types of epigenetic changes occur simultaneously within a single cell.

HOW DOES THE ENVIRONMENT CHANGE US?

In some cases, the cellular mechanisms by which our environment affects us are well understood. When we are infected by a virus, viral proteins bind to receptors on our cells, causing the receptors to change shape. This enables the virus to enter the cell. Once within the cell the virus binds to cellular proteins, causing certain genes to be transcribed while others are silenced.

The environment doesn't only have an impact on us when we are sick. Even something as mundane as the food we eat changes us at the cellular level. When we eat, our stomach first digests the food, and the resulting nutrient molecules are absorbed by our bloodstream. Receptors on the surfaces of cells then detect and utilize these nutrients. Glucose is combined with oxygen in our mitochondria to produce energy. Glucose also binds to receptors in our pancreas, causing the pancreas to make and secrete insulin. This insulin is detected by receptors in other cells, which activate or disable proteins that regulate a wide variety of cellular processes. It's amazing to realize that these cellular events are all initiated by something as basic as the food we eat.

HOW DO CELLS REMEMBER?

In contrast to the relatively simple examples described above, environmental factors affect our cells in other ways that are far less well understood. These include processes in which considerable time elapses between the environmental event and when the effects it induces are observed. For example, months or years may pass between the time when we read something and when we recall it from our memory. Similarly, it may be a long time from when we start exercising until we can observe any changes in our muscles or our athletic performance.

Protein shapes and mRNA concentrations tend to quickly return to their basal values after an environmental stimulus has been removed. Yet somehow cells are able to "remember" external events that occurred months or years in the past. Just how our cells accomplish this feat has long been a mystery. Recently, though, scientists have started to unravel the underlying biochemical mechanisms, and they're learning that one of the cell's key strategies is to alter its chromosomes.

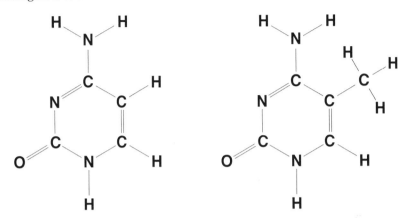

Figure 13.1. Schematic illustration of normal (left) and methylated (right) cytosine. The structures shown are oversimplified, as the actual structures are three-dimensional. N = nitrogen, O = oxygen, C = carbon, H = hydrogen. Lines = chemical bonds. Double lines = chemical bonds in which two electrons are shared.

You'll recall that chromosomes consist of DNA, which is packaged with proteins to support and protect the DNA. The sequence of DNA letters in our chromosomes rarely changes, and if it does, the result tends to be dramatic, often resulting in a tumor or cancer. Instead, our environment generally modifies our chromosomes in subtler ways that do not change their DNA sequence. In some cases the chromosome's DNA is chemically modified, though the sequence of DNA letters remains unchanged. The most common form of such DNA modification is called DNA methylation. In DNA methylation, special enzymes called DNA methyltransferases, or DNMTs, transfer a carbon atom and three hydrogen atoms (collectively called a methyl group) onto a DNA cytosine nucleotide (cytosine is one of the four DNA nucleotides). Figure 13.1 shows a two-dimensional schematic of the structures of methylated and normal, unmethylated cytosine.

Other chromosomal modifications involve chemical changes to the proteins that surround the DNA. None of these chromosomal modifications change the DNA recipes for the proteins our cells produce, but they can influence the rate at which genes are transcribed and the proteins that they encode are synthesized. As a result, such subtle chemical modifications to our chromosomes can have a profound effect on which proteins are produced by our cells.

CELLULAR ENVIRONMENT AND DEVELOPMENT

Nothing is more important for a cell to remember than what kind of cell it is. A blood cell has to remember to transport oxygen, and a pancreatic beta cell has to remember to secrete insulin. This phenomenon, which is called cell differentiation, may sound trivial, but it is really quite remarkable and profound, since every cell has the same sequence of DNA and is ultimately a descendent of the same fertilized egg cell.

How does an embryo orchestrate its development so that some of its cells become blood cells while others become skin cells or neurons? Scientists have learned that a key element of this process is the pattern of protein and DNA modifications that each cell stores in its chromosomes. These varying patterns of chromosomal alterations ensure that neurons synthesize only proteins required for nerve signaling, while muscle cells just make the proteins required for mechanical energy production.

A developing organism produces one set of chromosomal modifications within each of its blood cells and another set of chromosomal alterations in each of its skin cells. In fact, variations in chromosomal modifications between *types* of cells are greater than the chromosomal variations between individuals. In other words, the chromosomal modifications of your blood cells are more similar to the chromosomal modifications of my blood cells than they are to the modifications of your heart cells.

Ensuring that the appropriate genes are transcribed in each cell is critical for an organism. If this process goes awry, an undesired gene may be made into protein or a critical protein may not be synthesized. The consequences for the individual can be catastrophic. Such abnormal epigenetic modification may be caused by something in the external environment or by the presence of an unusual genetic variant. For example, aberrant epigenetic modification triggered by environmental factors often plays a key role in cancer, while epigenetic variations caused by a mutation in the *FMR1* gene leads to fragile X syndrome.

THE CHALLENGE OF EPIGENETICS

The complete collection of protein, messenger RNA and chromosomal modifications within a cell is called the epigenome. Since the epigenome stores all of the accumulated environmental influences on our cells, studying the epigenome is critical to understanding how our environment affects us at the cellular level. Scientists generally study the epigenome by experimenting with animals. Typically, they will first measure part of an animal's epigenome. They can then expose the animal to some environmental change and determine whether any epigenetic modifications resulted from the external event. Similar studies can sometimes be performed with human subjects, but human environments and genomes are much more diverse and complex than those of a laboratory animal. Consequently, determining just what caused human epigenetic modifications is much more difficult.

Even with laboratory animals, epigenetics is much more challenging to study than genetics. After all, essentially every cell in the body has the same DNA sequence, and that DNA sequence remains the same throughout the life of the animal. In contrast, the epigenome varies among different cell types and changes over time. Epigenetic modifications can be quite different in cells taken from a child at age 5 and the same type of cells taken from the same child at age 10. Consequently, no one has attempted to map out the *entire* human epigenome, and probably no one ever will.

Despite these challenges, studying small parts of the epigenome has already yielded clues about how the environment affects us at the cellular level. And as DNA sequencing costs plummet, larger epigenomic analyses are becoming feasible. In the early 1990s, the cost of DNA sequencing was approximately $1 per base. In 2001, the cost of sequencing a million bases of DNA had dropped to approximately $5000. By 2005, the cost of sequencing a million DNA bases had fallen to $800, and by 2008 to only $40. In 2011, sequencing the same million bases cost 12 cents, and sequencing projects, which once required months to complete, could be performed in hours. As a result, scientists are now able to perform complex epigenetic analyses of large numbers of genes in multiple cell types under varying environmental conditions.

LEARNING CHANGES OUR CHROMOSOMES

Among the most profound ways that our external environment affects our lives is through learning and memory. We read a "life-changing" book or see an "unforgettable" movie and still remember it years later. After decades of research, scientists are beginning to understand the brain processes by which this remarkable phenomenon occurs. Animal experiments are showing that we learn by creating and removing interconnections among our brain cells.

These interconnections are formed when two brain cells grow close enough to one another so that one cell can secrete a neurotransmitter that can be detected by a receptor on the neighboring cell.

Scientists now understand that these brain-cell interconnections play a critical role in learning and memory, yet much less is known about the molecular mechanisms by which these interconnections are created and destroyed. Although experiments over the last 30 years indicated that epigenetic modifications, such as DNA methylation, influence learning and memory, only recently have researchers begun to understand in detail the brain processes by which this occurs.

For example, in 2010 researchers at the University of California, Los Angeles, studied learning and memory in mice that lack DNA methyltransferases, the enzymes required for DNA methylation. Engineering DNMT knockout mice is challenging. Mice have two DNA methyltransferases, DNMT1 and DNMT3,* and the scientists had to design double knockout mice with both the *DNMT1* and *DNMT3* genes disabled. Since DNA methylation is critical for overall animal development, the scientists needed to engineer conditional knockout mice, in which *DNMT1* and *DNMT3* were nonfunctional only in the neurons of the mouse's forebrain, where learning was known to occur.

Once the researchers had produced their *DNMT1-DNMT3* conditional knockout mice, they were able to compare the knockout mouse's learning ability with that of a normal mouse of the same strain. To test mouse learning, the scientists used an apparatus called the Morris water maze. It consists of a water tank with a platform slightly below the water's surface. A substance such as powdered milk is added to the water so that it becomes opaque and the platform can't be seen. When a mouse is placed in the water maze, it typically swims as fast as it can to try to get out of the water. Initially, it will swim in random directions, but by placing marks on the sides of the tank, researchers can teach mice to swim directly to the hidden platform. Mazes can also be used to test animal memory: one trains a mouse to solve a maze and then waits, say a week or a month, and tests how well the mouse remembers the correct route. Figure 13.2 shows a schematic of a Morris water maze.

---

* Mouse DNA has another gene, which was originally called *DNMT2* because its DNA sequence looked similar to that of *DNMT1* and *DNMT3*. Subsequent experiments showed that the DNMT2 protein methylates RNA, rather than DNA.

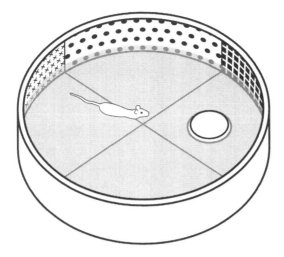

Figure 13.2. Morris water maze test. The mouse attempts to escape the water by finding the platform (which is shown as a double circle, but in the actual apparatus can't be seen through the turbid water). The four quadrants of the maze's tank walls have different patterns, so the mouse can learn in which quadrant the platform is located.

Using the water maze, the scientists observed that the *DNMT1-DNMT3* knockout mice were consistently worse learners than normal mice. After 12 days of training, the normal mice typically found the platform in less than 20 seconds; the mice with nonfunctional *DNMT* genes needed more than 30 seconds. The researchers also found that, after training, the DNA in the neurons of the normal mice had been methylated near genes involved in memory formation. Since the only known function of DNMTs is to methylate DNA, learning appears to involve DNA methylation (at least in mice).

CHROMOSOMES, LEARNING AND GETTING OLD

The link between learning and chromosomal modification is also supported by a striking experiment performed by scientists from the European Neuroscience Institute in Göttingen, Germany. The German researchers demonstrated that a different form of chromosomal modification, called lysine acetylation, also contributes to learning and memory. (Recall that lysine is one of the 20 amino acids that make up proteins and consequently is a component of chromosomal proteins.)

The Göttingen team used two different tests to assess mouse learning and memory: the water maze test and a fear-conditioning test. In fear conditioning, mice are presented with a light or other cue, followed by a mild electric shock. Mice learn to associate the cue with the shock and consequently display fear behavior (typically freezing in place) as soon as they see the cue. Fear conditioning can be used to assess mouse memory. Some mice show fear behavior from the visual cue 24 hours after training, while other mice have already forgotten the association between the cue and the shock by then. Figure 13.3 illustrates the fear-conditioning test.

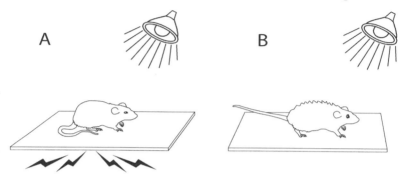

Figure 13.3. Fear conditioning. A: During training the animal is repeatedly exposed to a cue such as a bright light, which is then followed by a shock. B: Subsequently, the animal is exposed to only the cue, and the extent of the animal's fear response (typically freezing behavior) is measured. Fear conditioning is sometimes used to assess fear sensitivity and sometimes used to measure learning or memory.

Using these tests, the Göttingen scientists showed that young (3-month-old) mice were better learners than older (16-month-old) mice. After 10 days of training with the water maze, the 3-month-old mice found the hidden platform in approximately 12 seconds, while the older mice required twice as long. The scientists also found that younger mice were better learners in the fear-conditioning test. These observations, by themselves, weren't very exciting. Scientists already knew that, just like humans, aging mice aren't quite as sharp as young ones.

What was more interesting was that when the mice learned, several lysine amino acids in their brain's chromosomes became acetylated, but in the older mice, one of the acetylation reactions no longer occurred. Apparently, the brains of older mice are no longer able to carry out one of the chromosomal modifications that occur in mouse learning. To directly show that the loss of this lysine acetylation contributed to the impaired learning in the old mice, the researchers gave half of the old mice a drug that facilitates acetylation. Remarkably, the drug improved the memory of the older mice. After receiving the drug, the older mice on the drug froze in place 60% more often than the mice that hadn't received the drug – presumably because they still remembered to associate the visual cue with the shock.

NEUROTRANSMITTERS, RECEPTORS AND BRAIN DISEASE

In this chapter, we've had a glimpse of how the environment can change our cells and, in particular, how learning and forming memories modify the cells of our brain. Can we use this knowledge to gain any insight into what happened to Susannah Cahalan, the unfortunate young woman who suddenly found herself experiencing amnesia and paranoid delusions?

To explain what was happening to Cahalan we'll need a few more facts about how the brain functions. The first important fact we need to remember is that one of the most important neurotransmitters involved in learning and memory is a molecule called glutamate. You may recall that glutamate is the neurotransmitter that malfunctions in fragile X syndrome, which helps explains why people with the disease have intellectual disabilities.*

The second important piece of information you need to know is that a key nerve cell receptor of glutamate signals in the brain is a protein called the NMDA receptor. We'll explore the relationships among glutamate, NMDA receptor and brain function in more detail when we look at the biology of intelligence in chapter 18. For now, it's sufficient to know that if the NMDA-receptor proteins in someone's brain are impaired, it's going to have a severe impact on their brain's glutamate signaling and hence on their ability to learn and remember. Although most people's brains process glutamate signals properly, in a few unlucky individuals, something in the environment triggers the immune system to attack their own NMDA-receptor proteins. The result is the disease called anti-NMDA receptor encephalitis, in which malfunctioning in the brain's communication system leads to the psychological and memory-loss symptoms experienced by Cahalan.

Anti-NMDA receptor encephalitis was first reported in the medical literature only in 2007. Most likely the disease existed earlier, but nothing was known about it, and with no treatment, patients endured devastating psychiatric and neurological disturbances. Even now, scientists have only limited understanding of what environmental factors trigger the self-destruction of one's glutamate receptors. In many cases, the initial trigger

---

* By the way, glutamate is chemically similar to glutamic acid, one of the 20 fundamental amino acids we learned about in chapter 2, and it's also related to the familiar food additive MSG (monosodium glutamate). But that doesn't mean eating MSG will make you smarter.

is a cancer, typically an ovarian tumor in a young woman. The autoimmune response may also result from a viral infection. Often though, as in Susannah Cahalan's case, doctors are unable to identify an event that initiated the brain's autoimmune response. Moreover, even when the underlying trigger can be determined, how the immune system attacks the brain cell receptors is not well understood.

Despite our limited understanding of the details of anti-NMDA receptor encephalitis, scientists have made progress treating this dreadful disease using drugs that suppress the immune system. Today, approximately 75% of people with anti-NMDA receptor encephalitis fully recover and are able to resume normal healthy lives. Susannah Cahalan was one of that fortunate 75%. After a month of treatment, her psychotic symptoms and delusions disappeared, and her brain began to function normally again, though to this day she still can remember very little of what occurred during the month of her illness.

14

CAN WE LIVE TO 120 – BY EATING LESS?

When I was a boy, my mom would tell me stories from her childhood. One recurring character was her grandmother, Regina. Regina was a woman with no identifiable disease, but was always described as frail and old. My mother's parents told her that she had to be absolutely quiet and well behaved around Grandma Regina. Only years later did I realize that this "frail old lady" had actually been in her mid-50s.

Our perception of old age has changed dramatically since my mother's childhood. In 1910, human life expectancy was approximately 54 years. Today, women in industrialized countries often live more than 80 years, and medical treatments can cure many of the debilitating infectious diseases that crippled people a century ago. Changes in nutrition and lifestyle have also contributed to increasing human life expectancy. A 54-year-old woman is not likely to be described as frail or called an old lady anymore. Yet this begs the question of how much more can we increase human longevity. By further improving medical care and enhancing our environment, might people in the future lead healthy lives to the age of 120? To 150? Might we someday be able to create a world where we live, essentially, forever?

That our environment influences how long we live is hardly a surprise. Animals live longer when they are shielded from their predators, and humans live longer if they eat well, exercise and are protected from infectious diseases. Along with these obvious environmental influences, over the last century scientists have discovered more surprising factors that affect an animal's lifespan, and possibly the life expectancy of people as well. In this chapter, we'll look at some of these unexpected environmental influences on longevity and learn how extreme diets and other forms of environmental stress can dramatically increase an animal's lifespan. We'll even see evidence hinting that similar techniques for manipulating the environment may increase human longevity as well.

CALORIC RESTRICTION AND AGING

In the early 1930s, the United States was in the grip of the Great Depression. Chronic hunger was a serious and widespread problem, and many feared that in addition to causing obvious suffering, long-term hunger might also shorten life expectancy. To test whether semistarvation did shorten lifespan, in 1934, Clive McCay and Mary Crowell, researchers at Cornell University, performed a series of experiments with laboratory rats. In these experiments, rats were placed on calorically deficient diets for extended periods of time. (Just how restricted these diets were was not recorded in their paper, but typically in such experiments, animals consume 10% to 50% fewer calories than normal.) The researchers did give the rats all the nutrients they absolutely needed, but they made sure the rats ate less than they wanted.

The results of the experiment were as dramatic as they were unexpected. On average, the "starving rats" lived 35% *longer* than their litter mates who were allowed to eat normally. McCay and Crowell's discovery that caloric restriction leads to longer lifespan has subsequently been replicated in experiments on flies, worms, mice and other laboratory animals. The results nearly always show that caloric restriction increases life span, typically by around 30% to 40%. Occasionally, calorie-restriction experiments have had even more dramatic results. In some experiments, researchers have been able to double the life spans of laboratory mice, from three years to six.

CALORIC RESTRICTION AND LONGEVITY IN MONKEYS

Just because caloric restriction increases longevity in flies, worms and laboratory rodents doesn't prove that it also increases life span in people.* Humans are not laboratory animals that can be bred to have identical DNA sequences and forced to live in uniform environments, so controlled human longevity experiments are impossible. Consequently, to gain insight into the processes of human aging, scientists study aging in monkeys. Two long-term studies of aging and caloric restriction in rhesus monkeys were begun in the 1980s, one at the University of Wisconsin Primate Center and the other at the primate laboratories of the National Institute on Aging (NIA) in Bethesda, Maryland. These experiments had to be carried out over more than 20 years, since the average lifespan of a rhesus monkey in captivity is 27 years.

The Wisconsin researchers published their findings first, and their 2009 paper indicated that caloric restriction improves health and longevity in at least one species of primates. Monkeys allowed to eat as much as they wanted had three times the number of "age-related" deaths (that is, excluding deaths from, for example, injuries) as the calorically restricted animals. The calorically restricted monkeys were also healthier than the freely fed monkeys, using measures of monkey age-related diseases, such as cancer, diabetes and cardiovascular disease. Unfortunately, when the National Institute on Aging published its results in 2012, its conclusions regarding survival rates were quite different. The NIA experiment found that, although the calorically restricted monkeys had lower triglyceride and cholesterol levels and somewhat fewer age-related diseases, the two groups of monkeys had similar overall survival rates.

How could two highly respected research groups perform the same experiment and get such different results? The NIA researchers carefully compared their methods with those of the University of Wisconsin team and found subtle, but important, differences. First, in the Wisconsin experiment, the control-group monkeys were allowed to eat as much as they wanted; in contrast, the NIA control monkeys were fed a "standard," but not unlimited,

---

* We're not just talking about preventing *overeating;* scientific studies have consistently shown that obesity has a major negative influence on human health and life expectancy. Rather, the question is whether caloric restriction beyond what is typically considered a normal diet can increase human life expectancy.

diet. Second, the *compositions* of the monkey diets were different. The Wisconsin monkeys were fed a purified diet in which the composition of each nutrient was precisely known. The NIA monkeys were fed a more natural diet that included a greater variety of nutrients. Precisely how these methodological differences caused the differing experimental results is still not completely understood, but the NIA data showed that the beneficial effects of caloric restriction, at least in monkeys, were not as clear-cut as had been previously thought.

WILD MICE, LABORATORY MICE AND CALORIC RESTRICTION

The conflicting results in the two experiments were apparently caused by differences in the study design, in particular, in the ways that the monkeys were fed. For some scientists, this discovery did not come as a big surprise. These scientists, led by Steven Austad at the University of Texas, had already been skeptical of the caloric restriction experiments in mice and rats. While they hadn't questioned the dramatic lifespan increases in the calorically restricted rodents, the researchers had challenged the interpretation of these results.

Austad's main point was that laboratory animals are extremely different from normal, wild animals. Laboratory animals are highly inbred, get little exercise and live on different, and often less healthy diets from animals in the wild. In fact, laboratory mice have often been inadvertently *bred* to overeat, as any genetic variants that cause rapid growth and reproduction lead to more offspring in a laboratory setting. In contrast, genetic variants causing mice to eat less and exercise more are advantageous in the wild: animals with these genes are better able to evade predators and hence are more likely to reproduce. These healthy genetic variants tend to disappear from laboratory mouse DNA after many generations of such "unnatural selection." In other words, Austad argued that laboratory mice have artificially *short* life expectancy and that calorie-restriction experiments just enable laboratory mice to regain some of their natural lifespans.

To test such criticisms, scientists should ideally test mouse (or monkey) anti-aging interventions in the wild. Studying animal longevity in the wild is difficult, though; animals, and especially mice, typically get eaten long before they get old. So the University of Texas team proposed to study longevity in the laboratory, but with the offspring* of mice that had been captured in the wild. The Texas researchers found that the average lifespan of wild-offspring mice was approximately 20% longer than that of inbred

---

* Actually, they used the second-generation "grand-offspring" of the wild mice.

laboratory mice, whether or not the lab mice were calorically restricted. They also found that the *average* lifespan of the wild-offspring mice was the same whether or not they were calorically restricted. While more of the calorically restricted wild-offspring mice survived to mouse old age, they were also more likely to die when they were very young, compared to standard lab mice. So whether caloric restriction was a useful preventative for aging in wild mice remained controversial.*

CALORIC RESTRICTION, STRESS AND LONGEVITY IN HUMANS

The caloric restriction experiments in monkeys and wild-offspring mice dampened the earlier hopes that caloric restriction would significantly increase human life expectancy. Nevertheless, the animal results have been encouraging enough to motivate scientists to search for evidence that might show whether caloric restriction influences human longevity. One approach has been to study "natural experiments," in which people have been subjected to caloric restriction for reasons unrelated to any scientific experiment. Throughout the world, millions of impoverished people live with too little food all of their lives. Sadly, the overwhelming majority of these people also face poor nutrition and lack of adequate health care, resulting in low life expectancies.

One group of people, though, the Okinawans, lived for many years with restricted diets because of poverty, while having adequate basic nutrition and health care. Between 1945 and 1960, the average Okinawan consumed approximately 11% fewer calories than what is normally considered necessary for maintaining body weight. (By 1960, economic conditions on Okinawa had improved, so caloric intake was comparable to that of people from neighboring Japan.)

What makes the caloric data about Okinawans intriguing is that they include a high number of individuals who live to a very old age. While in most developed countries only 10 to 20 out of every 100,000 people become centenarians (that is, live to the age of 100), the rate for Okinawans is 40 to 50 per 100,000. These Okinawan centenarians were young adults during a period of caloric restriction on the island. Is it possible that their longevity is partly the result of their restricted diets 50 or 60 years earlier? Well,

---

* Of course, it's also possible that many people today who are overfed and don't get enough exercise are biologically more similar to laboratory mice than to wild mice. If so, then using lab mice rather than wild mice as models for human aging may not be a bad idea at all.

maybe. But the Okinawan studies also showed that many of the Okinawan centenarians belonged to the same families, suggesting that genetics also plays a role in helping Okinawans live long lives. As is so often the case when studying people, disentangling the effects of genetics and the environment isn't easy.

LIFE EXTENSION BY VOLUNTARY (SEMI-)STARVATION?

Despite the limited evidence that caloric restriction extends human lifespan, some people are willing to go hungry in the hope of having a longer, healthier life. One such person is Brian Delaney, a Swedish-American man of about 50. He's almost 6 feet tall and weighs approximately 135 pounds. Delaney eats only two meals per day, restricting his daily caloric intake to 1500 calories. (The normal daily caloric intake for a man his age is considered to be approximately 2400 calories.)

Delaney has been limiting his diet for 20 years, hoping that it will bring him health and long life, just as it has for rats and mice. So far, Delaney is healthy, and he says that continually feeling hungry doesn't bother him anymore. According to Delaney, having people always tell him he looks too skinny is more annoying – but not annoying enough to stop him from trying to live longer by eating less.

Brian Delaney isn't alone. Delaney estimates that at least a few thousand people worldwide are using caloric restriction in hopes of increasing their lifespan. Delaney should know. He is the founder and president of Caloric Restriction Society International (CRSI), an organization that provides support to people experimenting with caloric restriction and that helps foster research on its impact. CRSI is less than 20 years old, and whether its members will in fact live longer than other people is still unknown. In addition, CRSI members may have other habits that are important to life expectancy: Do they exercise more? Take more nutritional supplements? Go to the doctor more often? So, even if CRSI members do live longer, attributing their longevity to their diet will be difficult. Nevertheless, studying CRSI members might provide hints to links between human health, longevity and caloric intake.

With this in mind, a 2004 study by scientists from Washington University Medical School compared physiological measures linked to health and longevity, such as cholesterol levels and blood pressure, between CRSI members and people who did not limit their calories. Since few people have been practicing long-term caloric restriction (the average time of caloric restriction of individuals in the CRSI study was six years), only 18

CRSI subjects were available for the study. With so few research subjects, the study's findings need to be viewed with caution. That said, the results were quite dramatic. On average, CRSI members were in the lowest 10% of people their age for both overall cholesterol and LDL cholesterol levels ("bad" cholesterol). Their average blood pressure, approximately 100/60 mmHg, was unusually low, also a sign of cardiovascular health. Medical records available for 12 of the CRSI subjects showed that before starting caloric restriction, they had average blood pressures of 132/80 mmHg, suggesting that the low blood pressure was a consequence of caloric restriction. Since only 18 people participated in the CRSI study, and they aren't old yet, one can't draw any conclusions from this experiment on the longevity benefits of caloric restriction, but the CRSI members do seem to be a very healthy group of people.

To learn more about caloric restriction and human longevity, scientists from Pennington Biomedical Research Center and Duke, Tufts and Washington universities have begun a larger trial, called the Comprehensive Assessment of Long-term Effects of Reducing Intake of Energy Project, or Calerie Project. This study will be the first large-scale attempt to monitor the effects of human caloric restriction. The project has recruited 225 volunteers, all making a two-year commitment to consume 25% fewer calories than their "personal caloric baseline."* The study excludes obese individuals, and though some are slightly overweight, many of the study participants are already in what doctors consider the healthy weight range. Although a two-year study can't demonstrate that caloric restriction increases human lifespan, if the volunteers show changes in cholesterol levels and blood pressure similar to those of the CRSI members, the results will suggest that caloric restriction does lead to a healthier, if not necessarily longer, life.

WHAT YOU EAT VS. HOW MUCH YOU EAT

Laboratory experiments have shown that caloric restriction can slow down aging in some experimental animals. They also demonstrated that the composition of the animals' food can influence the aging process as much as the quantity of food. Learning that *what* you eat is as important as how much you eat is hardly a surprise. After all we are continually being told that we should eat a nutritious balanced diet and not just "empty calories."

---

* Determining this personal baseline is important to ensure that participants receive sufficient nutrients, but it is not easy to measure, requiring sophisticated and expensive metabolic testing.

Experiments with mice and rats have shown, though, that the calories to restrict are not necessarily what you may have thought, especially if you've been trying to maintain your health with a high-protein diet. The animal experiments demonstrated that decreasing protein consumption, especially proteins with large amounts of the amino acid methionine, was the most important aspect of caloric restriction. In fact reducing methionine consumption was as effective as restricting total calories in increasing lifespan. In experiments performed back in 1993, rats were fed rat chow with only 20% of the usual amount of methionine.* Since methionine is important for growth, the methionine-deprived rats grew to only one-quarter the size and weight of normal laboratory rats (100 grams versus the normal lab rat weight of 400 grams). But they also lived 30% longer than their litter mates. They ate as much as they wanted and still lived long lives, as long as their methionine consumption was restricted.

How does restricting methionine intake slow aging? Scientists don't have the complete answer. In part, the reason is that methionine is an essential amino acid, meaning that animals can't synthesize methionine and must obtain it from their food. Animals lacking methionine in their diets have difficulty growing and reproducing. Rodents on diets lacking other essential amino acids also live longer than normal, though the lifespan increases are less dramatic than for a methionine-restricted diet. Beyond being an essential amino acid, methionine is unique in that, as you may remember from chapter 2, *every* protein must start with a methionine amino acid; consequently, with limited methionine, cells have difficulty producing any protein. A methionine-limited animal will have even more difficulty growing and reproducing than an animal that has any other essential amino acid restricted.

Recently, there have even been hints that lowering protein consumption may increase longevity in humans. In 2014, researchers from the University of Southern California reported an 18-year study of health and longevity of over 6000 individuals for whom they had data on their relative consumption of protein, carbohydrates and fat. They found that people who in their 50s and early 60s consumed a lower ratio of protein to carbohydrates, and especially a lower rate of animal-derived protein to carbohydrates, had significantly lower mortality rates. To be sure, they also found that individuals over age 65 who were consuming lower amounts of protein had slightly *higher*

---

* When the amount of methionine was reduced even further, the rats started dying at a younger age, showing that a small amount of methionine is essential for survival.

mortality rates – so clearly there are many factors that still need to be sorted out before one can say whether and how low-protein diets affect longevity.

That said, this early human study, along with the studies on rodents, might lead one to hope that methionine or overall protein restriction might benefit humans too. If this is true, the impact on human health could be substantial. Designing a low-methionine and low-protein diet for people is not that difficult; most vegan diets are low in protein and methionine. Most fruits, vegetables and grains contain relatively little protein per calorie, and whatever protein there is turns out to have lower methionine content than animal protein. And a vegan diet, though perhaps not the most appealing to some, definitely beats the perpetual hunger of total caloric restriction.

STRESS, REPRODUCTION AND LIFESPAN

Caloric restriction is not the only unexpected environmental stress that can increase an animal's life expectancy. Researchers have shown that varying an animal's core body temperature or exposing it to limited oxidative damage (for example, by placing a worm in a chamber with very high air pressure) can also increase longevity. Recent experiments on worms sent to the International Space Station suggest that even zero gravity can slow the aging process – at least in worms. These stresses need to be introduced carefully (not for too long or at too extreme a level); otherwise the animals die. The lifespan increases are generally modest, typically 10% to 20%, but the experimental results do show that mild environmental stressors other than restricted food intake can also lead to longer life.

Another unexpected environment influence on life expectancy is that (at least among experimental animals) the potential of having future offspring can lead to increased longevity. When researchers at University of California, San Francisco, removed the sperm and egg cells* from nematode worms while keeping their reproductive systems intact, the worms' life expectancy increased by approximately 60%. In contrast, if the scientists removed the worm's entire reproductive system (so that the worms could *never* again produce sperm and egg cells) the worm did *not* live longer. In other words, the worm's reproductive tissues were somehow able to communicate to the rest of its body, leading to increased lifespan, and apparently the worm's longevity-enhancing trigger signal is the combination of not being able to reproduce now yet being able to reproduce in the future.

---

* Nematode worms are hermaphrodites, meaning they are part male and part female, as each worm produces both sperm and egg cells.

The potential for future reproduction may also be linked with longevity in mammals. In 2003, a team from University of California, Davis, transplanted ovaries from young mice into old ones (female mice beyond the normal mouse breeding age). The mice with the transplanted ovaries consistently lived longer than old female mice that had not received young ovaries. The scientists also observed that the mice with transplanted ovaries had hormone concentrations typical of younger mice, suggesting that altered hormone levels might have contributed to their increased longevity.

REPRODUCTIVE STATUS AND HUMAN LIFE EXPECTANCY

Might reproductive potential affect longevity in humans as well? Two recent studies hint that this could be the case, at least for women. The first study assessed the impact of oophorectomy (surgical removal of the ovaries) on mortality. The researchers compared long-term survival of women who had their ovaries removed at the time they had a hysterectomy (surgical removal of the uterus) with those who had not.[*] The scientists found that the women with oophorectomies had lower rates of ovarian (and breast) cancer than those who had not undergone oophorectomy. This was not surprising; oophorectomy is known to be an effective preventative measure against these cancers. Remarkably, though, the *overall* mortality rates (including noncancer deaths) were *higher* among women who had oophorectomies, suggesting that hormones or other molecules produced by the ovaries may be creating signals in a woman's body that promote longevity.

The second study that suggested a link between a woman's reproductive status and her life expectancy analyzed oral contraceptive use and longevity. Scientists compared mortality rates among 1700 women, some of whom had never used oral contraceptives; the others had used them for at least eight years. The data confirmed that cervical cancer mortality was higher among women who had used oral contraceptives. This had long been known and was the concern that had originally prompted the study. More interesting was the study's second finding: women who regularly used oral contraception had *lower overall* mortality rates. In other words, women taking contraceptives appear to be subjected to two effects: a bad one, increasing chance of cervical cancer, and a mysterious positive one that increases their overall health. The difference in mortality rates between the two groups of women was small,

---

[*] Oophorectomy is sometimes performed, as a preventative measure, at the time of hysterectomy in women who are considered to be at higher risk for ovarian or breast cancer.

yet again it appears that a pro-longevity signal may be coming from the reproductive tissue. As has been shown in worms and mice, the potential for giving birth in the future may also increase human lifespan.

It is important *not* to conclude from these studies that a woman will live longer if she stays on birth control pills and refuses to have her ovaries removed under any circumstances. The measured survival differences were small, and the only reason they could be detected at all was that the studies averaged survival rates among large numbers of women. In fact, there are often important medical reasons for oophorectomy or avoidance of oral contraception, especially among women in families with histories of certain cancers. Rather than asserting that oral contraceptive use or ovary conservation increases a woman's life expectancy, the experimental results simply suggested that the interaction between longevity and reproductive potential, which had been observed in worms and mice, may have subtle parallels in people as well.

Before leaving the topic of longevity and reproductive status, I should probably mention a recent Korean study that compared the longevity of Korean eunuchs who lived between the 15th and 19th centuries and other Korean men from that time period. The researchers found that among the 81 eunuchs for whom they had longevity data, the average lifespan was approximately 70 years. In contrast, in a comparable group of non-eunuch Korean men during this time period the average lifespan was less than 55 years. Remarkably 3 of the 81 eunuchs lived to 100 years, while generally only 1 person in 3000 to 4000 reaches 100 years.

Lest one think that the increased eunuch lifespan came from living privileged palace lives, the authors pointed out that the eunuchs usually lived outside the palace and that even the kings and other male members of the Korean royal family generally lived less than 50 years. Admittedly, with 81 subjects, the eunuch study's sample size is not especially large. Nevertheless, the increased average lifespan of the eunuchs is striking. That said, even if the extended life expectancy among eunuchs turns out to be a real and reproducible effect, it seems unlikely that being a eunuch will become a popular method for enhancing male life expectancy.

CAN WE INCREASE OUR LIFE EXPECTANCY AFTER ALL?

There is now little doubt that at least in the laboratory, animals live longer, healthier lives if they experience mild stresses such as limited caloric restriction. We've even seen hints that similar processes occur in people. Yet by just looking at environmental influences, we've limited our understanding of how we age and how one might be able to slow the aging process.

Aging is not only the result of our environment but also of inherited genetic variants. By analyzing the interactions between genes and environment, scientists are learning how aging occurs within our cells and are developing drugs that combat aging by mimicking the effects of caloric restriction. Only by looking at the combined genetic and environmental influences on our proteins and DNA will it be possible to understand complex biological phenomena, such as human aging. It is to this interplay of genetics and epigenetics that we'll turn to next.

PART V

NATURE AND NURTURE: DISENTANGLING THE INFLUENCES OF HEREDITY AND ENVIRONMENT

15

CAN WE LIVE TO 120 –
AND STILL EAT CAKE?

On February 18, 2005, Leonore Kahn Reichert, longtime resident of New Rochelle, New York, and an active supporter of the Girl Scouts, died at the age of 101. Considering her age, Ms. Reichert's being survived by nine grandchildren, 11 great-grandchildren and one great-great-grandson was not especially surprising. More remarkable was that her three *siblings* also all outlived her. They became centenarians as well and were healthier than most people 20 or 30 years younger.

The Kahns are not the only family with several members who've lived to 100. In fact, a woman with a brother or sister who lived to 100 has an eight times higher chance of making it to 100 herself than a woman who doesn't have a centenarian sibling. For men, the difference is even starker: a man with a centenarian sibling has *17* times the chance of reaching 100 as other men.

The Kahns didn't spend their lives on calorie-restricted diets. They weren't unusually conscientious about exercising or maintaining a healthy lifestyle, either. Their living environments were quite diverse since their childhoods, and that was a *very* long time ago. Rather, they simply appear to have been very lucky.

When several people from the same family have luck with longevity, scientists become suspicious that genetic factors are playing a role. Of course, some genetic variants clearly affect a person's life span. Inheriting a mutation that causes a fatal childhood disease, such as Tay-Sachs or Niemann-Pick disease, will certainly decrease one's life expectancy. What scientists are really interested in are genetic variations linked with the *overall* aging process. They want to find DNA variants that lead to the well-known, general symptoms of aging, including loss of muscle- and bone-mass, diminished endurance, decreased skin elasticity (wrinkling), slower healing and diminished immune response.

METHUSELAHS AMONG THE ANIMALS

Evidence that genetic variation contributes to the aging process comes not only from families like the Kahns but also from the animal kingdom. The wide range of life spans among animal species seems too great to be simply the result of living in different environments. Some species live for a very long time, indeed. Certain carp, such as koi, have been reported to live to 200. Tortoises can live 150 years or more. And, of course, tortoises and koi accomplish these feats of longevity without any of the benefits of modern medicine.

Even among our closest animal relatives, the mammals, we find species with unexpectedly long life spans. Perhaps the most remarkable and intensely studied is an East African rodent called the naked mole rat. These little fellows are about the size of a mouse, 3 to 4 inches in length, and weigh only a bit more than an ounce. They don't see very well, but they don't need to, since they live underground. In fact, their bodies are well-designed for underground life. Within their elaborate tunnel systems, mole rats can move backward just as quickly as they move forward.

Naked mole rats are very unusual animals. They adjust their internal body temperature to their environment rather than keeping their body at a fixed temperature like other mammals (including humans). Mole rats also have an unusual social system. Like bees and ants, but unlike other mammals, naked mole rats live in colonies of 70 to 80 animals in which only a single female, the queen, and a small number of males reproduce. All the other rats in the colony are sterile.

What interests longevity researchers about naked mole rats is that they live long and healthy lives. Mole rats live up to 28 years, while similar rodent species live only a fraction of that time. Laboratory mice rarely live more than 2 to 3 years. In the wild, mouse life expectancy is even shorter, since mice generally are killed before they grow old. Not only do naked mole rats live longer than mice, they are healthier as well. Unlike many mammals, mole rats never get cancer; they don't seem to be vulnerable to tumors of any kind. A naked mole rat at age 24 still has the muscle and bone composition of a young animal. Twenty-year-old mole rats are just like young animals by almost every physiological test, and elderly female naked mole rats continue to be as fertile as young ones.

Until quite recently, the mechanisms that enable naked mole rats to live longer than other rodents were completely unknown. Then in September 2013, scientists from the University of Rochester, in New York, discovered that one of the components of the ribosome (the enzyme that translates messenger RNA into protein), was synthesized in a highly unusual manner in the cells of the naked mole rat. The researchers also observed that the mole rat's ribosomes made far fewer mistakes in translating RNA messages into proteins than ribosomes in other animal species. Quite possibly, this more accurate conversion of the information from genes to proteins might contribute to the mole rat's long and healthy life.

The secrets of tortoise, carp or mole rat longevity may not be relevant to people. But, mole rats are not *that* different from us. They have bodies made of cells. Those cells are composed of proteins that are encoded in the mole rat's DNA. Of course, human and mole rat DNA sequences are different, and those differences cause a fertilized mole rat egg to become a mole rat whereas a human egg becomes a person. But the underlying biological circuitry is similar, and there is every reason to believe that longevity in animals can teach us about human aging.

LONGEVITY INVOLVES BOTH GOOD GENES AND THE RIGHT ENVIRONMENT

Looking at families such as the Kahns, as well as comparing mole rats and mice, suggests that inherited factors contribute to long life. The caloric restriction experiments described in the last chapter, along with studies linking good nutrition, exercise and availability of health care with human life expectancy, show that one's environment also contributes to longevity.

In earlier chapters, we've generally looked at either how our genes affect us or how our external environment does. I chose to focus separately on genetic and environmental influences primarily to keep the descriptions simple. In fact, our destinies are rarely affected simply by our genes or our environment; rather, our lives are shaped by a complex interplay of both genetic and environmental influences. And it is to these interactions of our genes and our environment on our proteins, our cells and our lives that we turn in the remaining parts of the book.

We begin by looking at how genes and environment affect human aging. We'll first find out how scientists identified the key cellular proteins that underlie the aging process. Then we'll learn how researchers can alter genes or the environment so that an animal has a longer, healthier life. Finally, we'll take a look at how these molecular insights are being used to develop drugs that may increase human longevity.

CHROMOSOMES DEGRADE WITH AGE

Over the last 50 years scientists have learned that an animal's aging is closely linked to the aging of its cells. One key way that a cell ages involves changes to its chromosomes. Over time, a cell's chromosomes degrade. Every time a cell divides its chromosomes lose a small amount of DNA from their tips (called telomeres). This occurs because cell division requires each chromosome to be duplicated, and chromosome duplication results in a small loss of DNA. When an animal is young and growing, its cells produce plentiful amounts of an enzyme called telomerase, which restores the telomeres to their original lengths. Since adult animal cells usually don't produce telomerase, each cell division leads to slightly shorter telomeres. Eventually, after approximately 50 cell divisions, the telomeres become too short for the chromosomes to function properly.

By measuring telomere lengths, scientists can correlate the aging of an animal with the aging of the animal's cells. Studies of animals ranging from zebra finches to dogs indicate that animals whose chromosomes have longer telomeres generally live longer and healthier lives than animals with shorter telomeres. Is telomere length also linked to health and aging in people? Many scientists think so. In the last five years researchers have found evidence linking shortened telomeres with depression, stress and a history of childhood abuse. Some data even suggests that maternal stress during pregnancy can lead to babies being born with shorter telomeres. Although to date studies associating telomere length and human health have been carried out only on small numbers of research subjects, this situation will soon change. Researchers at the University of California, San Francisco, and Kaiser Permanente (the largest private health insurance and medical services provider in the United States) are measuring telomere lengths in

100,000 people. This project is still in its initial stages, but preliminary results suggest that telomere length is indeed correlated with health and life expectancy in people.*

Loss of telomere DNA is not the only way that chromosomes slowly degrade. Chromosomes also deteriorate as a consequence of the same energy-generation processes that cells need to survive. Recall that cells produce energy through biochemical reactions in cellular substructures called mitochondria. Mitochondria convert food and oxygen into the energy-containing molecule ATP. This energy generation is essential for the cell, but it also leads to the production of toxic by-products: highly reactive, oxygen-based electrically charged molecules, or ions. These ions damage mitochondria, and ultimately the rest of the cell, by a process known as oxidative damage. To minimize oxidative damage, cells synthesize special enzymes that neutralize oxygen-carrying ions, as well as proteins that can repair the damage. Nevertheless, even with protective enzymes, oxidative damage eventually degrades the cell's DNA and proteins.

ALTERED CHROMOSOMES AND SENESCENT CELLS

Chromosomal damage is not the only way cells show their age. Cells also keep a record of how old they are by gradual (nondestructive) alteration of their chromosomes. These are the same type of epigenetic modifications as those by which a cell "remembers" whether it is a blood cell or a muscle cell. Over time, such epigenetic changes to chromosomes cause some proteins to be produced in greater quantities while others are synthesized in smaller amounts. Eventually, these epigenetic alterations, along with the chromosomal degradation caused by telomere loss and oxidative damage, signal to the cell that it should stop growing and dividing. At this point, the cell may enter a process of self-destruction (called programmed cell death) in which enzymes degrade the cell's substructures and its constituent proteins are recycled by the body for use in healthier cells. Sometimes, though, aging cells are not removed and recycled. They stop growing and dividing but don't die. Scientists call these senescent cells.†

---

* Because telomere degradation is linked to aging, scientists initially considered administering telomerase as a way to slow aging. Unfortunately, telomerase also makes cells more susceptible to cancer and consequently is unlikely to be useful as an anti-aging agent.

† Some types of cells don't grow and divide when they are healthy, but unlike senescent cells they can be *forced* to divide, for example, under cancerous conditions. In contrast, senescent cells *never* divide.

Despite the fact that senescent cells no longer grow or divide, they continue to perform important cellular functions. They activate cellular self-preservation networks. They also secrete molecules that signal to other cells that they are under stress of telomere loss and oxidative damage. For example, when an animal's tissue is wounded, some cells become senescent and send inflammatory signals that stimulate neighboring cells' self-preservation pathways. However, during the aging process this regulatory network can get out of control, with constant inflammatory signals leading to accelerated aging in a vicious circle.

Scientists initially suspected that senescent cell signaling could contribute to aging because some cells, which don't usually grow or divide, often display dramatic symptoms of aging. Such nondividing cells, such as heart cells and brain cells, should be immune from the chromosomal damage associated with cell division and growth. Yet somehow these cells also "think" they are getting old. One possibility was that they receive aging signals from senescent cells in other types of cells. In 2011, a research team from the Mayo Clinic demonstrated that signaling from senescent cells does in fact contribute to the aging process. The experiments used transgenic mice that produced a protein known as FKBP-cap6 exclusively inside their senescent cells. Since a drug called AP20187 selectively kills cells having FKBP-cap6, the Mayo Clinic scientists could kill all of a mouse's senescent cells while sparing cells that were not senescent.

With these transgenic mice the scientists could perform two dramatic experiments. In the first, the scientists administered AP20187 to the transgenic mice every three days starting when the mice were just three weeks old. The drug killed all the mouse's senescent cells before they could secrete any signaling molecules. The scientists observed that these mice displayed fewer symptoms of mouse old age: less bone loss, fewer cataracts in their eyes and fewer age-related spinal problems. Apparently senescent cells transmit signals that cause the animal's tissues to age, and by removing the senescent cells, the aging process was slowed.

The second experiment tested whether killing senescent cells late in an animal's life could slow down, or even reverse, the aging process. This time, the researchers gave the transgenic mice AP20187 only *after* the mice started showing symptoms of old age. Under these conditions, the anti-aging effects of AP20187 treatment were more limited. Some ailments of old age, such as cataracts, are irreversible and couldn't be cured by killing the senescent cells, but other symptoms, including muscle loss and decreased endurance, did improve. These experiments suggested that selectively killing senescent cells could be an effective anti-aging therapy – assuming that senescent cells can somehow be targeted in normal (not transgenic) animals. As a result, several research teams are actively searching for drugs that could selectively kill senescent cells in humans.

THE BIOCHEMICAL PATHWAY OF CELLULAR AGING

The molecular pathways that regulate cell growth, aging and energy generation are closely linked together. Over the last 40 years scientists have been able to unravel many of the secrets of these important biochemical networks and have identified many of the proteins that regulate how cells convert food into energy and, in the process, grow old. Many of the original experiments were carried out on a humble roundworm called the nematode. Worms are excellent organisms for studying aging. Nematode worms are easy to cultivate and they reproduce rapidly. They are relatively simple animals with only 959 cells, and after 50 years of extensive study, the relationships among the nematode's cells have been deciphered with extraordinary precision. Since nematodes normally live only two to three weeks, they can easily be used in experiments that study lifespan.

Countless biological insights that were first discovered in worms were subsequently reproduced in flies, mice, rats, monkeys and humans. Largely from worm experiments, scientists discovered two complex and interconnected biochemical pathways, involving hundreds of proteins, by which cells react to caloric restriction. These signaling networks, which each play a critical role the cell's detection of nutrient availability, are referred to in the scientific literature as the TOR pathway and the IIS (or insulin/insulin-like growth factor-like signaling") pathway. Cells with nutrient sensors (typically involving glucose receptors) monitor changes in nutrient levels and trigger the TOR/IIS pathway by releasing signaling molecules. Among the most important is a hormone called insulin-like growth factor 1 (IGF-1). IGF-1 travels through the blood and is detected via IGF-1 cell surface receptors. Once bound with IGF-1, the IGF-1 receptors become chemically active and initiate cell growth involving many regulatory proteins with names like TOR, AKT and FOXO (see figure 15.1).

The proteins in the TOR/IIS pathway perform many important biochemical functions within the cell. Some alter the cell's chromosomal modifications that mark the aging of the cell. Others trigger the production of enzymes that mitigate oxidative damage. Remarkably, the cellular pathways that regulate cell growth and energy have also been shown to regulate longevity in virtually every species tested: when a cell's environment pushes it toward fast growth and high energy/nutrient use, aging is accelerated. In contrast, if a cell's environmental sensors tell the cell to slow down its growth, they also trigger self-maintenance functions that lead to longer cell life.

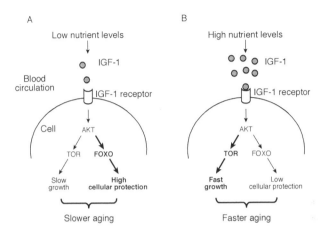

Figure 15.1. Nutrient sensing, aging, and the TOR/IIS pathway. A: Simplified illustration of the pathway when local nutrient levels are low, leading to slow cell growth and increased cell self-preservation. B: Changes in the pathway when nutrient levels increase; IGF-1, AKT, TOR and FOXO are among the important proteins in this pathway. Many others, including PI3K, SIRT1 and AMPK (described later in the book), have been omitted for simplicity.

CALORIC RESTRICTION AND THE TOR/IIS PATHWAY

By studying worms and mice, scientists have learned how the TOR/IIS pathway is altered during caloric restriction. Animal experiments reveal that caloric restriction activates a class of enzymes that protect the cell's chromosomes from oxidative damage and other enzymes that reverse the chromosomal modifications that occur during aging. Researchers have also learned how to manipulate the genes involved in the TOR/IIS pathway so that an animal's cells act *as if* caloric restriction were occurring. For example, scientists were able to engineer transgenic worms with a version of the worm's IGF-1 receptor protein that always acted as if there were a glucose shortage. These worms lived twice as long as normal worms, even when fed normal diets. In later experiments, researchers demonstrated that similar alterations of IGF-1 receptor genes could also increases longevity in flies and mice.

Modifying the IGF-1 receptor gene is just one way to genetically engineer a worm or a mouse with increased longevity. Twenty years of experimenting has shown that modifying other genes in the TOR/IIS pathway can also create an animal with an extended lifespan. Different genetic changes increase lifespan by varying amounts, and combining mutations can produce a transgenic animal with even greater longevity. The current record holder is a strain of worms with mutations in both copies of a gene called *PI3K*, which encodes another important TOR/IIS pathway protein. This particular mutation leads to a worm that is a bit finicky and needs to be raised in an environment where the temperature is carefully regulated. If these conditions are met, the result is truly a "Methuselah" worm. *PI3K* transgenic worms live on average 150 to 180 days, which is approximately *10 times* the lifetime of your everyday nematode worm.

AGING, GROWTH AND SIZE

Since animals need calories to grow, mice and rats on caloric restriction are smaller than their litter mates. IGF-1, which plays a central role in sensing nutrient and caloric levels, is also an important regulator of size. In the following chapter, we'll see that a genetic variation near the *IGF-1* gene is the key genetic difference between large dogs like Great Danes and small ones, like Chihuahuas. Since worms and mice with disabled IGF-1 receptors live longer than normal animals, IGF-1 may also have a darker side. It may not be a coincidence that the little Chihuahua lives twice as long as the Great Dane. Being bigger may not always be better.*

A joint study by scientists from University of Southern California (USC) and the Institute of Endocrinology, Metabolism and Reproduction in Ecuador has even found evidence of a trade-off between size and healthy lifespan in humans. The team studied age-related diseases among 99 Ecuadorian individuals with a condition called growth hormone receptor deficiency (GHRD). People with GHRD have a DNA mutation leading to nonfunctional growth-hormone receptor proteins. Consequently, their cells can't detect growth hormone, resulting in extremely short stature. (Individuals with GHRD are generally less than 5 feet tall.)

The Ecuador-USC team found that the incidence of age-related diseases, such as cancer and adult-onset diabetes, were significantly lower among individuals with GHRD than among their relatives who did not

---

* Smaller animals generally outlive larger ones, but only within a single species. Elephants and rhinos do live longer than flies and hummingbirds.

have GHRD. Because a larger number of the individuals with GHRD, in this small study, had died of non-age-related causes, such as alcohol abuse and accidents, the researchers were unable to conclude that GHRD could be linked to increased human lifespan. Nevertheless, the results suggested that short stature – at least in the case of individuals with rare growth-hormone-receptor mutations – might be associated with slower rates of human aging.

There is even evidence that slower growth rates and healthier aging might be linked among people of normal stature. In 2008, researchers from University of California, Los Angeles and the Einstein College of Medicine in New York found that mutations in the gene for the receptor for IGF-1 were significantly more common in the DNA of centenarians than in other people. To be sure, their results need to be viewed cautiously, since only a small number of research subjects were available. (IGF-1 receptor mutations were found in nine centenarians and in a single control subject.) Still, the results did again suggest that factors that regulate cell growth might also affect the rate at which people age.

LONGEVITY AND HUMAN GENETIC VARIATION

The evidence that modifying genes can mimic the effects of caloric restriction, and thereby increase animal life span, is compelling. How relevant such experiments are to human aging is less clear. For one thing, genetic tests of centenarians have found only limited evidence that the genes central to mouse or worm longevity are also important in human lifespan. In addition, even if genes are identified that have comparable effects on human longevity as *PI3K* or *IGF-1* receptor genes do in other animals, we still can't modify them in people the way we can in transgenic animals.

Instead, scientists are searching for drugs that reproduce the effects of caloric restriction or genetic manipulation. The first step in such a strategy is to identify potential drug targets. Along with the human analogs of longevity-related proteins in laboratory animals such potential targets include proteins that cause premature-aging diseases, called progerias. You may be familiar with fictional versions of progeria, which afflicted the character of J. F. Sebastian in the movie *Blade Runner*, as well as the title character in the movie *Jack*.

The most common of these rare diseases is called Werner syndrome; it occurs in 1 in every 100,000 children. Children with Werner syndrome develop normally until puberty, at which point they stop growing and begin to develop symptoms characteristic of older people. Their hair becomes grey, their skin wrinkles, and they develop diseases such as cataracts, diabetes and heart disease. Typically they die by about the age of 50. In 1996, Scientists identified the gene (called *WRN*) that is mutated in Werner syndrome. *WRN* codes for a protein that unwinds the double helix of DNA. WRN appears to assist telomerase in chromosome repair, and the cells of individuals with Werner syndrome have similar telomere damage to that found in the cells of old people.

Finding mutations that lead to premature aging is a boon for scientists searching for proteins that regulate aging. Equally important would be discovering DNA variants, such as those presumably located somewhere in the genomes of the Kahn siblings, that help people live long healthy lives. Unfortunately, there aren't many families with multiple centenarians. Instead, scientists search for gene variants commonly found among unrelated centenarians. These studies haven't yet had much success. Although some genetic variants are consistently found in centenarians, these variants are generally associated with well-known diseases of the elderly. For example, one variant of a gene called *APOE* is often found among centenarians. This variant has been associated with decreased risk of Alzheimer's disease and cardiovascular disease. Consequently, finding this *APOE* variant among centenarians doesn't shed much light on the general process of aging; it just confirms that this *APOE* variant helps protect people from two specific diseases.

Discovering genetic variants linked to healthy old age and not associated with common aging diseases has been more challenging. Initial searches for genetic variants associated with human aging were hampered by the high cost of testing multiple genomic markers in large numbers of people. As a result, only small numbers of centenarians were screened, and their DNA was tested at only a limited number of marker locations. To address these limitations, in 2013, researchers from Boston University carried out a meta-analysis* of five genetic studies of healthy individuals who had lived beyond the age of 90. Over 2500 nonagenarians and centenarians were included, and approximately 500,000 genomic marker locations were tested. Even with this relatively large sample, the *APOE* variant linked to decreased risk of Alzheimer's disease was again the only variant found often enough for scientists to be confident that it was not identified by chance. Nevertheless,

---

* A meta-analysis is an aggregation and review of multiple related scientific reports, typically using sophisticated statistical techniques to combine the results of studies with widely varying numbers of research subjects.

variants found in two other genes, which were among the 20 highest hits throughout the genome, were intriguing. One of these variants occurred in *WRN* (the gene that is disabled in Werner syndrome). The other was found in a gene called *LMNA1*, which is mutated in another rare premature-aging disease called Hutchinson-Gilford progeria syndrome. Finding atypical variations of *WRN* and *LMNA1* among healthy centenarians suggests that these genes may not only contribute to diseases of premature aging but also provide clues about mechanisms that underlie healthy aging.

RAPAMYCIN, CALORIC RESTRICTION AND ANTI-AGING INTERVENTION

To increase healthy human lifespan without genetic manipulation or caloric restriction, scientists search for drugs that mimic the biological effects of these anti-aging procedures. This effort has largely focused on drugs targeting proteins activated during caloric restriction. One such drug is rapamycin. Although rapamycin is more commonly used as an immunosuppressant to prevent rejection of transplanted organs, rapamycin is also a candidate for anti-aging therapy because it blocks the TOR enzyme of the TOR/IIS pathway. (In fact, TOR originally got its name because scientists observed that TOR was a "target of rapamycin.")

Initial rapamycin experiments with mice were encouraging. Mice that were given rapamycin when they were 600 days old (which for a mouse is equivalent to being a 60-year-old person) lived longer than untreated mice. The male mice lived 9% longer and females 14% longer than untreated mice. Unfortunately, rapamycin also has nasty side effects. Being an immunosuppressant, rapamycin makes a person more vulnerable to infections and infectious diseases. In mice, long-term rapamycin administration also affects the insulin regulation system, which can lead to diabetes. Despite these problems, there may still be hope for a rapamycin-like drug in anti-aging therapy. Recent experiments indicate that TOR's role in the longevity pathway is separate from its action in insulin regulation, suggesting that it may be possible to find an alternative drug with rapamycin's anti-aging activity, but without its effects on insulin regulation. If such a drug also lacked rapamycin's immunosuppressive effects it would be a promising anti-aging drug.

METFORMIN, RESVERATROL AND SIRT1-ACTIVATING DRUGS

Metformin is another drug that targets the TOR/IIS pathway. Metformin decreases blood glucose, insulin and IGF-1 concentrations. It is widely prescribed for treating diabetes and is being investigated as a treatment for breast cancer as well. The first clue that metformin might have anti-aging properties came from cell culture experiments indicating that metformin blocked S6K1, an important TOR/IIS pathway protein. In 2011, a Russian research group demonstrated that giving female mice metformin starting at an age of three months increased their average lifespan by 14%. For unclear reasons, metformin didn't increase the lifespans of male mice. It is also unknown whether metformin slows aging in animals other than rodents, and whether it is effective only if it is administered from an early age. Despite these caveats, the mouse data has been encouraging enough for metformin to be actively investigated as a human anti-aging agent.

Resveratrol, a naturally occurring compound found in fruits and plants, has also been widely tested as an anti-aging agent. Mice on resveratrol don't live longer, but they do display fewer symptoms of old age, such as cataracts, bone density loss and diminished motor coordination. Resveratrol's action at the molecular level is not yet clearly understood, but it appears to target the TOR/IIS pathway through a protein called SIRT1, which activates enzymes that protect the cell from oxidative damage. Unfortunately, large doses of resveratrol are needed to produce the mild anti-aging effects in mice. Comparable doses in humans are unrealistically high. As a result, at least one pharmaceutical company, Sirtris tried developing synthetic compounds that mimic resveratrol's effects on SIRT1.* The initial studies indicated that its compounds had 1000 times resveratrol's potency, and they showed promise in treating obesity and combating aging (at least in elderly mice). Although these encouraging initial results led GlaxoSmithKline to acquire Sirtris for over $700 million in 2008, more recent studies have failed to replicate the beneficial effects of the SIRT1-targeting drugs, and their efficacy in slowing human aging remains uncertain. In fact, in 2013, GlaxoSmithKline closed down its Sirtris subsidiary, although the drug company asserts that the Sirtris research strategy is promising and that it is still pursuing it.

LIVING LONGER WITH ANTIOXIDANTS?

Antioxidants are another class of compounds that have long been touted as slowing the aging process. Antioxidants include a wide range of substances

---

* Another reason pharmaceutical companies want to develop synthetic versions of resveratrol is that resveratrol is a natural compound, making it difficult to patent.

ranging from vitamins A, C and E to turmeric and ginkgo biloba. By definition, antioxidants neutralize the electrically charged oxygen-carrying ions that cause oxidative damage, so using them to combat aging makes intuitive sense. Still, there are problems with this simple idea. The required antioxidant dose would be large. In addition, animal studies have generally failed to demonstrate that oxidative damage is the biological cause of aging. In fact, experiments with worms have even suggested that limited oxidative damage may actually be helpful in combating aging, since the oxidative damage can trigger the self-protective mechanisms of the TOR/IIS pathway.

Nevertheless, since oxidative damage was long believed to be the main cause of aging, many research groups have studied the effect of antioxidants on human aging. The results of these studies have not been consistent, in part because they tested different antioxidants or used varying medical criteria to assess health benefits. In addition, some investigations only studied a small number of people or were not carried out in scientifically rigorous ways (for example, the studies were not "blind").* On balance, scientists and physicians reviewing multiple studies have concluded that antioxidants have little effect on slowing human aging. For example, researchers from St. Louis University, writing in the respected *Cleveland Clinic Journal of Medicine,* concluded that "Antioxidants have little effect on cerebrovascular and cardiovascular diseases and in fact may even increase overall mortality."

HORMONE REPLACEMENT ANTI-AGING THERAPIES

Hormones, including growth hormone, testosterone and dehydroepiandrosterone (DHEA), are also often proposed as treatment for the effects of aging. The rationale is that as we age, our bodies produce smaller amounts of these important hormones. So it sounds plausible that receiving growth hormone or testosterone supplements to restore our hormone levels to what they were in our 20s might also restore our strength, endurance and vitality.

Unfortunately scientific studies in elderly people receiving hormone replacement therapy show few anti-aging benefits. Older people taking growth hormones do have increased lean muscle mass and lower fat, but they generally aren't stronger or have more endurance. As with the

---

* A blind medical study is one in which one group of subjects receives the treatment, while remaining subjects are given a pill without any medicine. None of the subjects knows whether they are receiving the real medicine. Blind studies help quantify the effect in which people sometimes see their medical condition improve simply because they *think* they are being treated (the placebo effect).

antioxidants investigations, the results of the hormone replacement therapy studies don't always agree, so scientists need to analyze multiple studies. The same St. Louis University researchers who analyzed the antioxidant trials also reviewed numerous studies on the anti-aging effects of hormones. They concluded that

> None of these [testosterone] studies showed an improvement in measures of muscle strength, overall physical performance, energy, or sexual function. ... Routine replacement of DHEA in older adults provides no meaningful benefit. ... [And] although initial studies of growth hormone looked promising, according to the available research, the risk of therapy in people who are not growth-hormone-deficient outweighs the benefit.

In addition, earlier in the chapter we saw that some experiments suggest defects in growth-hormone receptors may actually help prevent age-related diseases. As a result, the case for the use of growth hormones as an anti-aging intervention is far from proven.

FINAL CAVEATS

There is no question that over the last two decades extraordinary progress has been made in understanding the aging process. Recent animal experiments with rapamycin, metformin and other potential anti-aging drugs make many scientists optimistic that this knowledge may soon lead to drugs that enable people to live longer healthy lives.

Before we start making plans to attend our great-grandchildren's weddings, though, it's worth remembering that all new drugs may have unexpected side effects, and the side effects of the proposed anti-aging drugs remain unknown. Beyond concerns about side effects, we shouldn't forget that there is still a lot we don't understand about human aging.

A case in point is the sad story of Brooke Greenberg. Brooke Greenberg was born in Baltimore, Maryland, in 1993. She was born a month prematurely, weighing just four pounds at birth. That Brooke was extremely abnormal became quickly apparent. Not only did she have serious illnesses, including stroke, seizures and ulcers, but she also didn't get older. Some biological switch in Brooke's body was not being thrown. Although doctors and scientists studied Brooke for years, just what was happening, or not happening, in her body was completely unknown. What was known was that Brooke remained an infant. In the last published scientific report, when she was 15 years old, Brooke weighed only 15 pounds and was 28 inches tall. She was described as having a mental age of less than nine months, and her doctors were unable to predict whether she would eventually start to grow up and age, whether she would die as an infant, or whether she would remain in her current state indefinitely. Ultimately, Brooke died of a lung defect on October 24, 2013, at the age of 20, although physically and mentally she was still barely a toddler.

Besides being both amazing and tragic, Brooke Greenberg's story reminds us that there are still large gaps in our understanding of human aging. In fact, her illness was so strange and mysterious that many aging scientists question whether it had any relevance to normal human aging at all. That said, it still seems that as long as our knowledge of human aging is limited, we would be well advised to be careful about what we wish for and consider carefully what we are willing to risk to make those wishes come true.

16

WHAT MAKES A SUPER-ATHLETE?
PART 1: SIZE MATTERS

⌒⌒⌒

The doctors were astonished. When the infant we'll call Heinrich was born in a Berlin hospital in 1999, they realized immediately that he was an extraordinary little boy. Heinrich was unlike any of the other babies in the hospital – indeed he was different from just about every other child. To protect the privacy of the boy and his family, only limited amounts of information have been published about him, but what is known is remarkable. When Heinrich was just six days old, ultrasound examinations showed that he was far more muscular than other babies. By the time he was four years old he was lifting 3 kilogram dumbbells (about 6 1/2 pounds), a feat far beyond the abilities of other youngsters his age. The question already seemed to be *which* sport Heinrich would ultimately excel at, not whether he would be a top performer at any sport.

Although children born with almost superhuman strength like Heinrich are extremely rare, children with some special athletic gifts are much more common. When we were young, we all knew children who were natural athletes. Maybe you were one of them. I wasn't. These kids ate the same foods as I did. They didn't exercise any more than I did (at least not at an early age). Yet they could run faster or throw a ball farther or lift more weight than other children their age.

Athletic excellence often runs in families. Think of Venus and Serena Williams, sisters who dominated women's tennis for over a decade, or the Manning family, Archie and his sons Peyton and Eli: all stars of American football. Bobby Hull was an ice hockey legend, while his brother Dennis and son Brett were outstanding hockey players as well. And the list goes on.

Training and coaching are also important in the making of athletes, and just because super-athletes cluster in families doesn't mean the reason is genetic. Super-athlete fathers may have trained their sons, and elite athlete sisters may have shared the same top coach. Nevertheless, the fact that athletic excellence often runs in families suggests hereditary factors are important.

The evidence that inherited DNA contributes to athletic excellence is not limited to observation of families with star athletes. Over the last 40 years, numerous studies of athletic ability in twins and among family members have shown that inherited genetic variants are responsible for approximately half of the variation in physical attributes such as strength and endurance. In this chapter and the following one we'll explore many of the factors, both genetic and nongenetic, that make an elite athlete. We'll see how genes, exercise and performance-enhancing drugs all affect the proteins in our cells and thereby contribute to athletic performance.

ELITE ATHLETES ARE OFTEN TALL

In many sports few things are more important than being tall. Nearly all top men's basketball players are well over 6 feet tall, and most leading tennis and volleyball players are tall as well. There are exceptions, say, if you dream of becoming a champion jockey, coxswain or gymnast. But even then size matters, though in those cases, it's often better to be small.

Height also runs in families. Most of us have known families where just about everyone was quite tall (or short). Comparisons of identical and nonidentical twins have shown that families with many tall individuals don't merely share a similar diet; there is a genetic component as well, and scientists have long wanted to find the causal genetic variants in families with tall people, just as they identified *BRCA1* gene variants in families with recurring breast cancer.

Some people are *so* tall, or so short, that scientists are able to identify the underlying biological causes. Often the trail leads to a protein called (appropriately enough) human growth hormone (hGH), or to a protein that regulates hGH or is regulated by it. Human growth hormone is one of the critical signaling molecules that tell our cells when to grow and divide. Needless to say, the production and secretion of hGH needs to be tightly regulated to ensure that our bodies grow enough, but not too much. Figure 16.1 depicts the hGH network, including a few of the important proteins, such as insulin-like growth factor 1 (IGF-1), that interact with hGH.

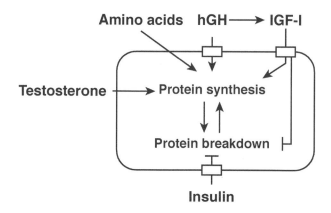

Figure 16.1. Growth hormone signaling pathways. Important growth-signaling molecules include hGH, IGF-1, the sex steroids (including testosterone) and insulin. Arrows indicate activating signals, while lines with T ends indicate inhibitory regulators. The large rectangle with rounded corners represents the outer membrane of a cell, while the small rectangles represent receptors in the cell membrane that bind the hormones shown adjacent to them (e.g., hGH, IGF-1 and insulin). Testosterone and amino acids enter the cell directly, rather than by binding to cell membrane receptors.

If a child's growth-hormone system isn't functioning properly, there is a good chance they will become very tall or very short. Children who produce too little hGH have a condition called growth hormone deficiency. Unless they are given synthetic growth hormones, they wind up very short, usually 4 to 5 feet tall.* Other children produce too much hGH, often because of a tumor in the pituitary gland, which produces hGH. In extreme cases, a pituitary tumor may lead to a condition called gigantism. And people with gigantism are tall indeed, often growing to be over 7 feet tall.

---

* Note that the symptoms of growth hormone deficiency are quite similar to those of growth hormone *receptor* deficiency, which was described in the preceding chapter.

In some cases, a pituitary tumor leading to gigantism is caused by a mutation in a gene called *AIP*. Perhaps, the most famous individual who had an *AIP* mutation was the Irishman Charles Byrne. Byrne, who lived in the 1700s and was known as the Irish Giant, grew to be 7 feet, 7 inches tall. Because of his great height, Byrne was a sideshow celebrity and became wealthy at an early age. Unfortunately, Byrne also had a fondness for alcohol, and he died in his early twenties largely because of his alcoholism.

After Byrne died, his skeleton was preserved, and recently a team of British researchers extracted DNA from his bones. Byrne's DNA showed that he had an *AIP* gene variant often found in people with pituitary tumors. Remarkably, the London researchers found the same *AIP* gene variant in the DNA of four different families of tall individuals living in Ireland today. Almost certainly, Charles Byrne and these four families are distantly related, descended from some other Irish giant who had the original mutation hundreds of years ago.

Although people with gigantism are very tall, they're rarely successful athletes. The same growth hormone misregulation that makes them tall makes them awkward and uncoordinated, not helpful traits in a competitive athlete. In contrast to gigantism, there is anecdotal evidence that a related, but less severe, condition called acromegaly may sometimes help people become elite athletes. People with acromegaly also have pituitary glands that secrete excess hGH, but in this case the pituitary only begins to secrete extra hormone late in puberty or in adulthood. As a result, they do not become as tall as individuals with gigantism, though they are generally quite tall and have relatively large hands and feet. However, acromegaly is not a good way to become an elite athlete, as acromegaly often leads to heart disease, diabetes and other diseases later in life.

People with a different rare medical condition, Marfan syndrome, are also tall. They typically have very long arms and sometimes they do become top athletes. Several outstanding athletes, including Flo Hyman, the great American women's volleyball player, who was 6 feet, 5 inches tall, and Chris Patton, a 6-foot, 9-inch basketball player, both had Marfan syndrome. Unfortunately, Marfan syndrome can also come with serious side effects. People with Marfan syndrome often have an awkward way of walking and may suffer from arthritis or cardiovascular problems. Both Flo Hyman and Chris Patton died from cardiac complications of Marfan syndrome. Hyman was 31 years old when she died suddenly; Patton was just 21. Sometimes people with Marfan syndrome don't appear to suffer any ill effects. They're just tall, and have long limbs and long fingers. For example, some scientists speculate that Abraham Lincoln, who was 6 foot, 4 inches tall, had Marfan syndrome.

Genetic studies of families with Marfan syndrome were able to track the causal variants of Marfan syndrome to chromosome 15, near a gene called *FBN1* or *fibrillin-1*. The fibrillin-1 protein, which is coded by this gene, is an important component of connective tissue such as tendons and ligaments. Subsequently, 600 variants in the *FBN1* gene have been detected among individuals with Marfan syndrome, each variant leading to a somewhat different set of symptoms of the disease.

In contrast to gigantism and Marfan syndrome, some genetic variants cause people to be short. One dramatic example is achondroplastic dwarfism. Men with achondroplastic dwarfism are typically just over 4 feet tall; women are even shorter. The proportions of their bodies are also unusual; individuals with achondroplastic dwarfism have relatively normal upper bodies, but have very short legs. Achondroplastic dwarfism is rare, occurring in only 1 out of every 25,000 people. The condition runs in families, and studying affected families revealed the causal gene variant was in *FGFR3*, the gene for fibroblast growth factor receptor 3 protein, an important regulator of bone growth. The variant in people with achondroplastic dwarfism leads to nonfunctional FGFR3 protein, causing severely shortened bones and abnormal cartilage.

Another gene with a dramatic effect on height is *SHOX*. The *SHOX* gene encodes a regulator protein, one that controls when other genes are converted into proteins. Just how *SHOX*'s regulation of protein production affects human growth is poorly understood. The *SHOX* gene is one of the few genes found on both the X chromosome and the Y chromosome Although nearly all of us have two X chromosomes or one X and one Y, some people don't. As we saw in chapter 11, women with Turner syndrome have only one X chromosome per cell. They have only one copy of the *SHOX* gene per cell and are very short. In contrast, men with Klinefelter syndrome have two X chromosomes in addition to their Y chromosome. Their cells produce excess SHOX protein, and these men are usually tall. Though height variation due to the *SHOX* gene is more common than achondroplastic dwarfism or Marfan syndrome, it is still uncommon, occurring in approximately 1 in 1000–2000 people.

COMMON GENETIC VARIANTS AND HEIGHT

Gigantism and Marfan, Klinefelter, and Turner syndromes are all rare. They don't explain why most tall people are tall, and with the exception of a few, mostly tragic examples like Flo Hyman and Chris Patton, they don't explain the height of elite athletes. Tall athletes generally have different gene variants to thank, but identifying those variants and understanding how they function have been difficult tasks. The reason is that these genetic variants each have only a small influence on how tall we are, making it impossible to identify them using the relatively small number of research subjects available in family-based genetic studies. So scientists instead genotype large numbers of unrelated people who are tall (say, over 6 feet tall) or short people (less than, for example, 5 feet, 6 inches), and then look for genetic variants consistently found more often among the tall people or the short people. One important difference between such investigations, called genome-wide association studies, or simply GWAS, and family-based studies is that GWAS studies are huge, often involving tens of thousands of unrelated individuals. Using large numbers of people should make genetic variations unrelated to height "average out," enabling the detection of DNA variants with more subtle effects.

As mentioned in chapter 13, although DNA testing was previously too expensive to consider genotyping thousands of people, the cost of genetic testing has been dropping at a rapid rate. Large GWAS studies have now become feasible, and in 2007, a GWAS study of almost 5000 people identified the first genetic variant with a statistically significant association* with human height. The variant was near a gene called *HMGA2*. Although the link between the *HMGA2* variant and human height was statistically significant, the variant's influence was small. Having two copies of the variant accounted for only 1 centimeter of height difference. The function of the HMGA2 protein is still not known, but its association with height was not a complete surprise. *HMGA2* knockout mice are small (they are known as pygmy mice), while transgenic mice with an unusual variant of the *HMGA2* gene are large. In addition, a rare mutation in *HMGA2* had been discovered in an eight-year-old boy who was exceptionally tall for his age.

By 2010, scientists had assembled enormous data sets in attempts to link genetic variants with height. Several GWAS studies, involving over 100,000 people, were analyzed in a search for genetic variants at more than two million marker locations. Although no single genetic variant with a large effect on a person's height was found, 180 genetic markers that each contributed to a small variation in human height were identified. The researchers found that if they considered all 180 marker locations, they could

---

* Statistically significant means that enough people were tested to be confident that the association was not caused by chance.

account for much of the height variability among the study participants. They also discovered that the 180 locations were not randomly distributed throughout the genome; many were located near genes known to code for proteins involved in skeletal growth. Clearly, identifying the causal variant(s) at each of these 180 locations and understanding how they influence human growth remains extremely challenging, but at least there is now a start to finding the reasons some people are taller than others.

DOGS, SIZE AND GENETICS

Since tracking down the genetic causes of human height variation is difficult, some scientists have been looking at size variations in animals. The hope is that the genetic picture will be simpler and that the insights gained will be relevant to understanding human growth. Perhaps no animal's genetics has been studied more closely than the dog's. As described in chapter 10, the reason is that most dog breeds have been developed only over the last few hundred years. By the standards of evolution, this is a very short time. Consequently, relatively few mutations have occurred and fewer irrelevant genetic variations need to be sifted through to find causal variants underlying a trait of interest.

Looking for gene variants related to body size in dogs has another advantage; the range of dog sizes is huge. A Great Dane is five times as tall as a Chihuahua. Compared to the difference between a Chihuahua and a Great Dane, the height difference between Charles Byrne, the Irish Giant, and a 4-foot-tall person with achondroplastic dwarfism is pretty small. Yet the Dane and the Chihuahua do belong to the same species. They had a common ancestor a relatively short time ago, and in principle they are able to mate and have viable offspring (though there would be logistical challenges).

In the early 2000s, a scientific team from the National Human Genome Research Institute began a series of genetic studies of dog size. The results were dramatic. While in humans, 180 genetic markers are needed to explain about half of height variability, in dogs 5 or 6 markers can predict most variation in dog size, and a single marker, located near the gene for IGF-1, can explain a significant portion of variability in dog size. Nearly all small-dog breeds have the same variant at the *IGF-1* gene. You'll recall that in people the IGF-1 protein is a central component of the growth hormone regulatory network. A similar relationship exists between IGF-1, growth hormone and growth in dogs. So in a sense, Great Danes could be called the Charles Byrnes of the dog world.

In a related study, the same research group searched for genetic variants in dogs like dachshunds and basset hounds that have short legs. They found that short-legged dogs have an unusual variant in a gene called *FGF4* (for fibroblastic growth factor 4). This name might sound familiar. *FGF4* is related to *FGFR3*, the gene with an unusual variant in people with very short legs due to achondroplastic dwarfism. Once again, the results demonstrated how similar we humans are to other animals.

STRONG MUSCLES MAKE TOP ATHLETES

Along with being tall, top athletes are strong. Although strength depends on training and exercise, there is strong evidence that genetics also plays a role. In extreme cases, genetic differences in muscle strength are painfully obvious. Childhood muscular dystrophies can cause muscle-protein deficiency, or the production of malformed proteins, leading to muscle loss. In these genetic diseases, the regulatory system that usually ensures that we make enough muscle has gone seriously wrong.

For most of us, the muscular development system works fine. We don't make too little muscle and we don't make too much. Of course, this raises the question of just how much muscle is too much. Perhaps our regulatory system is a little too strictly adjusted, and we might live healthier lives, and be stronger, if our bodies made a bit more muscle. Animal experiments suggest this may be true.

Scientists began to understand how genes regulate muscular development by studying mice. In 1997, researchers at Johns Hopkins University were systematically searching for mouse genes suspected of being involved in muscle growth because their DNA sequences were similar to those of known muscle regulatory genes. They identified one such gene, which they called myostatin, and created a knockout mouse in which the myostatin gene was truncated and nonfunctional. Surprisingly, mice with the nonfunctional gene had muscles twice the size of other mice. Although one might not expect that a nonfunctional gene would lead to bigger muscles, the scientists discovered that myostatin protein acts as a *brake* on muscle growth. Consequently, disabling myostatin leads to increased muscle growth. The myostatin knockout mice appeared to be generally healthy, so perhaps, at least in mice, removing the myostatin brake isn't a bad thing.

Figure 16.2. Normal mouse and myostatin knockout mouse.

Mice aren't the only animals that have taught us about myostatin. We've also learned from a dog breed called a whippet, which has been bred over centuries to be racers. Some whippets also have unusual variants in their myostatin genes. Although the whippet myostatin mutation is different from the one in the knockout mice, the end result is the same: nonfunctional myostatin protein. Occasionally, a whippet inherits a nonfunctional myostatin gene from both parents. These dogs have no functional myostatin protein whatsoever and are known in the dog-breeding world as bully whippets. They have a muscle-bound appearance and a noticeable overbite. As a result, they are often euthanized by breeders, but those that are allowed to live are quite healthy. In contrast to bully whippets, some other whippets inherit just one nonfunctional myostatin gene, along with one normal myostatin gene from their other parent. These dogs produce functioning myostatin protein, but only about half as much as in a normal whippet. They appear normal and are healthy, and they generally run faster than normal whippets.

After the nonfunctional myostatin gene variant was found in whippets, scientists began looking for nonfunctional myostatin variants in other dog breeds. Surprisingly, no such variant has been found, not even in greyhounds, a related racing-dog breed. Could a nonfunctional myostatin gene be more advantageous to a whippet than to a greyhound? Actually it could, because myostatin has a second function in regulating muscle development: myostatin also plays a role in skeletal muscle-fiber *differentiation*.

Figure 16.3. Normal whippet (left) and bully whippet (right).

Skeletal muscles, the muscles that we use in athletics, consist of long thin cells that form muscle fibers. Dogs (and people) have two main types of muscle fibers: "slow twitch" fibers and "fast twitch" fibers. Both are important. Fast twitch muscle fibers, as you might guess, fire more rapidly. They generate more force than slow twitch fibers, and they do it quickly. Sprinters, for example, rely on fast twitch muscle fibers for quick bursts of speed. Slow twitch muscle fibers generate less force per fiber, but they generate their force more efficiently. Marathon runners and endurance athletes rely primarily on slow twitch muscle fibers.

Myostatin causes muscles to make relatively more slow twitch than fast twitch fibers. This fact was also discovered in experiments with myostatin knockout mice. Not only did these mice have more muscles than normal mice but their muscles had a higher proportion of fast twitch to slow twitch fibers. Since slow twitch muscles are important for endurance sports, scientists predicted that the myostatin knockout mice would have less of an advantage in endurance tests than in activities requiring short bursts of strength. In 2009, researchers from the U.S. National Institutes of Health compared the endurance performance of myostatin knockout mice with unaltered laboratory mice. Endurance performance was determined by putting the mice on an inclined treadmill and measuring how long the mice ran before they got exhausted and stopped running.* The researchers found that despite the fact that the knockout mice had bigger muscles, they performed 40% *worse* on the treadmill endurance test than normal mice.

How do slow twitch fibers and fast twitch fibers relate to whippets and greyhounds? Since myostatin stimulates slow twitch muscle fiber production, a disabling myostatin mutation should help whippets (who run short races of approximately 200 to 300 meters) more than greyhounds (who

---

* Most mice like to run on treadmills, but not all of them. In this experiment, mice that wouldn't run until exhaustion were encouraged to keep running by either having an airstream blow on their hind feet or (no joke) having their hind feet "tickled."

run 500 meters or more). Even if greyhounds did once have a nonfunctional *myostatin* mutation, they probably lost it over generations of breeding, since the mutation wouldn't have provided any competitive advantage.

Myostatin gene mutations have even been found in cows. Belgian Blues are a breed of cattle prized for their lean musculature. Not long after the discovery of myostatin, scientists tested the DNA of Belgian Blue cattle and discovered that they too carried nonfunctional variants of the myostatin gene. Six different variants of the myostatin gene have been found in Belgian Blue cattle. They all lead to nonfunctional myostatin protein and cows with lots of muscles and little fat.

SUPER-MICE AND SUPER STRONG PEOPLE

Myostatin knockout mice, bully whippets and Belgian Blue cattle all demonstrate that disrupting the myostatin function leads to very muscular animals. Before the birth of Heinrich, though, it wasn't known whether disabling myostatin gene variants occur in people as well. Because of Heinrich's remarkable physical abilities, his DNA was tested for unusual genetic variants. The tests revealed that both copies of Heinrich's myostatin gene produced nonfunctional myostatin protein. Heinrich's variants are different from those found in myostatin knockout mice, fast-running whippets or Belgian Blue cows, but the end result was the same. Neither of Heinrich's myostatin genes works properly, so he doesn't produce any functional myostatin protein.

Since both copies of Heinrich's myostatin gene are nonfunctional, most likely his mother and his father also have a nonfunctional copy of the myostatin gene.* Heinrich's mother was an Olympic sprint swimmer, and producing only half the usual amount of myostatin might have contributed to her exceptional athletic abilities. Although no information regarding Heinrich's father is available, several of Heinrich's other relatives were described as unusually strong.

How many top athletes have nonfunctional myostatin? No one knows. Before Heinrich's birth in 1999, human myostatin mutations had never been found, but then again, before Heinrich was born, no one had been looking for myostatin mutations in people. It's possible that people with genetic variants in proteins that interact with myostatin also exist. One such person may be a boy from Michigan, named Liam Hoekstra. Hoekstra hasn't been

---

* There is also a small possibility that one or both of Heinrich's gene variants were de novo, meaning the mutation was new and not inherited from either parent.

studied in a published scientific investigation, but since his birth in 2005, he has often been described in the popular press. Generally portrayed as a "super-boy," young Liam is clearly very strong and muscular, though not quite as strong as Heinrich.

Why Liam is strong is not known. His myostatin genes appear to be normal. Most likely Liam has a mutation in a different gene that also leads to myostatin deficiency or that perturbs another biochemical pathway that regulates muscular development. One possibility would be a gene in the insulin growth factor (IGF-1) pathway. We've already learned about IGF-1, the hormone that works with human growth hormone to regulate how rapidly we grow. *IGF-1* is also the gene closest to the genetic marker variants that differ between Great Danes and Chihuahuas. So variants in *IGF-1* might well increase muscle growth. In fact, when scientists engineered transgenic mice experiments to produce increased IGF-1, the mice had 25% more muscle mass and were stronger than their nonaltered littermates.

Another possibility is that Liam's strength comes from an unusual variant in the gene for follistatin, a protein that limits myostatin production. Scientists have been studying follistatin in hopes that its ability to block myostatin production and promote muscle growth might help treat muscular dystrophy. In 2007, the Johns Hopkins team that had initially discovered the myostatin gene also developed a strain of transgenic mice that produced increased amounts of follistatin in their muscle cells. These mice were truly "Schwarzenegger mice." While the original myostatin knockout mice had muscles roughly twice the size as those of normal mice, the follistatin transgenic mice had muscles that were *four* times normal size. Apparently follistatin has other, as yet unknown functions in stimulating muscle growth, beyond inhibiting myostatin production.

As important as they are, muscle size and strength are not all there is to a super-athlete. In many sports, being able to generate large amounts of energy quickly is just as important. Efficient energy production requires different proteins from those involved in building strong muscles. How much of these proteins our muscles produce depends on genetic variants in our DNA as well as on the epigenetic alterations of the chromosomes in our muscle and cardiovascular cells. While some elite athletes benefit from outsized muscles, others gain their edge by being better able to harness the energy stored in their muscles, and one such super-athlete was an Olympic ski champion named Eero Mäntyranta, whom we'll meet in the next chapter.

17

WHAT MAKES A SUPER-ATHLETE?
PART 2: ENERGY MATTERS TOO

〜

Nineteen sixty-four was a good year for the Finnish Winter Olympics Team. The Finns won 10 medals including 3 golds. Not bad for a country with only four million inhabitants, but then again winter sports are a national pastime in Finland. A single athlete, Eero Antero Mäntyranta, won three of Finland's medals in 1964, including two of its gold medals. Perhaps Finland's greatest skier of all time, Mäntyranta won seven Olympic medals between 1960 and 1968. Like all champion cross-country skiers, Mäntyranta exercised vigorously and had spent years since childhood mastering the skills of his sport. Yet, Mäntyranta was different from his competitors in one important way. He had inherited an extremely rare genetic variant for a protein called the erythropoietin receptor (EpoR). As a result, the oxygen-carrying capacity of his blood was almost 50% higher than normal.

Did this rare gene variant turn Mäntyranta into an Olympic ski champion? Is *EpoR* a "champion skiing gene"? Of course not. Indeed, DNA testing of 200 of Mäntyranta's relatives showed that 29 of them shared Mäntyranta's rare genetic variant. Although physiological testing showed that Mäntyranta's relatives transported oxygen more efficiently than most people, none were ski champions. No, there isn't any Olympic skier gene.

That said, if Mäntyranta's relatives had devoted themselves to endurance sports, their *EpoR* gene variants would definitely have given them an edge.

Figure 17.1. Eero Antero Mäntyranta, Finnish Olympic ski champion.

In this chapter we'll see how a variant in a gene for an erythropoietin receptor, as well as other genetic variants, can lead to improved physical ability. We'll also find out how nongenetic factors contribute to athletic excellence. We'll see how exercising alters our chromosomes and how performance-enhancing drugs can change our athletic ability, no matter what genes we have inherited. But first we need to go over a few basic facts about how cells produce the energy that powers our muscles.

Strong muscles are just part of being a top athlete. Those muscles need energy, in the form of adenosine triphosphate (ATP). To produce ATP, muscles require an energy source, which ultimately comes from the food we eat. Our muscles use two biochemical pathways to produce ATP, each involving scores of chemical reactions and enzymes. The first pathway, called anaerobic energy generation, has the advantage of not requiring oxygen. Skeletal muscle cells are the only type of cell that can produce ATP without oxygen, and they do so when they need energy more quickly than the blood can supply them with oxygen. The second way that muscles convert food molecules into ATP, called aerobic or oxidative energy generation, requires oxygen (from the air we breathe). Aerobic energy generation is more efficient than anaerobic ATP production and is the only form of energy production available to cells other than skeletal muscle cells. Figure 17.2 presents an overview of key characteristics of the energy-generating processes we'll be looking at in this chapter.

Energy-generation processes	Anaerobic	Short-term aerobic	Long-term aerobic
Oxygen requirements	None	High amount for short time	Moderate amount for extended time
Muscle fiber type	Fast twitch	Slow twitch	Slow twitch
Energy efficiency	Low	High	High
Red blood cell requirement	None	Higher	Lower
Blood pressure requirement	None	Higher	Lower
Key proteins	ACTN3, less myostatin	ACE, EpoR	Bradykinin receptor, EpoR, HIF2α, PPARδ, AMPK
Typical sport	Sprinting	Intermediate distance running	Marathons

Figure 17.2. Overview of aerobic and anaerobic energy-generation processes.

MUSCLE FIBERS AND ATHLETIC PERFORMANCE

In the last chapter we saw that our muscles contain two types of muscle fibers: fast twitch and slow twitch fibers. The fast twitch fibers, used by sprinters, generate more force but are less efficient than slow twitch fibers, which marathon runners or other endurance athletes rely on. One reason for these differences is that fast twitch muscle fibers primarily use anaerobic energy generation, while slow twitch fibers produce ATP using oxygen.

We've already encountered one protein, myostatin, that regulates the relative amounts of fast and slow twitch muscle fibers in our muscles. You'll recall that myostatin knockout mice not only have bigger muscles than normal mice but they also have more fast twitch muscle fibers. As a result, the myostatin knockout mice excel on strength tests but don't perform well on treadmill endurance runs.

The alpha-actinin proteins are also key factors in fast twitch and slow twitch muscle fiber development. Two principal forms of actinin are found in our muscles, α-actinin-2 (ACTN2) and α-actinin-3 (ACTN3). ACTN2 is found in all muscle fibers, whereas ACTN3 is used only by fast twitch fibers. The relative concentrations of ACTN2 and ACTN3 influence whether muscles produce more fast twitch or slow twitch fibers, with decreased ACTN3 production leading to muscles with more slow twitch fibers.

To study the role of ACTN3 in physical performance, scientists in Australia engineered *ACTN3* knockout mice. These mice had a high ratio of slow twitch to fast twitch muscle fibers, and, sure enough, the mice performed poorly on athletic tasks that required short bursts of energy, while they outperformed normal mice on endurance tests.

ACTN3 also affects human athletic performance. Some people have an *ACTN3* variant, which results in nonfunctional ACTN3 protein. Approximately 18% of Europeans have two copies of this nonfunctional gene variant, while 42% have one nonfunctional variant and produce only half the usual amount of ACTN3 protein. The *ACTN3* knockout mouse experiments suggested that people lacking a functional *ACTN3* gene would be worse sprinters and better marathoners, and several investigations of DNA variations in human athletes support this hypothesis. World-class endurance athletes have two nonfunctional *ACTN3* variants more often than sprinters or nonathletes. Similarly, top sprinters are less likely to have two nonfunctional *ACTN3* variants than endurance athletes or nonathletes.

ATHLETIC PERFORMANCE AND BLOOD PRESSURE

Our body also adapts to differing energy requirements by adjusting our muscles' oxygen supply. By regulating oxygen transport, our blood system can minimize the amount of inefficient, anaerobic energy production in our muscles. Oxygen delivery is controlled, in part, by adjusting our blood pressure. For sprinting, we need high blood flow, so our blood pressure increases. For endurance sports, we generally need to lower our blood pressure so that we can maintain our energy level for a longer time.

To regulate blood pressure, the cardiovascular system uses signaling hormones, including bradykinin and angiotensin. Bradykinin signals blood vessels to dilate, and therefore lower blood pressure, while angiotensin causes blood vessels to constrict. Consequently, variants in the genes for these hormones or their receptors could affect athletic performance. The bradykinin gene itself has no known human variants, but a common DNA variant exists *near* the gene for the bradykinin receptor. One variant, which is missing nine base pairs and is called the −9 variant, leads to higher bradykinin receptor protein synthesis than the other gene variant, the +9 variant.

To test whether bradykinin receptor variants are linked to athletic performance, scientists tested muscle efficiency and bradykinin receptor genes in 81 British Olympic sprint and long-distance runners (muscle efficiency was measured on a stationary bicycle energy meter). The researchers found that long-distance runners were more likely to have two copies of the −9 bradykinin-receptor variant than sprinters, and that athletes with two copies of the −9 variant scored higher on the muscular efficiency tests than those with the +9 variant.

Angiotensin also regulates blood pressure. Although there are no common genetic variants for either angiotensin or its receptor, two common variants do exist near the gene for angiotensin converting enzyme (ACE). ACE converts angiotensin from its inactive to its active form and also degrades bradykinin, so increasing ACE leads to higher blood pressure. That's why ACE inhibitors, drugs that block the activity of ACE, are often prescribed for people with high blood pressure.

The two *ACE* gene variants differ by an insertion of 287 base pairs. The longer variant is associated with decreased ACE activity, so it was predicted to be more common in endurance athletes. Several studies have compared *ACE* variants among top athletes in different sports, and have found that endurance athletes generally do have the longer variant, while the shorter variant is more common in sprinters and athletes in other power-oriented events. For example, in a study of mountaineers who had climbed to higher than 8000 meters without supplemental oxygen, each had at least one long *ACE* variant, and generally the climbers with the best records had the long variant in both of their *ACE* genes.

OXYGEN TRANSPORT, ENDURANCE SPORTS AND LIFE IN THE HIMALAYAS

The cardiovascular system also adjusts oxygen transport by controlling the production of the oxygen-carrying red blood cells. Red blood cell production is regulated by means of a signaling molecule called erythropoietin (Epo). When Epo binds to Epo receptors (EpoR) on the surface of bone marrow cells, the bone marrow is stimulated to produce red blood cells.

There are no known variants of the *Epo* gene leading to improved athletic capacity, but a rare variant in the *EpoR* gene does influence physical ability. This is the DNA variant that gave Eero Mäntyranta his competitive edge. This variant leads to an erythropoietin receptor that is 70 amino acids shorter than the EpoR proteins found in most people. Those 70 extra amino acids form part of a control system that limits the body's production of red blood cells. Such red blood cell regulation is generally beneficial; people with high red blood cell levels are often susceptible to weakness, headaches and fatigue. Yet as Eero Mäntyranta has shown, occasionally having this extra EpoR regulation is neither necessary nor advantageous.

Some people with exceptional cardiovascular capabilities have different unusual genetic variants to thank. These individuals are not from Finland, but from Tibet. In Tibet, cardiovascular performance is very important. Tibet's capital, Lhasa, is 11,450 feet above sea level, and much of the rest of Tibet is even higher. Just breathing and extracting enough oxygen from the rarefied air is challenging at these elevations.

Tibet interests geneticists because two distinct groups of people live there: the historical Tibetans (we'll simply call them Tibetans) and the Han Chinese. Archeological and genetic data indicate that the Tibetans have lived in the mountains for approximately 5000 years. Five thousand years is enough time for evolution to have produced genetic variants that facilitate breathing when oxygen is in short supply. As a result, Tibetans excel at strenuous activity in the thin mountain air, that is, under what are called hypoxic conditions. In contrast, the Han Chinese have only lived in Tibet for the last 100 years, not long enough for genetic adaption to enable advantageous DNA variants to become widespread. So Han Chinese, even those who were born and raised in Tibet, generally show no more endurance under hypoxic conditions than people from lower altitudes (once they've had time to acclimate).

GENETIC VARIATION IN HYPOXIA GENES

Scientists have long suspected that Tibetans and Han Chinese have genetic variants that underlie their different endurance levels, and between 2005 and 2010, seven genetic studies were carried out in Tibet looking for differences in the DNA of Tibetans and Han Chinese. In some investigations, the scientists simply looked for genomic locations with consistent differences between Tibetan and Han Chinese DNA. In other studies, the researchers searched for positive selection, the DNA pattern described in chapter 10 that characterizes genetic regions containing a recently acquired, advantageous DNA variant. In chapter 10, the DNA variants being sought contributed to gentleness in the wolves that humans were breeding to become dogs. Here, the researchers were looking for genetic variants that help people be more physically fit in the mountains.

Each of the seven Tibetan–Han Chinese comparative genetic studies yielded several candidate chromosomal regions predicted to contain genetic variants contributing to improved physical ability under hypoxic conditions.

The various studies often identified different candidate regions, in part because the research teams studied different Tibetan subpopulations. Yet one genetic region on chromosome 2 turned up in every study. This region included a genetic marker where 91% of the Han Chinese had one genetic variant while 87% of the Tibetans had the other variant. The marker is near a gene known either as *EPAS1* or *HIF2α*. The name *HIF2α* is derived from the fact that the HIF2α protein is a hypoxia-induced factor (HIF). As you might guess, cells produce hypoxia-induced factors when they need more oxygen, and one function of the HIFs is to signal the kidneys to produce erythropoietin, the hormone that stimulates red blood cell production.

When the researchers examined the *HIF2α* gene itself, they found that Tibetans and Han Chinese had identical DNA sequences and consequently produce identical HIF2α proteins. In fact, scientists still don't completely understand the biological consequences of the variation at the genetic marker near the *HIF2α* gene. Perhaps the Tibetan variant causes different amounts of HIF2α protein to be produced in times of oxygen shortage. Possibly the variation doesn't have any biological consequences; it may just be a genetic marker, inherited along with some nearby, as yet undiscovered causal variant.

In any case, *something* important for human endurance is coded in our DNA close to the *HIF2α* gene. The scientists found that at low elevations, the 13% of Tibetans with the "Chinese variant" had somewhat lower levels of hemoglobin, the molecule red blood cells use to bind oxygen, than the Tibetans with the "Tibetan variant." In contrast, at 4000 meters, people with the Chinese variant had greatly increased blood hemoglobin levels (even higher than level of Tibetans with the Tibetan genetic variant), while the hemoglobin level among the people with the Tibetan variant was unchanged. Apparently, people with the Tibetan variant always produce enough hemoglobin for their blood to function properly, while individuals with the Chinese variant don't naturally produce enough hemoglobin for hypoxic conditions. So when they travel to high elevations, their blood cells start synthesizing additional red blood cells, which leads to excessively high hemoglobin levels, as well as the symptoms of mountain sickness.

ENERGY, METABOLISM AND SUPER-MICE

Altering an animal's metabolism can also dramatically affect its athletic performance. PEPCK is an enzyme that controls an important step in the synthesis of glucose, the principal energy source used by our cells to make

ATP. In the early 2000s, a group of scientists at Case Western Reserve University wanted to determine what happens if an animal produces excess amounts of PEPCK in its muscles. Previous experiments had shown that an animal with excess PEPCK throughout its entire body is susceptible to diabetes. Consequently, the scientists added a specially modified *PEPCK* gene to mouse DNA so that the mice produced extra PEPCK only in their muscles. As a result, the mice didn't contract diabetes; they were actually quite healthy.

In fact, the mice were not just healthy; they were super-healthy. At three months of age, these mice could run 4 kilometers at 20 meters per minute, while normal three-month-old mice were exhausted after running less than half a kilometer at that speed. The *PEPCK* transgenic mice were accomplishing these athletic feats by generating energy just as an elite human athlete would, by primarily burning fatty acids. As a result, the transgenic mice ate 60% more than their litter mates, yet they weighed half as much. They could run for four or five hours straight without needing to stop to eat or drink, and on average they lived twice as long as their normal litter mates. These *PEPCK* transgenic mice really were "super-mice."

ATHLETIC EXCELLENCE IS NOT JUST GENETIC

Genetically modified super-mice demonstrate that a DNA variant can lead to an animal with strength or endurance far beyond that of normal animals. People like Liam Hoekstra (whom we met in chapter 16) and Eero Mäntyranta teach us that humans, too, can have extraordinary athletic capabilities if they are lucky enough to win in the genetic lottery, while the studies linking athletic performance and common variants in *ACTN3*, *ACE* and bradykinin receptor genes suggest that inherited genetic variation influence everyone's athletic abilities, and not just the performance of elite athletes with rare genetic variants.

That said, athletic performance is clearly not just about genetics. Exercise and diet also have an important impact on our physical abilities. Not only do we know this intuitively, but scientists are discovering how exercise affects our athletic abilities at a cellular level. One way that exercising directly alters our muscle cells are via the fast and slow twitch muscle fibers. We've learned that certain genetic variants can induce muscle cells to make relatively more fast twitch or slow twitch fibers. Although genetics is the key factor in the relative amount of fast and slow twitch muscle fibers we are born with, muscle fiber concentrations can change with exercise. If we train for an endurance sport our muscles convert some fast twitch muscle fibers into slow twitch, while when we train for sprinting, the muscle fiber conversion occurs in the opposite direction.

In 2004, scientists from the Salk Institute in California showed that this muscle-fiber conversion is regulated by a protein called PPARδ. The Salk researchers were able to mimic the effects of endurance-sport training by creating transgenic mice that produce higher concentrations of activated PPARδ protein. These mice had increased slow twitch muscle fibers and were true mouse endurance athletes. They could run for up to two and a half hours on the treadmill, while normal mice became exhausted after just an hour and a half. During each running session, these transgenic mice ran approximately 1800 meters on the treadmill, while normal mice ran just 900 meters. In short, the transgenic *PPARδ* mice earned their media titles of "marathon mice."

Exercise changes the concentrations of proteins, such as PPARδ, by altering the amounts of the messenger RNAs that are transcribed from the muscle cell's DNA. Scientists have long known that exercising alters the muscle-cell mRNA concentrations of many proteins involved in protein synthesis, mitochondrial production and conversion of food into energy. What had been more difficult to understand was how these transient changes in messenger RNA transcription led to the long-term changes in muscles that result from exercising. Messenger RNA concentrations generally return to basal levels soon after exercising stops; yet the cellular benefits of exercise, such as the increased production of mitochondria, often are only apparent after weeks or months of training.

Although scientists still have only a limited understanding of how exercise changes muscle cells, recent data suggests that DNA methylation and other chromosomal modifications are involved. In 2012, scientists from the Karolinska Institute in Stockholm, Sweden, compared DNA methylation patterns in biopsied skeletal muscle cells obtained from healthy volunteers before and after short sessions of strenuous exercise. The scientists measured DNA methylation changes at eight genes important in cellular energy production and found that five of these genes were significantly less methylated after exercising. Since DNA methylation typically decreases gene transcription, the results suggested that muscle cells decreased DNA methylation during exercise in order to produce more of the proteins required for energy generation.

Recent experiments are also beginning to elucidate how long-term exercise modifies our chromosomes and thereby affects our physical condition. In 2013, a research group from Lund University in Sweden showed that individuals tested six months after joining an exercise program had changes in the DNA methylation patterns of the chromosomes in their fat cells. These DNA methylation changes led to alterations in the amounts of DNA that was being transcribed into mRNA and subsequently into proteins in their fat cells. Ultimately, this was leading to the observed physiological changes that are associated with the benefits of long-term exercise.

USING DRUGS TO IMPROVE ONE'S GENES

Although exercise and good nutrition can help compensate for unfortunate genetic variants, there are limits to their effectiveness. Individuals with genetic variants causing dwarfism or muscular dystrophy can't make themselves tall or strong simply by exercising more or eating more nutritious foods. Nor can people with genetic variants leading to anemia gain normal levels of endurance just by participating in a sports training program.

Nevertheless, understanding how the environment changes cells can help scientists develop drugs that can alter muscle or blood cells to compensate for disabling genetic mutations. Doctors can treat people suffering from severe anemia by giving them synthetic erythropoietin, which helps their endurance just as extra erythropoietin receptors helped Eero Mäntyranta. Children who are short because of insufficient growth hormone can receive injections of synthetic growth hormones to substitute for the hormones their own bodies are unable to produce. If such hormone replacement therapy is started at an early age and continued for several years, height gains of 8 inches or more can be achieved. Even for muscular dystrophy, which in the past has been largely resistant to medical intervention, there is now hope that follistatin or some other myostatin inhibitor may prevent the muscular underdevelopment and muscle wasting of the disease.

Whether drugs also provide benefits to healthy people who just want to be stronger or have more endurance is less clear. Some endurance athletes who were not lucky enough to have Eero Mäntyranta's *EpoR* genetic variant have used synthetic Epo to give themselves a competitive edge. At least they used to before unannounced testing for synthetic Epo was instituted at the 2002 Winter Olympics (resulting in three cross-country skiers being forced to relinquish their medals*). Other attempts to improve athletic performance have focused on drugs that mimic testosterone. Since testosterone stimulates muscle growth, administering extra testosterone might be expected to enhance muscle development and increase strength. Injected testosterone is degraded quickly in the liver, though, so competitive athletes more often use anabolic steroids, synthetic drugs that act like testosterone but aren't destroyed by the liver.

It is a curious fact that much of the scientific literature claims that such drugs provide little benefit in athletic performance, whereas many competitive athletes remain convinced that these drugs do work. Olympic champions such as Ben Johnson and Marion Jones have admitted to taking performance-enhancing drugs, and countless other top athletes including baseball superstars Mark McGwire and Barry Bonds and bicycling champion Lance Armstrong are either known or believed to have used them. According to a study published in the late 1990s by *Sports Illustrated* magazine, more than 50% of competitive athletes have used steroids, even though sports federations have banned steroid use since the 1970s.

Recent evidence suggests that, in this debate between the scientists and the athletes, the athletes are at least partly correct. A 2009 article by University of Southampton professor Richard Holt, a leading authority on performance enhancing drugs in sports, compared winning women's Olympic shot put distances from 1948 to 2008 (see figure 17.3). The graph shows that each time a policy was instituted to combat drug use by athletes – for example by introducing a new steroid drug test – the winning Olympic shot put distance decreased or stayed relatively unchanged. In subsequent years, presumably as more athletes found ways to evade the drug tests, the winning Olympic shot put distant started edging up again. Holt argues that the opinions of the scientists and the athletes regarding performance-enhancing drugs differ because they're using different criteria. While scientific investigations emphasize that the performance benefits of drugs are small, from the elite athlete's perspective, a 1% performance improvement might be the difference between an Olympic Gold medal and going home empty handed.

---

* Before Epo injections for athletes became available (if not legal), some athletes engaged in an even more questionable practice to transiently increase their blood Epo levels. In this procedure, called blood doping, some of the athlete's own blood would be removed and stored and then infused back into their bloodstream immediately prior to a competition.

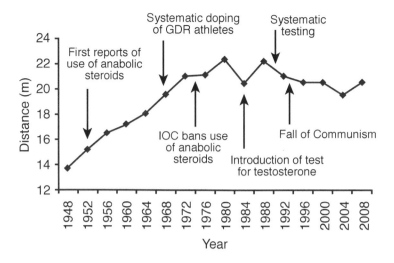

Figure 17.3. Winning women's Olympic shot put distances and milestones in testing for drug use in sports. GDR = German Democratic Republic (the former East Germany), IOC = International Olympic Committee.

DRUGS FOR COUCH POTATOES?

Most of us neither suffer from anemia nor compete in endurance sports at the Olympic level. But eventually we do all get old, and many of us stop exercising regularly even earlier. As a result, our muscles make fewer slow twitch muscle fibers and our endurance capacity drops. Consequently, some scientists are searching for drugs that can trick our muscles into thinking we have been exercising more, in a similar way to how genetic modification of the *PPARδ* gene increased the endurance of the transgenic marathon mice.

These drugs typically target either PPARδ itself or AMPK, a protein that interacts with PPARδ. One PPARδ-stimulating drug, called GW1516, was tested on normal (that is, not transgenic) mice to see if it improved mouse endurance. The experiments showed that after five weeks of GW1516 treatment, mice were able to run 70% further than they had been able to previously. Unfortunately for those of us who don't exercise regularly, the researchers discovered that GW1516 didn't *automatically* improve the endurance of the mice. In order for the GW1516 treatment to work, the mice also had to be on a regular exercise routine.

But there may still be hope for mice – or people – who can't seem to stay on an exercise regime. A new AMPK-stimulating drug called AICAR also increases endurance in mice, and it works on both exercising and so-called sedentary mice (mice for which running on treadmills is just not their thing). With AICAR, even sedentary mice are able to run more than 40% farther than if they hadn't been given the drug. So perhaps AICAR will even enable couch potatoes to improve their endurance. In fact, AICAR has been unofficially referred to as the couch potato drug.

Resveratrol is also being studied for its potential to improve human endurance. You'll recall that resveratrol is a natural substance found in fruits, plants and even red wine, that is being actively studied for its anti-aging effects. Resveratrol also improves endurance, at least in mice. Just how resveratrol works isn't completely understood, but it appears to stimulate the conversion between fast twitch and slow twitch muscle and hence between aerobic and anaerobic energy generation. What is known is that after receiving large daily amounts of resveratrol (0.2 to 0.4 milligrams per gram of mouse weight), mice are able to run approximately twice as far as mice that haven't received resveratrol. Unfortunately, a person would need to consume an enormous amount of resveratrol to receive a comparable dose (per weight) as the mice in the endurance experiments. If you attempted to consume that much resveratrol from red wine (which typically contains about 2 to 4 milligrams of resveratrol per liter) you wouldn't be running far at all, since you'd have to drink over 100 liters of wine.*

DANGERS OF PERFORMANCE-ENHANCING DRUGS

All performance-enhancing drugs have serious potential side effects, especially at the high doses often taken by competitive athletes. Quantifying those risks is difficult, since asking participants in a scientific study to use these drugs would be unethical. What is known is that blood doping or injection of synthetic Epo can increase one's risk of stroke or heart disease. Anabolic steroids can also have undesirable side effects, including heart and circulatory problems, liver abnormalities, increased aggressive behavior, and baldness. Excess growth hormone, if caused by an overactive pituitary gland, carries increased risk of cardiovascular disease, diabetes and cancer, though it is not yet known whether elevated growth hormone levels from drug use have similar health consequences. Meanwhile, drugs like GW1516 and AICAR are so new that their long-term effects are only starting to be understood in mice, let alone in humans.

---

* A gram is approximately 1/3 of an ounce and a milligram is 1/1000 of a gram. Also, a liter is a little more than a quart, so 100 liters is about 25 gallons.

Possibly no form of artificial athletic enhancement illustrates both its benefits and potential dangers more dramatically than the use of PEPCK. As we've learned, genetically modified mice that produce excess PEPCK in their muscles are stronger than normal mice, have increased endurance and live twice as long. Sounds pretty good. In fact, as the PEPCK mouse experiments were performed over 10 years ago, you might be wondering whether this research has led to similar treatments for people. The answer is no, in part because pills or injections elevate PEPCK levels throughout the body, thereby increasing diabetes risk, while genetically administering PEPCK just to muscles (as was done with the super-mice) is extremely difficult.

Beyond these technical challenges, PEPCK-based physical enhancement faces a more fundamental problem, which had been completely unexpected. It turns out that besides being long-lived super-athletes, the *PEPCK* transgenic mice were all extremely aggressive. So even if a PEPCK-based performance-enhancing drug could be made safe, the people taking it, though possibly strong and healthy, might be quite unpleasant.

18

WHY ARE SOME PEOPLE SO SMART?

Doogie M. is one smart fellow. On just about any intelligence test, Doogie scores significantly higher than his peers, including his identical twin brother, Dewey. Actually, Doogie's and Dewey's genomes are not *completely* identical. Doogie has one more gene than his brother. Could this single gene be the reason that Doogie is so smart? Could this bizarre story possibly be true?

Remarkably, the story of Doogie is true. Well, mainly true. As you may have guessed, Doogie isn't a person but a mouse. Doogie belongs to a strain of transgenic mice originally developed by Joe Tsien, a professor at Georgia Regents University. Doogie's DNA includes an additional copy of a single gene called *NR2B*. And Doogie *is* smarter on mouse intelligence tests than mice that don't have an extra copy of *NR2B*.

WHAT IS INTELLIGENCE?

Wait a second, what does it mean for a mouse to be "smarter" on a "mouse intelligence test"? For that matter, what does it mean for one person to be smarter than another person? Was Einstein smarter than Shakespeare or da Vinci, or did he just have a different kind of intelligence? Actually, for what we'll be talking about in this chapter, we don't need to agree precisely on what intelligence is. All we need to accept is that Einstein was *different* from you and me, and that understanding how Einstein's biology, his genetic makeup and his environment made him different would be pretty interesting. In other words, rather than tackling the thorny question of defining intelligence, I'll just paraphrase Supreme Court Justice Potter Stewart: Intelligence may be difficult to define, but you know it when you see it.*

If a child can solve logic or math problems that baffle most adults, it's hard not to conclude that something's unusual about him. And few children were more unusual in their problem-solving ability than young Carl Friedrich Gauss. Gauss was born in 1777 in Braunschweig, Germany, to poor, uneducated parents. Eventually, he became one of the greatest mathematicians of all time. Here, though, we're interested in Gauss when he was a boy. As told by Wolfgang Sartorius, Gauss' colleague at the University of Göttingen, when Gauss started elementary school, his teacher was a Mr. Büttner, someone who enjoyed tormenting his students with long, tedious arithmetic problems. One day Büttner told his students to add up all the numbers from 1 to 100. As the other students started writing and adding numbers furiously on their writing slates, young Gauss immediately wrote a single number on his slate (5050) and handed the correct answer to the teacher.†

Even if we can't define intelligence, to study how biology affects it, scientists need to be able to measure it or at least to measure *something* that seems related to intelligence. Usually intelligence is measured using problem-solving and memory tests, as well as tests of language ability and reading skills. In the next chapter we'll be looking at language, so here we'll focus on problem-solving and memory tests. Such tests may not be the perfect criteria for intelligence, but they give us a place to start, and if nothing else, they've been shown to be consistent. In other words, any one of these tests might not determine who is the most "intelligent," but at least it would predict who would score highest on a different "intelligence test"!

One advantage of problem-solving and memory tests is that they can be adapted to studying intelligence in animals. For example, one can set up a maze with food at the end and measure how long it takes a hungry laboratory mouse to find the food. In practice, a different kind of maze, the Morris water maze, in which mice are trained to escape to a dry platform,

---

* Actually, Justice Stewart was talking about pornography – but you get the point.
† Don't worry. If you don't know the mathematical trick that Gauss used, we'll return to this problem a little later in the chapter. For now you might take a minute to see if you can guess how Gauss solved the problem. Trying to solve this arithmetic problem may help you appreciate the genius of seven-year-old Gauss.

is more commonly used to measure mouse intelligence. (You'll recall from chapter 13 that mice don't like being immersed in the water maze.)

But why should we study intelligence in mice, if we are interested in human intelligence? The answer is that studying intelligence in animals, and especially laboratory animals, is vastly easier than studying human intelligence. If a person scores well on an intelligence test, determining whether their high score resulted from genetics or a great education or simply a better childhood environment is difficult. With laboratory animals, scientists can control environmental conditions. All animals can be given identical training and exactly the same living conditions and amount and type of food. By using transgenic animals, scientists can more easily study the effects of genetic variation on intelligence. In contrast, any two humans (except identical twins) differ by millions of genetic variations. Even if one could identify every genetic variant in Einstein's DNA, determining which variants (if any) made Einstein so smart would still be beyond science's current capabilities. Lastly, studying the mouse brain has the advantage that, despite the similarities between mouse and human brains, ultimately humans are smarter than your average mouse, or even an exceptional mouse like Doogie. So studying a mouse brain should be less difficult than studying the human brain. Of course, any insights that scientists gain from studying animal intelligence won't necessarily be valid for people. With any luck though, they'll provide a good starting point for looking for the origins of human intelligence.

INTELLIGENCE, GLUTAMATE AND NR2B

Even a mouse brain is extremely complex, involving millions of nerve cells with approximately 10^{11} interconnections among them (that's a million, million interconnections!). In previous chapters, you've learned that neurotransmitters and their receptor proteins are used by nerve cells to communicate and that one of the brain's key neurotransmitters is glutamate. NR2B is part of an important glutamate receptor, known as the NMDA receptor. So it shouldn't be surprising that a mouse with an extra copy of the *NR2B* gene is a smart mouse.

By producing more NR2B, Doogie's NMDA receptors can stay open and detect glutamate for almost twice as long as those in a normal mouse. Although it is not understood how having glutamate receptors stay open longer enhances learning, many experiments have shown that animals with glutamate receptors that do so learn and remember better. For example, young animals have more NR2B protein in their glutamate receptors than older animals, which appears to be why younger animals learn and remember more easily than older ones (a fact that readers of a certain age can probably relate to). And it's because his glutamate receptors remain open longer that Doogie is a faster learner than a normal mouse. After three training sessions, Doogie and the other *NR2B* transgenic mice were able to solve the Morris maze in an average of 25 seconds, while mice without the extra *NR2B* gene required 35 seconds. In tests assessing whether what a mouse has learned is still retained one to three days after training, the "Doogie mice" again consistently outperformed the normal mice.

Mice are not the only animals that learn quicker with added NR2B protein. Demonstrating this has been challenging, though, because genetically modifying animals other than mice is difficult. While the paper describing Doogie and his transgenic brethren appeared in 1999, it wasn't until 2009 that scientists were able to develop transgenic rats that overexpress NR2B. (Though we sometimes lump mice and rats together, they are quite different animals. Rats are more difficult to genetically manipulate than mice, but in many ways, they are more closely related to humans.) In any case, once the technical hurdles of adding *NR2B* to rats were overcome, the researchers found that transgenic *NR2B* rats also surpassed normal rats in learning and memory tests. Scientists are also actively attempting to develop strains of transgenic monkeys. If transgenic monkeys that overexpress NR2B can be engineered, it will be interesting to see whether they too are faster learners that their normal brethren.

Remarkable as the learning accomplishments of *NR2B* transgenic animals are, their limitations are also revealing. For example, after three training sessions the *NR2B* transgenic mice solved the test maze in 25 seconds, while the normal mice required 35 seconds. Yet after just three more training sessions, *both* sets of mice solved the maze in approximately 20 seconds. In other words, the transgenic mice were faster learners, but ultimately all the mice did equally well. Similarly, although the transgenic mice did remember the maze better for up to three days after training, within a week, both groups' performance had fallen to the same level. So while genetics does influence (at least rodent) intellectual ability, the environment, in the form of good schooling and continued practice, appears to be just as important. And to the extent that mouse intelligence is similar to human intelligence, this is most likely true of human learning as well.

LEARNING AND OTHER GLUTAMATE PATHWAY PROTEINS

The brain's glutamate signaling networks involve hundreds of proteins, not just NR2B. Variations in the genes for any of these proteins are likely to influence brain performance just as variations in *NR2B* do. For example, the FMR1 protein regulates glutamate signaling, and, as we saw in chapter 7, a genetic variation near the *FMR1* gene can lead to FMR1 deficiency and fragile X intellectual disability. KIF-17 is another important protein in the brain's glutamate signaling pathway. KIF-17 is a kinesin, that is, a protein whose structure creates electrical fields that can move other proteins. Such transport proteins are especially important in neurons, because neurons can be very large cells. To transport newly synthesized neurotransmitters and receptors to the parts of the cell where they are needed for brain signaling, neurons use an elaborate transport system of microtubules. The kinesin proteins, including KIF-17, are important components in this transportation network. In experimental cell culture systems, one can actually observe (under a microscope) KIF-17 transporting the NR2B protein, the very component of the glutamate receptor whose overexpression made Doogie so smart. Although NR2B transport via KIF-17 has not yet been directly observed in the brains of living mice, most likely the same transport mechanism is used in the living animal as in cell cultures.

In 2002, scientists succeeded in genetically engineering mice that overexpress KIF-17 in their brain cells. Presumably their brain cells transport NR2B more efficiently than the neurons of normal mice. In any case, when given maze-learning tasks similar to those given to the *NR2B* transgenic mice, *KIF-17* transgenic mice also master them faster than normal mice.

CDK5 is another protein that affects learning via the glutamate signaling network. CDK5 is a versatile protein that plays an important role in many processes in cell development. One of CDK5's key roles is to help degrade NR2B protein in neurons. (Neurons need to synthesize and degrade receptors whenever new information is learned or committed to memory.) Because CDK5 degrades NR2B, scientists guessed that disabling CDK5 in a mouse's brain cells would *increase* the amount of NR2B, and thereby make the mice smarter. However, CDK5 protein is required in many essential biological processes, so completely knocking out the *CDK5* gene leads to death in utero or shortly after birth. Consequently, in 2007 a research team at the University of Texas at Dallas engineered conditional *CDK5* knockout mice, in which the *CDK5* gene was disabled only in the brain cells of adult mice. The scientists observed that these mice had increased NR2B, as well as higher levels of glutamate transmission, in their brains. They also excelled on the "mouse IQ tests."

INTELLECTUAL ABILITY IN HUMANS

I think you'll agree that improving intelligence (even in a mouse) by manipulating genes is remarkable. That said, do these experiments on laboratory animals tell us anything about human intelligence? In other words, does our innate biology affect who will struggle in school and who will be a budding Einstein? Or is our intelligence determined by factors in the environment, such as how intellectually engaged we were as children or what schools we attended?

To address such questions, scientists often examine genetic and environmental variations among people with differing intellectual abilities. Especially when looking at young children who have not yet had extensive schooling, it seems likely that unusual cleverness is at least partly genetic. In fact, the extent to which we accept that intelligence is innate is reflected by our referring to such children as "gifted." You've probably read about such children or seen one on television. Perhaps you've known one, or maybe you were one yourself.

Carl Friedrich Gauss was such an extraordinarily gifted child. Even at the age of seven, Gauss had the intelligence and creativity to realize that if you need to add the numbers from 1 to 100, you don't have to perform the sum in the obvious way of:

Total = 1 + 2 + 3 + 4 .... + 49 + 50 + 51 + ... + 98 + 99 + 100.

Instead, you could get the same answer by *reordering* the numbers and adding them this way:

Total = (1 + 99) + (2 + 98) + (3 + 97) .... + (49 + 51) + (100) + 50.

In this reordering, the parentheses don't change the answer, they just tell you to perform the additions within the parentheses before you do the other additions. If you do reorder the numbers this way, something remarkable happens: each partial sum within parentheses equals 100. Since there are exactly 50 of these sums within parentheses, if you add them together you get 50 × 100 = 5000. The only number between 1 and 100 that's not inside one of these pairs of parentheses is the final 50, so the total sum of the numbers from 1 to 100 = 50 × 100 + 50 = 5050, which is the answer that seven-year old Gauss wrote on his slate.*

Gauss was not the first person to discover how to add quickly, in one's head. Professional mathematicians have known these techniques for hundreds of years. You may have even learned such a method for doing sums in high school or college math class. But for an unschooled seven-year-old to figure out how do this on his own is pretty dramatic, if anecdotal, evidence that intelligence can come to some people almost automatically, without any teaching or education. Did Gauss have some unusual genetic variant that made him so smart? It would be interesting to study Gauss' DNA to try to find out, though considering the millions of genetic variants we all have, finding the causal variant(s) would still be challenging.

GENES AND INTELLECTUAL DISABILITY

It's often easier to learn how something works by finding situations where it's broken. For learning about the biology of intelligence that means studying people who suffer from intellectual disability. Intellectual disability may stem from genetic or environmental factors; the three most common causes are Down syndrome and fragile X disease, which are genetic, and fetal alcohol syndrome, which is environmental. Overall, scientists estimate that roughly 1 out of 250 people are affected by severe intellectual disability (as measured by an IQ of less than 50), or over 20 million people throughout the world.

Among genetic causes, over 400 genes have been linked to intellectual disability, and even this list is almost certainly incomplete. Current scientific estimates indicate that variants in as many as 2000 genes may contribute to intellectual disability. That so many genetic variants causing severe intellectual disability exist might seem surprising. After all, intellectual disability is a serious condition. People suffering from it rarely have children, so one might guess that such mutations would disappear from the population. This is in fact true. But it is also true that new mutations causing intellectual disability keep recurring in the human population. These de novo mutations aren't in either the mother's or the father's DNA; they occur during the production of the sperm or (less frequently) the egg. Scientists have discovered that most children have 50 to 100 de novo mutations, and, on average, one of these

---

* Sartorius' account of young Gauss doesn't say whether Gauss used this particular trick to solve the addition problem. Other, similar techniques also work. That said, since Gauss got the correct answer so quickly and effortlessly, it's almost certain he used some such method.

mutations leads to an amino acid change in some protein. For children with intellectual disabilities, the number of protein-altering de novo mutations is often much higher. When researchers recently compared the genomes of 10 children with severe intellectual disability with their parents' DNA, they found an average of five protein-changing de novo mutations. In 6 of these children, one of the de novo mutations was in a gene linked to the brain structure or function, and was most likely at the root of their cognitive problems.

Although sorting out the influences of thousands of genes linked to intellectual ability is daunting, the task is made easier because many of these genes code for proteins, which belong to just a few brain-signaling pathways. Researchers have identified over 220 proteins in the glutamate signaling pathway alone, including over 180 involved in NMDA-receptor glutamate signaling. By studying just a few brain-signaling pathways, scientists can learn how hundreds of genes affect learning and intelligence. For example, when the DNA of people with previously unexplained intellectual disability was tested recently, rare mutations in six genes in the brain's glutamate signaling pathways were found. In fact, it seems that each time scientists sequence the DNA of individuals with intellectual disability new mutations in genes for brain signaling proteins are discovered.

TWIN STUDIES, GENETIC VARIANTS AND INTELLECTUAL ABILITY

The evidence that genetic variants can lead to human intellectual disability is overwhelming. But do genetic variations influence intelligence in people of average or above-average intelligence, as well? Stories of extraordinary children like the young Gauss suggest the answer is yes, but anecdotal evidence (and especially 200-year-old anecdotal evidence) should be viewed with caution.

To determine the extent that our intellectual gifts come from inheriting lucky genetic variants, scientists study twins. Over the last 30 years, multiple studies have shown that genetic variation contributes to intelligence,* or at least to whatever "IQ" tests measure. For example, researchers at Vrije University in Amsterdam, tested the IQ of 125 monozygotic (MZ) twin pairs and 112 dizygotic (DZ) twin pairs, first at age 5 and then again at age 12. They found that the IQ scores of the MZ twins were had a 68% correlation,

---

* It's worth emphasizing that linking individual genetic variation to intelligence is very different from linking race to intelligence, something for which there is no scientific evidence. In fact, simply defining the concept of race is difficult in the framework of modern genetics.

while the scores of the DZ twins had only 54% correlation. When these twins were retested seven years later, not only were the MZ twin scores again more highly correlated than the DZ twin scores, but, for unclear reasons,* the difference between the monozygotic and dizygotic twins had even become greater: the MZ score correlation was 81%, while for DZ twins the correlation was only 43%.

Since the Dutch report included only 237 twin pairs, a recent Ohio State University–University of London study surveyed 1279 monozygotic twin pairs and 1155 same-sex dizygotic twin pairs. These twins were all tested at the age of 10 with numerical and problem-solving tasks that measure mathematics ability. The children's performance was rated as "high performance" if they scored in the top 15%, and the study compared the concordance† for high performance between MZ twins and DZ twins. Once again, the MZ twins were significantly more similar to each other (59% concordance between the boy twins and 58% among the girls) than the DZ twins were (40% concordance for the boys and 44% among the girls).

Finally, let's look at a twin study that separated genetic and environmental effects in a different way. Over the last 30 years, the Minnesota Study of Twins Reared Apart has collected data on over 100 monozygotic twins who, for various reasons, were raised separately. By comparing these twins, the genetic similarities shared by monozygotic twins can be isolated from the influences of a common childhood environment that twins also usually share. The researchers found that even when raised in different environments, MZ twin IQ scores had a correlation of 69%. Although the correlation between the test scores of MZ twins raised together was even higher (88%), the fact that MZ twins, even when raised apart, had highly correlated test scores confirmed that genetics plays an important role in intelligence.

Twin studies demonstrate that human intellectual ability has a significant genetic component, but they don't indicate *which* genetic variants make a difference. To date, genetic association studies have consistently failed to find individual genetic variants with large effects on IQ. Intelligence

---

* Possibly some genetic factors only become important later in childhood. Since the number of twin pairs was small, one shouldn't take the precise values of the correlations as seriously as the clear and consistent differences between the MZ and DZ twin pairs.

† Twin concordance measures the extent that twins share a trait or phenotype. Generally, scientists use concordance to assess twin similarity in discrete categories, such as "high math performer" or not, and use correlations to measure similarities of quantities (such as raw test scores) which can accommodate a wide range of values.

appears to be like height, which as we learned in chapter 16, is influenced by hundreds of genetic variants, each having only a tiny effect. In one of the largest attempts to identify genetic variants associated with high intellectual ability, in 2011 a team of over 30 scientists from a dozen research centers tested the IQ of 3500 subjects, as well as their DNA at over half a million genetic marker locations. Yet even this large study was unable to detect a single specific variant that could be confidently linked to IQ.

BRAIN SIZE AND INTELLECTUAL ABILITY

Because genetic association studies have failed to detect variants directly linked to intellectual ability, scientists have begun to search instead for genetic variants linked to brain size or structure. The rationale behind this strategy is that brain structure shouldn't be influenced by variations in education and environment. Looking at brain variation only makes sense, though, if brain size or structure is actually correlated with intelligence. While extremely small brain size (a medical condition known as microcephaly) is associated with intellectual disability, whether brain size in the normal range is correlated with intelligence is less clear. Evidence from animals indicates that although brain size may be linked to intelligence, at best, it's only a piece of the puzzle. After all, the brains of elephants and whales, which weigh approximately 11 and 18 pounds respectively, are significantly bigger than the 3-pound human brain. Yet elephants and whales are not smarter than people (at least not according to any intelligence test that humans have come up with). If you think the criterion should be brain size *relative* to body size, then your average songbird would be smarter than a person. While some individuals do have a head or brain that is disproportionally large (conditions known as macrocephaly or megalencephaly), they generally have average or below-average intelligence. No, any link between brain morphology and human intelligence is subtler than simply overall brain size or brain-to-body size ratio.

To find subtle brain differences that may be linked to intelligence, scientists use MRI (magnetic resonance imaging). MRI can measure brain gray matter or white matter volume, or the size of brain substructures, such as the hippocampus, which are believed to be linked to intelligence. Researchers can then determine whether differences in any of these brain structures are correlated with variations in IQ scores. In perhaps the most interesting such study, in 2012, a team of over 200 scientists combined brain imaging with testing genetic markers and identified one genetic location on chromosome 12 near a gene called *HMGA2*, where certain genetic variants were consistently associated with differences in brain size. By itself, linking *HMGA2* with brain size was not remarkable. As we learned in chapter 16, *HMGA2* variants had already been linked to human height variability in 2007, so learning that taller people might also have somewhat larger brains doesn't seem that surprising. What was more striking was that the researchers found that the *HMGA2* variant linked with larger brain size was also associated with a small increase in IQ scores. On average, about 2.5 IQ points difference between people with two copies of one variant versus people with the other variant. Although the IQ difference was small and the results need to be replicated with larger numbers of subjects, if this association is confirmed in future investigations, it will represent the first genetic variation directly linked to IQ found in people who do not suffer from intellectual disability.

HUMANS, APES AND INTELLECTUAL ABILITY

Scientists also search for genetic variations linked to intelligence by looking for variants that exist in human DNA but not in DNA of other primates. In 2012, a research team from the University of Colorado published the results of a search for segments of DNA that are duplicated multiple times in the human genome, but that are not duplicated (or are duplicated much less often) in other animals. Such genetic duplications, which occur periodically over the course of millions of years in the genomes of all animals, are interesting because they are believed to be one way that species acquire new traits and abilities. The idea is that after a genetic region is duplicated, the genes in one of the duplicated regions can gain new mutations that allow them to encode novel proteins, while the other copy of each gene remains in the genome so the animal hasn't lost its recipe for the original proteins.

In any case, the Colorado team found one segment of the human genome that was duplicated to a remarkable extent. This region on chromosome 1, called DUF1220, codes for an amino acid sequence that is approximately 65 amino acids in length (the length of the DUF1220 segments varies slightly) and is found about 270 times on chromosome 1; the exact number of copies varies among different people. Other mammals have far fewer DUF1220 regions in their genomes. Chimpanzees have 125 DUF1220 regions, gorillas have 99, orangutans 92, and macaques 35. Outside of the primates, animals have even fewer DUF1220 regions: for example, dolphins have 4, mice only 1, and non-mammals don't have any at all.

Although the function of the DUF1220 region is still unknown, recent evidence suggests that it is linked to brain size, at least in people suffering from intellectual disability. In a recent study of the DNA of 59 individuals with intellectual disability, the Colorado group observed that the subjects with unusually low numbers of DUF1220 copies typically had microcephaly and decreased gray-matter volume, while those with unusually large numbers of DUF1220 tended to have macrocephaly and increased gray-matter volume.

To determine whether DUF1220 copy number was correlated with brain size and gray-matter volume in people with normal intelligence as well, the team then measured brain structure and DUF1220 DNA variation in a group of healthy individuals. The researchers again found significant correlations between gray-matter volume and numbers of DUF1220 copies. The question remains whether DUF1220 copy number is just correlated with brain morphology or is also linked to intelligence. Studies investigating whether such a link with intelligence exists are currently being carried out.

SMART DRUGS

In some ways, our current knowledge of the biology of human intellectual variation is similar to the understanding of human *athletic* variation some 20 years ago. In the case of athletic performance, scientists have identified hundreds of performance-enhancing genetic variations that give some people an edge in athletics. Over the next 20 years scientists will probably begin to identify a similar list of cognitive-enhancing genetic variants leading to advantages in intellectual pursuits.

Simply identifying cognitive-enhancing genetic variants won't help those of us who don't have them. Yet it may be possible to use this genetic knowledge to increase our intellectual abilities in other ways. Modifying *NR2B* or other brain-signaling genes to improve human intelligence is not likely to be possible in the foreseeable future. Although the idea of using our emerging knowledge of brain pathways to inspire novel ways to learn more effectively is appealing, it is also still speculative. In the short term, the most promising approach of enhancing human intelligence is likely to be through the development of drugs that mimic the biochemical changes that occur in "super-smart" transgenic animals or genetically gifted people. In fact, developing drugs to enhance intellectual ability has become an active research field. So far, most of this research is focused on treating intellectual disabilities. Chapter 7 described drugs targeting glutamate receptors that treat fragile X learning disability in mice and are currently being tested in people. Drugs have also been developed to combat the memory loss of Alzheimer's disease and other forms of dementia, as well as the learning difficulties that accompany attention deficit hyperactivity disorder (ADHD).

Beyond treating intellectual disability, can drugs enhance the intellectual abilities of people with normal intelligence? Over the last 40 years, many research groups have searched for such "smart drugs" (also known as intelligence enhancers or cognitive enhancers). Some of these drugs target the brain's signaling pathways, including the glutamate pathway. For example, in the 1980s and 1990s, scientists at several pharmaceutical companies developed drugs called ampakines. These drugs stimulate the brain's AMPA receptors, which function alongside the NMDA receptors in glutamate signaling, so activating AMPA receptors was a reasonable strategy for improving learning and memory. Studies on rodents and humans in the early 2000s suggested that ampakines do facilitate learning and memory, so DARPA, the U.S. military's research arm, and several drug companies have spent years testing them as potential smart drugs, though with only limited success.

More recently, scientists from MIT and China's Tsinghua University have been developing magnesium compounds that target the brain's NMDA receptors. They observed that – at least in rats – extra magnesium stimulates a rat's brain to produce more NR2B protein. Since the rat brain (and the human brain) generally has difficulty absorbing magnesium, the MIT-Tsinghua researchers used a special magnesium compound, magnesium L-threonate, which can be more easily move from the bloodstream into the brain.

The researchers found that when rats are given magnesium L-threonate their performance improves on intelligence and memory tests, such as solving mazes, presumably because of changes to the brain's NMDA receptors. The scientists also observed that magnesium decreased memory loss in rats that had been genetically altered to have a rat version of Alzheimer's disease. These results have encouraged some people to start taking magnesium L-threonate as a food supplement, although there is still no evidence indicating that magnesium improves human learning and memory as it does in rats.

Other drugs that do not target the brain's glutamate pathway have also been investigated as potential smart drugs, because they are helpful in treating brain or cognitive illnesses. Probably the most widely used are the stimulants. These include Ritalin and the amphetamines, such as Adderall, which improve attention and concentration in children with ADHD. They are also often used as "study aids" by healthy students without any intellectual disability. Still other drugs, including Namenda and Aricept, have appeared promising as smart drugs because they can treat cognitive loss in Alzheimer's disease. Whether they enhance memory or learning in healthy people is less clear. In particular, since Namenda *blocks* NMDA glutamate signaling, it may well not be helpful for people with healthy brain functioning. Provigil is another proposed smart drug that is used in the treatment of cognitive problems. It improves memory and cognitive function in people with sleep deprivation and has been prescribed "off label" as a smart drug, even being touted as "Viagra for the brain," because of its claimed ability to aid memory and enhance cognitive ability in (non–sleep deprived) older people. As with the stimulants and the anti-Alzheimer's drugs, little is known at the cellular or molecular level about how Provigil improves memory or cognition.

ARE SMART DRUGS THE NEXT BIG THING?

Although the precise number of healthy people who are trying smart drugs isn't clear, there's no question that the number is high. The largest study was performed in 2001 by researchers from the University of Michigan and included almost 11,000 students from over 100 U.S. colleges and universities. The survey, which was limited to nonmedical use of prescription stimulants, such as Ritalin and Adderall, found that more than 4% of the students had used such stimulants without a prescription during the previous year. Other studies, including those most often quoted in the popular media, report much higher smart drug usage. A 2006 survey of almost 2000 students at a U.S. university, which was reported about on the *60 Minutes* television show, stated that 34% of students had used a stimulant as a smart drug. Among juniors and seniors, the fraction using smart drugs was almost 60%.

These surveys yielded inconsistent results, in part because they asked different questions. Some surveys asked whether subjects had taken the drug even once, while others asked if they taken it in the last year. Sometimes subjects were questioned whether they took the drug specifically for cognitive enhancement, while in others they were simply asked if they'd taken a stimulant without a prescription. In these studies, students acknowledged taking stimulants without a prescription for a wide variety of reasons, ranging from "getting high" to "losing weight," but the most frequent responses – "help in memorizing," "before exams" or "to get better grades" – were learning related.

IS IT SMART TO TAKE SMART DRUGS?

No matter which survey one relies on, it is clear that large numbers of students, and many nonstudents as well, are taking smart drugs. If even 4% of students are taking stimulants to improve their grades, that's half a million university students in the United States alone. And these figures don't include the use of Provigil or other nonstimulant drugs, which weren't included in the surveys.

Despite these striking statistics, we are still left with important, fundamental questions. Do these drugs really improve learning or memory in normal healthy people? Do they actually enhance intelligence? In the case of the ampakine drugs, the results of multiple clinical trials have been largely discouraging, and ampakine drug development has been largely abandoned. Studies of cognitive enhancement from other proposed smart drugs have been more encouraging, though the results have generally still been inconclusive. Ritalin definitely helps healthy rodents learn more effectively. Some studies indicated that Ritalin also improves memory and spatial-analysis problem solving in healthy human volunteers, but others didn't. Similarly, some reports claimed learning or memory enhancement in healthy people using Adderall, Provigil, Namenda or Aricept, while others found little or no improvement from these drugs.

To better understand these conflicting results, three large meta-analyses were published in 2010 and 2011, reviewing over 100 experimental studies of the effects of Ritalin, Adderall, Provigil, Namenda and Aricept, and even coffee and glucose, on learning in healthy individuals. The results of these meta-analyses indicated that Ritalin and Adderall do improve memory, though they showed little evidence of enhancement of other aspects of learning or intelligence. Even everyday stimulants such as coffee and glucose were shown to have some positive impact on attention and memory, but not on any other aspects of learning. The meta-analysis of Provigil concluded that Provigil improved alertness, even in well-rested people, but there was little consistent evidence suggesting Provigil otherwise enhances cognition or memory. For Namenda and Aricept, the reviewers determined that not enough studies have yet been carried out on healthy people to come to any conclusions.

In other words, hundreds of thousands of students, and most likely comparable numbers of nonstudents, are using drugs with questionable benefits. Even the most favorable studies suggest that the current crop of cognitive-enhancing drugs work primarily by increasing short-term memory or attention span. Essentially no studies report improvement in long-term memory or learning, let alone increases in intelligence. To put these modest successes of smart drugs in context, a review of studies evaluating nonpharmaceutical forms of cognitive enhancement noted that regular physical exercise and adequate amounts of sleep also improve memory and aid learning, without any of the possible side effects of drugs.

And the smart drugs do come with risks. Stimulants, such as Ritalin and Adderall, are addictive. A 2006 review of data from the U.S. National Survey on Drug Use and Health showed that almost 5% of ADHD-stimulant users met the medical criteria for drug dependence or abuse. The side effects of more recently developed drugs are unknown. There is already concern in the medical literature about possible long-term side effects on a generation of adults who regularly took stimulants as children. For children with ADHD this risk is at least balanced by demonstrated efficacy at treating a serious disorder. For healthy individuals the relative benefits of stimulants and other smart drugs are much more questionable. On balance, ambitious students would probably be well-advised to carefully weigh the risks before gambling on smart drugs to get them an A on the next midterm.

19

WHERE DOES LANGUAGE COME FROM?

The boy we'll call Will was 15 years old when the doctors at London's Hospital for Sick Children first examined him in 1988. He had been referred for genetic testing by the speech and language clinic where he was being treated for severe language impairment. Although Will's hearing was normal and he had above-average intelligence, his speech was largely unintelligible. His reading skills were those of someone half his age, and his speech comprehension was impaired. Will's brother, "Alan," who was nine years younger, also had normal hearing and vision. His mathematical and other nonverbal skills were characterized as average as well. Yet Alan also suffered from language disability and unintelligible speech. And it wasn't just Will and Alan. As the doctors at the hospital's genetics clinic studied what became known as the K. E. family in the scientific literature, they learned that 13 other family members suffered from language impairment. Clearly something severely disabling was being inherited within the K. E. family.

LANGUAGE, CULTURE AND ZEBRA FINCHES

Language is arguably the quintessential human characteristic. Although many animal species, from mice and songbirds to dolphins and elephants, communicate via vocalization, animal communication isn't comparable to human speech. Even the remarkable animals that after years of training have learned some human language have severely restricted vocabularies. Koko the gorilla allegedly mastered 1000 signs of American Sign Language, while "Betsy the Talking Dog," who appeared on a 2011 *Nova* television show, understood only 300 words. In contrast, an average six-year-old child knows around 14,000 words. Moreover, Koko and Betsy have very limited ability in combining words into sentences. For example, Koko rarely strings more than two or three words together in what could be considered a coherent sentence. In contrast, generally by the age of four children regularly create five- or six-word sentences. Even a diehard animal lover like myself has to admit that true speech and language appear to be uniquely human capabilities. For better or worse, language may well be the reason that humans dominate the planet.

Human language is the result of intrinsic biological capabilities as well as learning and culture. The relative degree to which language is a universal trait or evolves differently depending on the local culture has been debated for over half a century. What is clear is that both biological and cultural factors contribute and that studying how they interact can provide important insights into the mechanisms of language development. And some of the most striking insights have come from studying laboratory animals.

Using animals to study language may sound strange, but animal communication does share characteristics with human speech. It is also much easier to study. For example, male zebra finches communicate with females by singing to them (the females don't sing). In the wild, all male zebra finches' songs are similar in terms of rhythm, pitch and tempo. Male zebra finch chicks raised in a laboratory also sing similar songs, provided that an adult zebra finch teaches them how to sing. If a chick is isolated from male adults, though, it won't learn to sing properly. Its song will sound different in ways that are easy to both hear and measure quantitatively. In other words, as with human speech, zebra finch song appears to be a mix of inherited factors and learning.

These observations led zebra finch biologists at New York's City University to attempt to create a new zebra finch "culture" with a different zebra finch language (or at least a new zebra finch song). In 2009, the researchers carried out an ingenious experiment that employed an enclosure housing a male zebra finch chick with three females. When the chick matured, he sang a strange zebra finch song, as was expected since no adult male taught him when he was young. Nevertheless he was still able to mate with the females in the enclosure (they didn't have any other males to choose, after all). When the chicks hatched, they were kept in the enclosure as well, and the adult male began teaching the male chicks his (strange) song. Remarkably, the chicks *mostly* copied their father's song, but they also modified it. By measuring song parameters, the scientists observed that the chicks' songs were more like a wild male zebra finch's song than their father's song was, even though their father had taught them.

The researchers then repeated the experiment with the newly grown chicks. Again each male was placed in an enclosure with females. When these females had chicks, the males were again taught to sing by their father. The songs of these chicks were also similar to their father's, but were another step closer to a wild zebra finch song. After this procedure had been repeated for four generations, the songs were no longer distinguishable from those of a wild bird. Rather than create a new song, the isolated zebra finches were *recreating* their old song of the wild, which had been lost four generations previously.

UNIVERSAL HUMAN LANGUAGE STRUCTURE

The results of the zebra finch experiments demonstrated dramatic differences between zebra finch song development and human-language evolution. When groups of people separate, their languages diverge; they don't spontaneously return to some ancient language. Of course, human language is much more complex than zebra finch song, so perhaps some *elements* of human language, if not the details of the words and syntax, are universal, and maybe all languages do converge to them.[*]

How might one test such an idea? No ethical scientist would create an enclosure where human infants were isolated from adults who could teach them language. Yet on rare occasions, infants are naturally isolated from human language. One way this can happen is if a child is born completely deaf. People who are born deaf are not *necessarily* isolated from human communication. In most societies, deaf children meet older people (often also deaf) who teach them to communicate in sign language. However, this is not always the case. In some rural societies in underdeveloped countries, deaf children may grow up in isolation from other deaf people. They may grow up with no language at all, beyond a few basic gestures to crudely

---

[*] The theory that all human languages share a similar underlying structure has existed for several centuries but was developed in most detail by Noam Chomsky in the mid-20th century. It's widely, though by no means universally, accepted.

communicate with their (usually hearing and non–sign language capable) parents and family.

Such linguistic isolation had been the case for many deaf people in Nicaragua before the Sandinista Revolution of the late 1970s. After the revolution, the new government decided to create schools for the deaf. These schools brought together previously isolated deaf individuals from throughout the country. These people didn't share a sign language. In most cases, they didn't *have* a sign language, just the few gestures they used with their families, which were often unique to each family. But the remarkable fact was that over the subsequent 15 years, these individuals, and especially the children among them, began to create their own language. Since they had no adults to teach them sign language, the language they created, now called Nicaraguan Sign Language, was different from other sign languages, including those used in other Spanish-speaking countries. With each influx of young deaf Nicaraguan children, Nicaraguan Sign Language became less the gesturing of isolated deaf people without a language and more a traditional sign language with all of its linguistic structure.

You may well be asking what makes a collection of hand and arm gestures (which is, after all, what a sign language consists of) a "real" sign language. This is not an easy question. Analyzing the elements that make up a language is a complex science and describing it in detail would take us far afield. We'll content ourselves here with a single example to illustrate the difference between gesturing and human sign language. Say someone wants to describe an object rolling down a hill. A hearing person would typically accompany their words with a *single* gesture that was both downward and involved rotating either a hand or finger to illustrate that the motion involved rolling. In contrast, in a sign language, a signer indicates rolling downhill using at least two separate signs: one indicating downward motion followed by (or preceded by) a distinct sign indicating rolling. Often a third sign would be included to indicate that the rolling and the downward movement were simultaneous. Although less compact than using a single motion, by separating "down" and "rolling" into separate motions, a true sign language creates reusable language elements, which can be combined in different ways. The first Nicaraguan deaf children who came to the school generally used single gestures to indicate rolling downhill. By the mid-1990s, the children were describing rolling downhill as a combination of two or more signs, just as in a conventional sign language, even though no signing adult had ever taught the children sign language. Not unlike the untutored zebra finches, the children were creating the structure of their language by themselves.

SPEECH IMPAIRMENT AND FOXP2

The spontaneous emergence of a new sign language with the linguistic structures of traditional languages suggested that sign languages, and presumably spoken languages as well, share a universal structure, and may even be "hardwired" in our brains. Yet identifying the biological mechanisms that underlie this universal structure remained elusive. It was for this reason that the discovery of the K. E. family was a gold mine for scientists investigating the biology of speech. By studying the DNA of the affected K. E. family members, scientists were able to discover the critical genetic variant they shared and thereby obtain clues about the molecular mechanisms underlying their language impairments.

The medical condition that the K. E. family suffers from is called either developmental verbal dyspraxia (DVD) or childhood apraxia of speech. Individuals with DVD have difficulty coordinating their mouth movements, which reduces the intelligibility of their speech. They make inconsistent errors from one utterance to another, and their problems increase with longer and more complicated sentences. Often they also have other language deficits; they may not be able to break up words into their constituent phonemes* or they may not understand basic grammatical concepts, such as word inflection or syntax. What is particularly intriguing about verbal dyspraxia is that the afflicted individuals have normal hearing, and their vocal chords also appear normal. Though their nonverbal intelligence (as measured by IQ tests) is sometimes below average, their minor overall intellectual deficits are on a different scale from their severe speech and language disabilities.

With 21st-century DNA technology, the genetic variant causing the K. E. family's DVD could have been identified in a few months, possibly even in a few weeks. Twenty years ago, though, such gene hunting was a daunting scientific challenge, and identifying the casual genetic variant required over a decade. Researchers needed to carry out multiple DNA analyses of the K. E. family members with ever-increasing numbers of genetic markers. Eventually, a research team from Oxford University identified a genetic location on chromosome 7 where all the K. E. family members with DVD had one DNA variant, while all the family members with normal language skills had a different variant.

The difference was a single nucleotide within a gene called *FOXP2*, a gene that had not previously been associated with language problems – or anything else for that matter. At this location in the *FOXP2* gene, every K. E.

---

* A phoneme is the smallest unit of sound used in a spoken language. For example the word *phoneme* consists of five phonemes: f, o, n, ee and m.

family member with verbal dyspraxia had an adenosine nucleotide, while every healthy family member (as well as over 300 unrelated, normal people who were also tested) had a guanine nucleotide. As a result, the afflicted family members produced FOXP2 protein that differed slightly from the FOXP2 protein produced by the healthy people; at location 553 in the FOXP2 protein, an arginine amino acid was replaced by a histidine amino acid in the family members with the disease.

Any remaining doubts that *FOXP2* gene variants could cause DVD were dispelled with the subsequent identification of people with DVD who had *other* rare variants in their *FOXP2* genes. Although these discoveries demonstrated that modified FOXP2 protein could cause verbal dyspraxia, they didn't explain *how* a single amino acid change altered FOXP2 functioning, nor did they shed light on how FOXP2 was related to language development. That said, the discovery of FOXP2 has played an important role in guiding researchers toward the molecular mechanisms underlying language. From FOXP2's amino acid sequence and inferred protein structure, scientists realized that FOXP2 wasn't a "speech protein" at all. FOXP2 is a transcription factor, a protein that regulates when other genes are transcribed. They learned that FOXP2 is produced not only in the brain but also in the lungs, the heart and the gut, and that it regulates over 300 genes. Some of these genes undoubtedly play roles in the molecular pathways involved in language, but determining which genes were related to language, or how changing a single amino acid in a protein that regulates them alters the language-processing biochemical pathway, is still only partially understood.

FOXP2, MICE AND SONGBIRDS

After the importance of FOXP2 in human speech was discovered, some scientists expected that FOXP2 would turn out to be a uniquely human protein, or at least that human FOXP2 protein would differ dramatically from the versions of FOXP2 found in other animals. Consequently, researchers began to search for FOXP2 protein in a wide variety of animals, including chimpanzees, mice and songbirds. The results were surprising. What the researchers found was that FOXP2 is, in fact, one of the proteins that has changed the *least* during animal evolution. Human and chimpanzee FOXP2 differ by only 2 amino acids (out of 715), and human and mouse FOXP2 differ by only 3 amino acids.

Mouse and human FOXP2 are so similar that, in 2009, scientists from the Max Planck Institute in Germany were able to create transgenic mice in which one or both copies of their *FOXP2* genes were replaced by human *FOXP2* genes. OK, you may be asking (if you're a wise guy), can mice with human *FOXP2* genes recite Shakespeare or otherwise display human language skills? Of course, the answer is no. FOXP2 is just one component in a language-processing pathway that contains scores, if not hundreds, of proteins. As a matter of fact, the "speech" of the transgenic mice with human *FOXP2* genes, that is to say their vocalizations, was essentially unchanged. The average number of vocalizations was the same, while the pitch variations of their vocalizations changed only slightly. Transgenic mice with human *FOXP2* genes also appeared just like normal mice in a wide array of other tests of physiology and behavior. But there were a few intriguing differences. The nerve cells of the transgenic mice had longer dendrites (dendrites are the protrusions that nerve cells use to detect neurological signals). They also exhibited increased synaptic plasticity, a measure of nerve cell learning capacity.

The increased synaptic plasticity in the mice with normal humanized *FOXP2* was in clear contrast with what had been observed previously in transgenic mice engineered to have the *FOXP2* mutation found in the afflicted K. E. family members. Those mice displayed diminished motor-learning skills, and their neurons had decreased synaptic plasticity. So although the precise function of FOXP2 in the brain still wasn't clear, it was becoming increasingly apparent that exactly what genetic variants of FOXP2 a mouse (or a human) had was affecting *something* important in the brain.

Zebra finches also taught scientists about language and FOXP2. The first zebra finch FOXP2 experiments found that during the developmental stage, when young finches learn songs, key regions of the brain involved in song learning had increased concentrations of FOXP2. Later experiments showed that altering a finch's FOXP2 concentration affects its singing ability. In these experiments young zebra finches had their brain's FOXP2 concentration artificially decreased by one-half* (to mimic what presumably occurs in the brains of the afflicted K. E. family members). The researchers found that finches with reduced FOXP2 were unable to properly imitate the songs of the adults that were teaching them how to sing. With less FOXP2 in their brains, the young finches made more mistakes in their songs and sang them in a less consistent manner than zebra finches with normal amounts of FOXP2.

---

* This important and widely used method for manipulating protein concentration is called RNA interference or RNA silencing.

CNTNAP2 AND OTHER LANGUAGE-RELATED GENES

FOXP2 regulates the production of over 300 other proteins, most of which have nothing to do with language processing. Although determining which FOXP2-regulated proteins affect language remains challenging, some clues are emerging from individuals with a less severe language disability called specific language impairment (SLI). SLI has long been known to run in families. Although the overall prevalence of SLI is only 5% to 10%, approximately 40% of parents, siblings and children of individuals with SLI also suffer from the condition.

Despite the strong evidence that SLI is inherited, only recently have genes associated with SLI been identified. One is *CNTNAP2*. The initial clues that CNTNAP2 might be involved in language processing came from studies of inbred populations. Identifying causal genetic variants is easier in inbred populations, because affected individuals in inbred populations generally receive both copies of their disease-causing genetic variants from a single ancestor (who was an ancestor of both their mother and their father). Consequently, genetic searches can be restricted to chromosomal regions in which an individual's maternally and paternally inherited DNA is identical. Using this approach, in 2006, a research team led by scientists from the Clinic for Special Children in Strasburg, Pennsylvania, discovered a rare mutation in the *CNTNAP2* gene among Old Order Amish individuals that caused severe brain malfunctioning, including epileptic seizures and language impairment.

Two years later, the Oxford University team that had discovered the *FOXP2* mutation was also led to *CNTNAP2*, even though they used a very different approach from the Pennsylvania researchers. Since the Oxford scientists knew that transcription factor proteins, like FOXP2, bind to DNA adjacent to the genes that they regulate, they searched for DNA regions that were bound by FOXP2 protein. One chromosomal region they detected included the *CNTNAP2* gene. This result motivated them to test the DNA on chromosome 7 in 184 British families with histories of language impairment, and again *CNTNAP2* gene variants were found to be associated with language impairment.

CNTNAP2 has also been associated with language via autism studies. Approximately 40% of autistic individuals have severe speech impediments or are completely unable to speak. In the early 2000s, UCLA researchers investigated the DNA of autistic children with impaired communication skills, such as not speaking their first words until they are relatively old. They found that the affected children tended to have different genetic markers in the region of chromosome 7 that includes *CNTNAP2*. Although the causal variants in the *CNTNAP2* gene were not identified, the fact that three different types of studies have all pointed to CNTNAP2 strongly suggests that CNTNAP2 somehow affects our ability to speak.

Other genes have also been linked to language impairment through genetic association studies, though little is known about precisely how they affect language development. One example is *SRPX2*. *SRPX2* is regulated in the brain by FOXP2, and cell culture experiments show that the K. E. family *FOXP2* variant disrupts FOXP2 regulation of *SRPX2*. Several families with epilepsy that have been studied have rare variants of the *SRPX2* gene, and members of one of these families also suffer from verbal dyspraxia. SRPX2's function in the brain is still unknown, but SRPX2 (along with UPAR, a related protein of unknown function, also regulated by FOXP2) is being studied intensely in the hope that its role in language processing will soon be understood.

Two other candidate genes, *CMIP* and *ATP2C2*, are located near one another on chromosome 16. *CMIP* and *ATP2C2* were strongly associated with language impairment in a 2009 British study of 806 individuals from 211 families. These results were then replicated in an independent sample involving 490 additional cases. Again, only the genetic markers associated with language impairment were found; the specific DNA variation(s) that actually *cause* the problem have not been tracked down. That said, just as the FOXP2 variant that causes verbal dyspraxia in the K. E. family was eventually found, it is most likely only a matter of time before the causal variant(s) in the *CMIP* and *ATP2C2* genes are also discovered.

UNDERSTANDING DYSLEXIA

A few years after the K. E. family came to the attention of doctors at the Hospital for Sick Children in London, a different remarkable family appeared at the Children's Castle hospital in Helsinki, Finland. The initial patient was a six-year-old boy we'll call Matias. Matias had below-average verbal and nonverbal test scores, as well as reading difficulties. Although Matias' low test scores made his reading difficulties unsurprising, the doctors soon discovered that Matias had two sisters with reading difficulties, despite having normal intelligence and speaking ability. The Helsinki doctors then learned that the children's father also had a history of learning and reading problems, though the mother and a fourth sibling had no reading difficulties. Apparently something was consistently afflicting certain members of this family with reading problems.

Difficulty in learning to read, a condition called dyslexia, is not uncommon. Its prevalence varies among people from different population groups, and twice as many boys as girls are diagnosed with dyslexia. Unraveling the biology of dyslexia is challenging. Rodents don't read, so animal experimentation can only provide indirect evidence, for example, by determining whether a suspected dyslexia protein is involved in animal brain development. Despite these challenges, hints as to the biological basis of dyslexia are slowly being found. One clue is that dyslexia often runs in families. If a child has a dyslexic identical twin, then the odds that the child will him- or herself be dyslexic is 68%, while if the child has a dyslexic nonidentical twin, the odds are only 38%. In contrast, the overall prevalence of dyslexia is only about 4% to 10%. The fact that dyslexia affects twice as many boys as girls also suggests underlying genetic factors, though gender differences in education might also play a role.

Because twin studies show that dyslexia has a significant genetic component, scientists have compared DNA of young children with and without dyslexia. (The point of testing *young* children is to minimize the influences of differing school and learning environments.) One gene that has been identified in such studies is called *DYX1C1*, for dyslexia susceptibility 1 candidate 1. *DYX1C1* is located on chromosome 15 and was also the gene that was mutated in the dyslexic family from Finland. The affected family members shared a rare genetic variation in which DNA had been exchanged between chromosome 2 and chromosome 15. The location on chromosome 15 where the DNA exchange occurred was in the middle of the *DYX1C1* gene, disabling the gene. Meanwhile, genetic tests showed that the unaffected mother and child in the Finnish family did not have DNA exchanged between chromosome 2 and chromosome 15.

Aside from the mutation identified in the Finnish family, causal *DYX1C1* variants for dyslexia have not yet been found. Only genetic markers, that is, genetic variants consistently found in dyslexic individuals that have no identifiable biological role, have been found. The molecular function of the DYX1C1 protein is also still unknown, but animal experiments suggest that DYX1C1 helps ensure that nerve cells are correctly positioned in the animal's brain during prenatal development.

LANGUAGE AND READING ABILITY IN "NORMAL" PEOPLE

So far we've looked at studies of people with language impairment: people with dyspraxia, SLI or dyslexia. But what about language learning in everybody else? Do genetic variations explain the subtle differences in language ability among normal people? Might genetic variants contribute to the talents of individuals who excel at language, such as poets or prize-winning novelists?

To address these questions, scientists have again studied twins, and they've found that, even among children without any language impairment, speech and reading ability has a genetic component. For example, a 2005 British study tested 281 pairs of monozygotic twins and 275 pairs of same-sex dizygotic twins between four and five years old to determine whether monozygotic twins have more similar verbal test scores than dizygotic twins. The results showed that this is indeed the case; the scores of monozygotic twins had a 55% correlation, those of dizygotic twins had a correlation of only 35%.

The twin studies only demonstrated that some unknown genetic variants were influencing normal language ability, but they didn't provide any clues to which variants were responsible. One obvious place to look was near *FOXP2*, but so far no genetic variants affecting language skills have been found near *FOXP2* (except for the rare variants found in the K. E. family and other individuals with verbal dyspraxia). In contrast, scientists have recently found evidence that *CNTNAP2* variants may be linked to language ability, even in people with normal language skills.

In 2011, the Oxford University research team joined scientists from the University of Western Australia to search for associations between genetic variants in *CNTNAP2* and language skills in healthy two-year olds. Since testing language skills in a two-year-old isn't easy, the team assessed the children's language skills by asking parents how well their child could form simple linguistic structures, such as two- and three-word constructs. The results suggested that *CNTNAP2* influences language skills in healthy children, just as it does in children with language impairments. To understand how the researchers came to this conclusion, you need to know an important fact about genes that I've previously ignored: rather than being a single sequence of DNA, most genes are split into multiple pieces along a chromosome (see figure 19.1). Such genes are spliced together only after they have been transcribed into messenger RNA. Specialized enzymes look for short sequences of nucleotide letters to determine where the messenger

RNA should be spliced together before it is translated into a protein, and if a mutation causes the RNA to be spliced in the wrong place, the result can be a serious disease. For example, certain types of cancer or muscular dystrophy are caused by errors in RNA splicing.

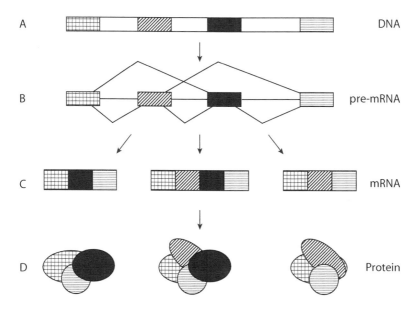

Figure 19.1. Gene splicing. A: Section of DNA of a hypothetical gene. The patterned regions of DNA, which will ultimately encode the protein's amino acids, are called exons. The white DNA regions, which are removed before protein synthesis, are called introns. B: Transcription. The DNA is transcribed into RNA (called pre-messenger RNA, or pre-mRNA, since it needs to go through the additional splicing step before it becomes mRNA). The introns are now shown just as lines. C: Alternative splicing. The introns (and depending on the cell type or cellular environment, one or more of the exons) are then removed from the pre-mRNA by means of special splicing enzymes. Three different ways that the four exons of the pre-mRNA might be spliced together are depicted. D: Translation. Each of the mRNAs can then be translated into a protein with a distinct shape. Gene splicing enables cells to synthesize differing, though related, proteins from a single region of DNA, depending on how the RNA is spliced before it is translated into protein.

The *CNTNAP2* gene is a particularly large and complex gene that is spread over two million base pairs and split into 25 pieces. These pieces must be precisely spliced together in order to produce a functional CNTNAP2 protein. The striking result of the Oxford study was that the genetic markers that were associated with language ability in healthy children were in the same single DNA piece (out of the 25 sections of the *CNTNAP2* gene) as the markers previously associated with language impairment. This finding suggested both that genetic variations also influence language ability in normal children and that these variants might be found by studying children with language disorders (where their effects are more extreme and easier to find). If this is correct, we might someday be able to test young children's DNA and get a hint of their future language ability. That said, the evidence linking genetic variation and language ability in the general population is still far from conclusive. The data need to be reproduced, the causal variants are yet to be identified, and the effects on the brain caused by these variants are still largely unknown. In other words, the day when we can look at a child's DNA and predict whether they are likely to become a talented writer or orator, if it ever comes, is still far in the future.

PART VI

THE MOLECULAR BIOLOGY OF BEHAVIOR AND EMOTIONS

20

WHAT IS GENDER?

⌒‿⌒

Back in the early 1980s, I became friends with a delightful, kind and compassionate man I'll call Lyle. Lyle was an engaging storyteller, his stories unfolding in a heavy Southern drawl, laced with gentle, self-deprecating humor. Lyle was also unabashedly gay, taking great joy in flaunting his sometimes outrageous feminine persona. At times, Lyle would talk affectionately of his twin brother, "Kyle." I didn't meet Kyle until much later. When we finally did meet, years after Lyle had died, another victim of the AIDS epidemic, it was as if Lyle had suddenly returned from the dead. Kyle looked just like Lyle and had the same gentle Southern charm and wonderful sense of humor. But there was a difference. Kyle is 100% heterosexual. He is no closeted gay. He is clearly "straight" and has been ever since he was a boy ... just as Lyle was thoroughly homosexual and couldn't remember a time he hadn't felt gay. That identical twins, with essentially identical genes and similar childhoods, can have such different sexual orientation is a challenging and still unsolved puzzle of biology.

Feelings of sexual attraction are among the most powerful human emotions and motivate some of our most striking behavior. In the final part of this book, we'll be looking at our behavior and our emotions to see how they are influenced by the genes and proteins in our brains. We'll find out how scientists are discovering that a small DNA variation or a subtle chromosomal alteration in a brain cell can make someone more vulnerable to depression, fear or violence. Neuropsychology, the scientific discipline that studies such phenomena, is a young one. Most of its data so far comes from either animal experiments or human studies involving small numbers of individuals. So we'll need to view our conclusions with caution. That said, a dramatic picture is already beginning to emerge of how the biology of our cells influences our behavior and emotions.

We'll begin by looking at what scientists call gender-specific behavior, behavior that typically differs between males and females. Gender-specific behavior takes many forms, ranging from how aggressive we are to how we nurture our young. (Some would even claim that compulsive shopping or the inability to ask for directions is gender-specific behavior, but we won't go there.) We'll focus on two striking manifestations of gender-specific behavior: sexual orientation and gender identity.* As we'll see, these distinct, yet related aspects of our personalities are affected by our genes and our environment in complex ways that are only now slowly beginning to be understood.

SEXUAL ORIENTATION IN ANIMALS

To learn about sexual orientation, scientists often turn to animals. Male animals are usually, but not always, sexually attracted to females, while females are generally attracted to males. Scientists have learned that an animal's sexual orientation is influenced by its sex hormones: testosterone and the estrogens, a group of related hormones including estrone, estradiol and estriol. The importance of sex hormones can be dramatically observed in transgenic animals. A male mouse genetically engineered to have nonfunctional androgen receptors can't detect testosterone and doesn't behave like a typical male mouse; it won't mount females and won't fight with other males. A mouse's sexual orientation can also be changed by nongenetic means. If a male mouse is castrated (its testes are surgically removed so the mouse can no longer produce testosterone) the mouse no longer acts sexually toward female

---

* Note that by sexual "orientation" I mean the sex or sexes an individual is attracted to, while sexual "preference" will refer to the sex someone is *most* attracted to. I'll also generally describe behavioral differences between males and females using the term *gender,* while referring to physical differences with the word *sex.*

mice.* If the castrated mouse is subsequently injected with testosterone, the mouse again displays typical male mouse behavior. Testosterone influences the behavior of female mice as well. After being given testosterone, female mice display "male" mouse behavior; they mount other females as readily as males will. At least in mice, changing testosterone levels can lead to what appears to be homosexual behavior.

Estrogens, the principal hormones produced in a female's ovaries, also influence rodent sexual behavior. Estrogens' influence on mouse sex is somewhat surprising, though. For unknown reasons, male mice and rats need estrogens, usually thought of as female hormones, to show typical male behavior. For example, estrogens restore a castrated mouse's male behavior almost as well as testosterone. Even male mice with normal concentrations of testosterone need estrogens to be "fully masculine" males. To demonstrate this, scientists engineered mice with disabled genes for aromatase, the enzyme that synthesizes estrogens from testosterone. These male aromatase knockout mice displayed unusually *low* levels of stereotypically male behavior. They fought less and mounted female mice less. Estrogens also affect the sexual behavior of female rodents in surprising ways. Newborn female mice given extra estrogen don't become "super-feminine" when they reach maturity. Instead, they behave like male mice; they are not sexually receptive of males, and they mark their territory and attack intruding males, just like males.

MOUSE SEXUAL ORIENTATION BEYOND THE SEX HORMONES

Testosterone and the estrogens do not act in isolation. They function as part of a complex regulatory system that includes many other proteins and signaling molecules. Genetic or environmental variations affecting any of these molecules can potentially influence an animal's sexual behavior. Recently, scientists have discovered how two such molecular changes, one involving an enzyme called fucose mutarotase and the other involving the neurotransmitter serotonin, can dramatically change sexual behavior in mice.

Fucose mutarotase is an enzyme that attaches a sugar molecule, called fucose, to proteins, a process known as fucosylation. One important fucosylated protein is alpha-fetoprotein, which is involved in the fetal regulation of estrogen. Since prenatal estrogen concentrations play an important role in

---

* In case you are wondering, sexual preference and orientation in mice are measured by their patterns of vocalizations and sniffing, as well as their mounting behavior.

sexual differentiation, some scientists wondered whether fucosylation might affect the development of animal sexual behavior. To test this idea, scientists from Korea's Advanced Institute of Science and Technology engineered fucose mutarotase knockout mice. Remarkably, the female knockout mice displayed sexual behavior normally associated with males. They mounted female mice and showed a preference for sniffing female urine. These female mice appeared quite healthy. They were fertile, and most of them even became pregnant, the pregnancies resulting from forced mating by males despite attempted rejections by the females. As far as anyone could tell, these knockout mice were 100% female. Physically they were not sex-reversed at all. Yet somehow their brain wiring had become homosexual.

Serotonin is a signaling molecule that can also turn a heterosexual mouse into a homosexual one. Serotonin signaling is used throughout the body. In the brain, serotonin plays an important role in mediating feelings such as pleasure, pain, fear and aggression. Scientists have long known that serotonin levels are also linked to sexual activity, though only recently has serotonin been shown to have a role in sexual orientation (at least in mice).

The experiments linking serotonin and sexual orientation were again performed with knockout mice. These experiments focused on a protein called TPH2, which the brain needs to synthesize serotonin. By engineering *TPH2* knockout mice, scientists can study the effects of serotonin deprivation. As would be expected, the neurons of *TPH2* knockout mice are unable to produce serotonin. More surprising was that male *TPH2* knockout mice showed only a slight sexual preference for female mice, while normal male mice are almost exclusively attracted to female mice. In subsequent experiments, the male *TPH2* knockout mice were injected with a chemical that enabled their brains to produce serotonin even without TPH2. Sure enough, after the injections, the male *TPH2* knockout mice preferred mounting and vocalizing to females, just like normal male mice.

SEXUAL ORIENTATION AND EPIGENETICS

Knocking out a gene, removing the testes, or administering drugs or hormones can alter an animal's sexual preferences. These observations led some scientists to wonder whether more subtle environmental influences could also change an animal's sexual orientation. For example, might it be possible to alter an animal's sexual behavior and preferences simply by changing the way it was reared as an infant? Although this question has not yet been conclusively answered, a series of experiments by researchers from the University of Wisconsin suggest that it may in fact be possible to alter an animal's sexual preferences simply by manipulating its early environment.

The University of Wisconsin experiments, published in 2010, again studied the effects of the estrogen-signaling pathway. The Wisconsin researchers first showed that an estrogen drug that affects mouse sexual behavior also alters estrogen-receptor concentrations and chromosomal modifications in mouse brain cells. These chromosomal modifications were similar to those that take place in muscle cells during long-term exercise or in brain cells involved in cognition during learning. In this case, though, the modifications occurred in brain cells known to be linked to sexual behavior.

The Wisconsin team then showed that they could induce the same chromosomal changes in a mouse's brain *without* drugs, just by changing how the mice were raised. The researchers knew that when mother mice groom their infants, male mice receive more grooming than females. So the scientists set up an artificial mouse-rearing environment in which they could control how much grooming each mouse received. The infant mice would be raised without their mothers. Instead of maternal grooming, the mice would be stroked with a small brush. The scientists increased the simulated grooming received by the infant female mice so they received as much grooming as would normally be experienced only by males. The researchers discovered that with the extra grooming, the brain cells of the infant females had the same chromosomal modifications normally found only in the brain cells of male mice. The female mice that received this extra grooming also produced less estrogen-receptor protein, just like normal male mice. In other words, the scientists were able to modify an infant female's estrogen receptor levels by means of altered grooming the same way they had by administering an estrogen drug. Although these experiments didn't specifically test sexual behavior, by showing that increased grooming led to the same chromosomal changes as a behavior-changing drug, they suggested that altered grooming might affect gender-specific behavior or sexual orientation. It will be interesting to see whether future experiments demonstrate that an animal's sexual preferences can in fact be understood simply in terms of the nurturing it receives as a newborn.

GENDER-SPECIFIC ANIMAL BEHAVIOR IN THE WILD

The research laboratory is not the only place where unusual sexual or gender-specific behavior can be observed in animals. In fact, nonheterosexual behavior is widespread throughout the animal kingdom and has been documented in hundreds of species. Nonheterosexual animal behavior in nature takes many forms. In the world of domestic sheep, for example, while most males are sexual only with females, approximately 10% are sexual exclusively with other males and 20% are sexual with both females and males. Among bonobos, close relatives of chimpanzees, approximately 60% of the females engage in sexual activities with other females.

Animals in the wild may also display nonsexual behavior that – if it occurred among humans – would be described as deviating from gender-specific norms. For example, among Laysan albatross, pairs of females may stay together as "socially monogamous" couples for over 10 years, during which time they return to their shared nests every year to raise their chicks. (These albatross females do mate with males to fertilize their eggs, but the male-female pairs don't stay together long.)

Initially, the discovery of these types of animal behavior in the wild was met with surprise and skepticism, even in the scientific community. Nonheterosexual behavior had been thought a "waste of energy," diminishing an animal's fitness. Some scientists believed that animal species that were not completely heterosexual would eventually disappear, but the continued occurrence of nonheterosexual behavior in many species has demonstrated that this is not the case.

What are the origins of natural homosexual behavior among animals? In the case of sheep, differences in sexual preference appear to be linked to variations in hormone levels. For other animal species, it's possible that homosexual behavior promotes cooperation and consequently improves the species' fitness. It may also be that homosexual behavior does diminish fitness, but not enough to drive the species to extinction. Whatever the reasons, homosexuality and bisexuality are common in the wild. The evidence is clear and well documented, even if unwelcome to evolutionary scientists with a specific ax to grind or religious fundamentalists who believe homosexual behavior is "unnatural."

SEXUAL ORIENTATION AND "CHOICE"

The evidence for homosexual and bisexual behavior in the wild is overwhelming, and laboratory experiments are beginning to identify the biochemical pathways that underlie animal sexual orientation. But are these animal studies relevant to human sexual orientation? Although human and animal sexual behavior are both regulated by the sex hormones, there are important differences. While sex hormones influence human sexual behavior, they don't appear to regulate human sexual *orientation*. Decreasing male testosterone levels (by castration or using drugs that inhibit testosterone production) does lower male sex drive, and administering testosterone and estrogen does help people who want to change their sexual appearance. Unlike in mice, though, there is no evidence that manipulating testosterone or estrogen levels influences whether a person is more sexually attracted to men or to women. Human and animal sexuality also differ in that mice and other animals communicate sexually via smell, using pheromones. Whether humans communicate using pheromones is still controversial. Even if we do, we must use different pheromone receptors, since the human version of a critical animal pheromone receptor gene is nonfunctional.

Since human and animal sexual signaling pathways are different, one might even question whether human sexual orientation is determined by a combination of genetics and environment, as it is in animals. In fact, some people have long maintained a contrary view, claiming that sexual orientation is simply a choice, and that some of us choose to be heterosexual while others choose to be homosexual or bisexual – just as some people choose to take piano lessons or to wear Hawaiian shirts.

Considerable evidence shows this explanation is almost certainly wrong. Most people, both gay and straight, become aware of their sexual orientation at an early age, either as children or when they reach puberty. In a 1990 survey, which asked self-identified homosexuals when they first became aware of their sexual preferences, the average age the men reported was 10 years old, while the women said 12. The actual age may well be even lower. Why? Because when the same question was asked of homosexuals in 1960, the answers were 14 years old for the men and 17 for the women (see figure 20.1). The change from 1960 to 1990 may well mainly reflect increased tolerance in society, which enabled young people to get in touch with their feelings at an earlier age. If this trend continues, in the future children may become aware of their sexual attractions at even younger ages.

That fact that adults rarely change sexual orientation also argues against sexual orientation being a choice. Determining how often people change their sexual orientation is challenging. Societal pressures may cause someone to change his or her sexual behavior, say by becoming celibate, without their sexual attractions having changed. Self-reported sexual preferences may also be unreliable; due to societal pressures, individuals may deliberately, or unconsciously, misrepresent their sexual orientation. For these reasons, few studies have attempted to determine how often adults actually have changed sexual orientation. Those that have often have been hampered by severe methodological weaknesses, such as relying on a psychotherapist's opinion, or interpreting a person's celibacy as proof that they had changed sexual orientation.

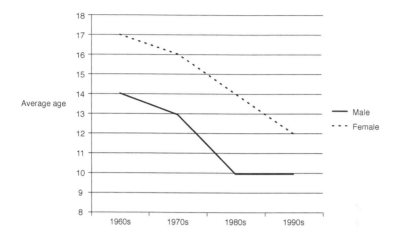

Figure 20.1. Average age of the first awareness of homosexual orientation among individuals who report same-sex attraction. Source: *Our Sexuality,* 10th edition, by Robert Crooks and Karla Bauer, Belmont, CA: Thomson Wadsworth, 2008, p. 234, fig. 9.1.

Only one reasonably objectively designed study, published by Stanton Jones of Wheaton College and Mark Yarhouse of Regent University in 2011, has reported that sexual orientation change is possible. This six-year-long study involved participants in a Christian "sexual-orientation conversion program." Admittedly, Jones and Yarhouse are not unbiased investigators. They are strong proponents of "conversion therapy," a controversial program that allegedly enables someone to change their sexual orientation. In any case, Jones and Yarhouse reported that out of 57 individuals* who had entered the program six years earlier, 5 had changed from primarily homosexual to heterosexual orientations. If these 5 people were not all lying or in denial, changing one's sexual orientation may be possible. That said, the fact that only 9% of a group of presumably motivated individuals

* The study also included individuals who had previously been in the program up to three years. Using data from this entire group led to a reported overall "success rate" of over 23%. As the authors themselves admit, using this larger experimental cohort biased the results, and when this and other biases were excluded, the "success rate" was approximately 9%.

changed their feelings after six years of intensive efforts is hardly evidence that sexual orientation is a choice. The results suggest instead that for a small percentage of the population, overcoming one's innate sexual orientation may, with great difficulty, be possible.

GENETICS, ENVIRONMENT AND GENDER-SPECIFIC BEHAVIOR

If human sexual orientation is not a choice, is it determined by one's genes, one's environment or some combination of the two? Until recently, gender-specific behavior, including sexual orientation, was believed to result almost exclusively from different child-rearing practices for boys and girls. Infants were believed to have brains that were "blank slates," and our gender-specific behavior was thought to come from the lessons of our parents, teachers and peers (except perhaps for a few biological effects resulting from our hormones).

Since then, overwhelming evidence from animal experiments, studies of twins, and studies of infants too young to have been influenced by child-rearing practices has shown that the blank-slate picture is wrong. Experiments show that from their first days of life, baby girls are more likely to gaze at human faces, while boys more often look at mechanical toys. Infants that are three to eight months old, just old enough to reach out and grab an object that interests them, also follow their traditional gender roles. When offered dolls or toy cars, the girls are more likely to choose dolls while boys more often pick the toy cars. To calm those who object that by three months, a child has already had gender-role training, similar experiments have been carried out with young vervet monkeys. With monkeys as well, the females tended to choose the dolls while the males chose balls or cars. Apparently something innate in infant female humans (and monkeys) generally attracts them to dolls or faces while males are more often drawn to objects that can be moved around and manipulated. That these behaviors could have been learned from adults in the first few days, or weeks, of life seems unlikely. Presumably these contrasting behaviors are hardwired in the infants' brains at birth.

TWIN AND FAMILY STUDIES OF HUMAN SEXUAL ORIENTATION

Evidence supporting a genetic component to sexual orientation also comes from studies of twins. Although such investigations have been hampered by the difficulty of finding enough homosexual twins, multiple studies consistently reported that, at least among males, homosexuality is more commonly shared by identical twins than by nonidentical twins. The largest such studies were carried out using the Australian and Swedish twin registries. The Australian study included almost 5000 twins, while the Swedish study involved approximately 7500 twins. Between the two groups, there were 110 pairs of male identical twins that included at least one twin who was gay. Of these 110 twin pairs, 16 pairs (15%) were concordant for homosexuality, meaning that both twins were gay. There were also 84 pairs of male nonidentical twins with at least one gay twin, but only 4 of those nonidentical twin pairs (5%) were concordant. In other words, the identical twin of a gay man was much more likely to be gay than the nonidentical twin of a gay man. Among female twins, the percentage of concordant identical and nonidentical twins was much closer. If one identical twin was lesbian, there was a 14% chance that her identical twin sister was also lesbian, while a nonidentical twin of a lesbian had an 11% chance of also being homosexual. Comparing these percentages with the overall prevalence of homosexuality in society would be interesting. Such comparisons are difficult to make, though, because the results often depend critically on how the study defined homosexuality. That said, according to a 2011 survey in the United States, approximately 3.5% of American adults identify themselves as lesbian, gay or bisexual, a figure that is significantly lower than the 14% to 15% prevalence among the identical siblings of homosexual twins.

Because of the small number of non-heterosexual twins, researchers have also studied nonheterosexual orientation among other relatives of gays and lesbians. The results of such family studies are more difficult to interpret than twin-based studies, since it's harder to distinguish the effects of shared genetic variants from those of a common childhood environment. Despite this limitation, family studies have provided at least one surprising clue pointing to genetic link to sexual orientation. Researchers have found that if a gay man has a gay uncle, the uncle is significantly more likely to be a maternal uncle (that is, the brother of the man's mother) than a paternal uncle. Similarly, if a gay man has a male first cousin who is homosexual, then most likely they are related on their maternal sides (that is, their mothers are sisters). These observations not only point toward the existence of genetic variant(s) predisposing some men to homosexuality but also suggest that these genetic variants may be located on the X chromosome, since the X is the only chromosome that men inherit exclusively from their mothers (see figure 20.2).

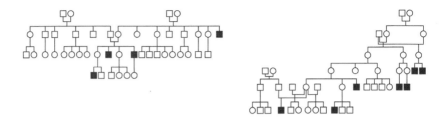

Figure 20.2. Evidence of maternal inheritance of male homosexuality.
Two family trees from the 1993 study that suggested that male homosexuality is
inherited through the maternal branch of the family tree. Males are depicted with
squares, females with circles. Self-described homosexual men are indicated by
solid squares. Note that when both a man and his uncle are homosexual, the uncle
is (at least in these families) always a maternal uncle – that is, the brother of the
man's mother and not the brother of his father – suggesting that the inherited
factor is being transmitted via the genes of the mother.

SEARCHING FOR GENE VARIANTS FOR HUMAN SEXUAL ORIENTATION

Since the twin studies indicated that human sexual orientation had at least
a small genetic component, beginning in the 1990s researchers attempted
to identify the causal genetic variants. The first studies tested for variants
in genes known to be involved in sexual development, for example, the
androgen- and estrogen-receptor genes or the aromatase gene. In a few
cases, genetic variants linked to human sexual orientation have been
found. Recall that congenital adrenal hyperplasia is a genetic condition
in which the adrenal glands produce far more testosterone than normal.
In the last 20 years, researchers have carried out numerous studies of
sexual orientation in women with elevated testosterone levels due to
congenital adrenal hyperplasia. A 2011 review[*] of 10 such studies reached
the "overwhelming conclusion" that women with congenital adrenal
hyperplasia were more likely to be homosexual or bisexual than other
women. So at least in the case of congenital adrenal hyperplasia, genetic
variation can influence sexual orientation.

Scientists have also focused on genetic variation on the X chromosome
because of the family studies suggesting that male sexual orientation
is primarily inherited maternally. A widely reported 1993 investigation

---

[*] A review article is a scientific publication that integrates the results of multiple
related experiments and studies.

led by Dean Hamer of the NIH, claimed to find genetic markers on the X chromosome that varied in a consistent manner between gay and straight men. Unfortunately, these early studies included only a small number of gay men. Since each of us has millions of genetic variants in our DNA, studies with small numbers of research subjects can easily find seemingly positive results by chance. For this reason, searches for genetic variants associated with heart disease or diabetes involve thousands of individuals, but such large-scale screens are expensive and have not yet been performed in genetic screens for DNA variants linked to sexual orientation.

As the initial investigations associating X chromosome genetic markers with sexual orientation only involved a few hundred subjects, it was not surprising that subsequent studies were unable to replicate their positive findings. Eventually even the NIH group that had made the original claims linking X chromosome markers and male sexual orientation acknowledged being unable to reproduce their results in larger follow-up studies.

Overall, the search for genetic variants linked to sexual orientation has been unsuccessful. Of course, not finding any genetic variants doesn't mean they aren't there. Twin studies still suggest that such genetic factors exist and would be found if larger numbers of subjects could be genetically tested. Consequently, in 2003 the NIH funded the Molecular Genetic Study of Sexual Orientation (also called the GayBro Project) to test the DNA from 1000 pairs of brothers who both identify as homosexuals. Finding a sufficient number of subjects has been challenging, though, and as of this writing the project still hadn't published any scientific papers. Most likely, as more families with multiple gay members are identified, scientists will finally be able to start identifying genetic variants that contribute to human sexual orientation.

PRENATAL ENVIRONMENT AND SEXUAL ORIENTATION

The twin studies have shown that inherited, genetic factors play a small role in determining sexual orientation, but only a small one. After all, in the Australian-Swedish study, more than three-quarters of monozygotic twins with at least one homosexual twin were found to be discordant for sexual orientation, that is, just like Lyle and Kyle, one twin was gay and the other was not.

If sexual orientation is not a choice, and the twin studies showed that genetics was not the main factor, what else might be determining human sexual orientation? Since a person's sexual orientation appears at an early age and rarely changes, presumably the answer had to be found somewhere in either the prenatal or early childhood environment. Since experiments on rats had shown that if pregnant rats are stressed, their male offspring often display stereotypically female behavior, several research groups tested whether human sexual orientation might also be linked to stress levels during pregnancy. Although an initial study of 200 men did report such an association, subsequent larger studies failed to find any link between homosexuality and stress during pregnancy.

Scientists also considered whether a drug called diethylstilbestrol (DES) might influence sexual orientation. DES is a synthetic form of estrogen, and for 20 years in the mid-20th century, DES was often prescribed to pregnant women at risk for miscarriage. (Later studies showed that DES didn't reduce miscarriages, but by then thousands of women had already taken it.) Since DES is chemically similar to estrogen, DES exposure in the womb might affect sexual development and influence the offspring's ultimate sexual orientation. An early report claimed that DES-exposed females were more likely to eventually become homosexual than women in the general population. As was true with the investigations of prenatal stress and sexual orientation, though, later studies couldn't replicate these findings, and at this point the scientific consensus is that prenatal DES exposure doesn't influence sexual orientation.

Yet another theory claimed that male homosexuality was the result of an immunological reaction toward the male fetus. The idea was that a male fetus induced an immunological response in the mother causing her to produce hypothetical "anti-male" antibodies. If the mother subsequently conceived another son, her anti-male antibodies would then somehow influence the sexual development of the fetus. This hypothesis was initially supported by data claiming that males with older biological (not adopted) brothers are more likely to be homosexual than men without older brothers. Once again, although early studies suggested each additional older brother increases a male's likelihood of being homosexual by approximately 30%, a recent meta-analysis that included 5000 men (five times the sample size used in prior studies), found no effect.

CHILDHOOD TRAUMA AND HOMOSEXUALITY?

If the prenatal environment isn't the critical factor underlying sexual orientation, perhaps early childhood experience is. Until the 1970s, the prevailing belief among psychologists was that heterosexuality was the norm and that homosexuality and bisexuality were abnormalities caused by childhood trauma. Freudian psychoanalysis asserted that overbearing mothers, or withdrawn or hostile fathers, were to blame. Other psychologists claimed that a homosexual childhood seduction by an adult or adolescent led to homosexuality.

These theories were typically presented with little or misleading supporting evidence. Early reports by psychotherapists described anecdotal reports of childhood mistreatment or abuse from their homosexual patients. To be sure, some nonheterosexual patients did experience childhood abuse, but these reports did not show that such negative childhood experiences were more common among homosexuals than heterosexuals. Later, more systematic studies sometimes also claimed to find associations between childhood abuse and adult nonheterosexual orientation. These studies have also had weaknesses. They included only a few dozen individuals who had suffered from child abuse and later identified as homosexuals. In addition, only a single study attempted to distinguish whether the childhood mistreatment was the *cause* of the nonheterosexual orientation or simply was the *consequence* of abusive adults responding to nonheterosexual behavior already present in the child. Although this one study also claimed that severe childhood abuse was more common among nonheterosexual adults, it still found that among the severely abused, approximately 80% of the women and 90% of the men described themselves as heterosexual. In other words, even if the results of this study are replicated, they would at most indicate that early child mistreatment is just one factor contributing to the development of sexual orientation.

COULD MAMA'S X CHROMOSOMES BE AFFECTING JUNIOR'S SEXUAL ORIENTATION?

Another possible factor shaping human sexual orientation is inherited chromosomal modifications. A 2006 study by UCLA researchers suggested that certain X chromosome modifications in women may contribute to sexual orientation.

Since women have two X chromosomes in each cell and men have only one, every cell in a women's body inactivates one of its X chromosomes to avoid getting a double dose of X chromosome genes. Whether any given cell silences the X chromosome the woman received from her mother or the one she received from her father appears to be largely random. Usually half of a woman's cells inactivate her maternally inherited X chromosome, while her other cells silence her paternal X chromosomes. Some women, though, have an extremely unbalanced or "skewed" distribution of inactivated X chromosomes, with more than 90% of their cells having active X chromosomes from the same parent.

What does silencing of a woman's X chromosomes have to do with sexual orientation? Perhaps nothing. Remarkably, though, the 2006 UCLA study reported that mothers of gay men were more likely to have skewed X chromosome inactivation. Among the mothers of gay men, 13% had extremely skewed distributions of X chromosome silencing, while only 4% of the mothers with no gay sons did. And 23% of the women with more than one gay son had skewed X chromosome silencing. Since only a single study with just 200 subjects was performed, the findings may well have been a coincidence. Even if the findings are confirmed, just how the skewing of silenced X chromosomes might affect sexual orientation is still unclear.

In the end, notwithstanding all the animal experiments and all the human studies, science still has only a very limited understanding of the biological mechanisms underlying human sexual orientation. We still don't understand how twins like Lyle and Kyle, with identical genomes and similar prenatal and childhood experiences, could grow into adults with such divergent sexual preferences. That said, at least there was never any question that both Lyle and Kyle were men. Not only did their bodies clearly have male anatomies, but each was clear that he considered himself a man. Yet for some people, identifying as a man or a woman is far from simple. Their bodies (and their chromosomes) may claim they are men, but in the core of their beings they are convinced that they are women. Or vice versa. Some feel that neither the female nor the male label fits them. In fact, there are thousands of people with a gender self-image that doesn't match their anatomy or their chromosomes. One of them is Lynn Conway.

LYNN CONWAY'S STORY

I first heard of Lynn Conway 30 years ago. I was studying electronic circuit design, and Conway, along with Carver Mead, was the author of one of my textbooks. I knew that Carver Mead was a famous Caltech professor and a pioneer in integrated circuit design, but I hadn't heard of Conway. I do remember thinking that she must be a remarkable person, since few women were leaders in the field of integrated circuit design. Little did I know just how remarkable a woman Lynn Conway really is.

Conway was born an apparently normal baby boy in White Plains, New York, in 1938. For reasons of family privacy, she has never revealed her birth name, so we'll call this baby boy Bob Smith. Bob was a precocious, intelligent boy who was healthy and normal in nearly all respects, except that from an early age, Bob wanted to play and dress like a girl. Since his parents strongly discouraged this unusual behavior, young Bob kept his feelings to himself. He graduated from MIT, became a leading computer designer at IBM, and married a coworker, becoming the father of two girls.

Deep within him, though, Bob's feelings that he was really a woman only grew stronger. Eventually, he could no longer keep these feelings to himself. He admitted to his wife that he wanted to become a woman. Soon thereafter, Bob Smith began the process of becoming Lynn Conway, a long, tortuous journey involving hormones and surgery. The hormones and the surgery turned out to be the easy parts of the journey. Shortly after becoming a woman, Conway was fired from IBM, saw her marriage dissolve, and was forbidden by the courts from having any contact with her children.

To avoid additional prejudice and discrimination, Conway decided to hide her past. Over the next 30 years she slowly reinvented herself as a female engineering research scientist, without a past. Only in 1998, at the age of 60, was she finally secure enough in her personal and professional life to publicly reveal her history. Since then she has become a leading spokeswoman for the rights of people who feel compelled to change their sex, while remaining a respected computer scientist.

GENDER IDENTITY AND GENDER REVERSAL

People who appear on the outside to be unambiguous members of one sex, while their brains tell them they belong to the opposite sex, are called transgender or transsexual. Their condition is called gender-identity reversal or gender dysphoria. Gender identity reversal and homosexuality are different. Some homosexual individuals (that is, people who are primarily or exclusively sexually attracted to individuals of the same sex) are gender-identity reversed, but most aren't. Some individuals with gender-identity reversal are attracted primarily to the opposite sex; others are not.

The prevalence of gender reversal is difficult to estimate because people with gender dysphoria are frequently reluctant to speak about it. Estimates of gender-identity reversal range from 1 or 2 per 1000 to 1 or 2 per 10,000, with considerable variation in different societies. Consistently higher estimates of transgenderism are found in surveys made by transgender activists than those by mainstream psychology organizations. Even using the most conservative estimates, some 30,000 transgender people are living in the United States, and an estimated 60,000 Americans have undergone gender-change surgery over the last 60 years.

By transgenders, I'm not referring to people who have played the role of the opposite gender to escape societal discrimination or to have more opportunities in life. Such individuals have usually been women; some, like the writer George Eliot, simply adopted male pen names, while others, such as the band leader Billy Tipton, took on complete male identities to increase their professional opportunities. To the extent that they were motivated by reasons other than an inner feeling that they were really male, these women don't fit our definition of transgender. Transgender people aren't seeking to avoid societal discrimination. On the contrary, they are willing to endure severe discrimination, just to be true to themselves. These are people like Christine Jorgensen, who underwent male-to-female surgery in the early 1950s, or Chaz Bono, born the daughter of the entertainers Sonny and Cher, who decided at the age of 39, that to be true to himself, he needed to begin the process of gender conversion to become a man.

NATURE OR NURTURE?

Until relatively recently, many scientists believed that a person's gender identity was largely, if not completely, determined by his or her childhood education and upbringing. If you were raised as a boy, you would see yourself as a boy, while if you were raised the way girls are traditionally raised, you'd develop a female self-image. This was just another aspect of the blank-slate view of human development.

Among the most influential scientists who held such views was John Money, a psychology professor and researcher at Johns Hopkins Medical School from 1951 to 2006. Money believed that biology only influenced gender identity via hormone secretion from the testes or ovaries, and that an "appropriate" child-rearing environment, supplemented by hormone treatment, could counteract these biological influences. In Money's words, from a paper he wrote in 1975, "Gender identity is sufficiently incompletely differentiated at birth as to permit successful assignment of a genetic male as a girl. Gender identity then differentiates in keeping with the experiences of rearing." We now know that Dr. Money was wrong. Genetics and prenatal biology play major roles in determining someone's sexual identity, and nowhere was the blank-slate theory of gender identity more effectively, or more tragically, debunked than in the story of David Reimer.

THE STORY OF DAVID REIMER

Bruce Reimer, along with his identical twin brother Brian were born normal and healthy boys in Winnipeg, Canada, in 1965. All went well for baby Bruce until he was diagnosed with a urinary tract problem at the age of six months. As part of Bruce's treatment, the doctors decided that he should be circumcised, normally a routine and minor surgical procedure. Sadly, the circumcision was performed in an unconventional manner and went terribly wrong. As a result, young Bruce's penis was irreparably damaged.

Bruce's parents sought medical advice to minimize the physical and psychological damage caused by his nonfunctioning penis. Ultimately, they were referred to Dr. Money, who persuaded the Reimers that the best option was to surgically remove Bruce's testes and to replace what was left of his penis with a vagina. The Reimers agreed, and the surgery was performed shortly before Bruce's second birthday. Bruce was renamed Brenda and raised as a girl. Dr. Money decided that Brenda should not learn about her early life history, and instead be told that she was a normal girl. Money was convinced that Brenda would become a happy, well-adjusted girl and woman, while her twin brother Brian, who shared the same genome, would be a healthy male, proving that childhood upbringing could change one's gender identity.

Unfortunately, Brenda Reimer's life did not unfold as planned. Although outwardly she appeared to be a normal girl, Brenda was a troubled child. She wasn't interested in dolls or activities that typically appeal to girls. Instead she was drawn to the kind of play that appeals to boys. But Brenda was not just a tomboy. Deep in her heart, Brenda Reimer felt she wasn't a girl at all. Brenda became confused and depressed, believing she was a complete misfit and that something was fundamentally wrong with her. Finally when she was an adolescent, Brenda's father told her how she had been born a boy and then "reassigned" to be a girl. In response, Brenda decided to reassign herself back to being a boy. Brenda gave herself the name David, and started receiving male hormones to replace those that would have normally been produced by the testes. She sought surgery to transform her body, as much as possible, back to being male. According to Reimer's biography, *As Nature Made Him: The Boy Who Was Raised as a Girl,* David's life improved greatly once he decided to live as a male. However, his story doesn't have a happy ending. After a number of years, he began to suffer from psychological problems, in part stemming from his childhood experiences as a girl, and at the age of 38, David Reimer committed suicide.

The importance of David Reimer's story rests on the sad irony that it powerfully demonstrated the opposite of Dr. Money's theory. Our genetic makeup and prenatal brain development do play critical roles in determining gender identity, and they can't be overcome via child rearing and hormones. Tragically, the case of David Reimer was not unique. Other infant boys with malformed or missing penises were subjected to gender reassignment during that time period, and the results were rarely more successful than David Reimer's. A scientific review published in 2002 found that among 16 infant boys who had been reassigned as girls at birth because of a missing penis, 8 eventually decided they wanted to change themselves back into males. And all but one made choices in their lives that largely fit typical male gender roles.

NATURAL GENDER REVERSAL

Although David Reimer's story played a key role in putting the blank-slate theory of sexual development to rest, it didn't shed any light on gender-reversal cases that were not the result of a tragic medical decision. Mostly, gender reversal remains a mystery. In a few cases, scientists have found underlying genetic factors. A small number of transgender people have unusual numbers of sex chromosomes. Some male-to-female transgenders have three X chromosomes, while a few female-to-male transgenders have one X and two Y chromosomes. Women with congenital adrenal hyperplasia, the condition in which the adrenal gland produces high concentrations of testosterone, are far more likely to identify as a man than other women are. Yet even among women with adrenal hyperplasia, gender reversal is rare, occurring only 1% to 3% of the time. Other genetic enzyme deficiencies can also lead to gender reversal. Recall that individuals who lack the enzymes that produce testosterone or its derivative, dihydrotestosterone, usually look like girls when they are born and are raised as girls, even though they have normal Y chromosomes. When these girls reach puberty, they often experience gender dysphoria. They begin to "feel male," and in a majority of cases choose to identify as men.

Along with genetic factors, there are also hints that one's childhood environment plays a role in shaping one's gender identity. In several countries of East Asia, such as Thailand and the Philippines, the frequency of gender dysphoria is significantly higher than in Western countries. Obtaining reliable data is challenging, but published estimates suggest that in Thailand alone, there are tens of thousands and possibly hundreds of thousands of male-to-female transgenders. Although it is possible that Thai transgenderism is the result of certain genetic variant(s) in the Thai population, it is tempting to speculate that the generally tolerant Thai attitude toward sexuality also contributes to the number of transgender individuals in Thai society.

Even with these clues, science still doesn't understand the determinants of gender identity any more than those underlying sexual orientation. The origins of human sexual orientation and gender identity remain among science's most frustrating and challenging puzzles. That said, although we may not yet understand the biological details of sexual orientation or gender dysphoria, biology has already taught us important lessons, and the most important is that people who are homosexual or bisexual or transsexual are in no way deviant or sick. Their challenges rarely stem from their intrinsic differences, but rather from the hostile attitudes and behavior they have been forced to face.

21

WHAT IS PAIN? WHAT IS PLEASURE?

S teven Pete works in the health and human services field and likes to drive the back roads of Canada and the U.S. Paul Waters is a British man who works in retail and enjoys web design. Superficially, Waters and Pete don't have much in common. Were it not for the one unusual trait that they do share, odds are their paths would have never crossed. But Waters and Pete are close friends and confidantes because they share a rare and remarkable trait: they live in a world without physical pain. Not that Waters and Pete are smiles all the time. Quite the contrary. Living with congenital insensitivity to pain, the medical name of their condition, is not easy. Our bodies and our brains have developed elaborate signaling networks for experiencing pain, as well as for enjoying pleasure, and if either doesn't work properly, the results can be disastrous.

PAIN SENSITIVITY AND GENETICS

Everyone knows what pain is. (Well, everyone except for a few people* like Steven Pete and Paul Waters.) You touch a hot iron or you fall and hit your head on the pavement, and you suddenly experience intense and unpleasant

---

* And a few fictional characters, such as Ronald Niedermann, from Stieg Larsson's Millennium series of thrillers *(The Girl with the Dragon Tattoo etc.)*.

sensations. Pain serves a vital function. It warns us to get our hand away from the hot iron and to be careful so we don't fall and hurt our head again. People who suffer from congenital insensitivity to pain, or CIP, must continually be on their guard to not inadvertently endanger their lives. Because of their lack of an early warning signal of pain, in the past, people with CIP rarely lived past their 20s.

Congenital insensitivity to pain is one example of how genetic variation can affect the experience of pain. The first evidence linking genetics with pain came from experiments in the 1980s with laboratory rats. Marshall Devor, a researcher at the Hebrew University in Israel, discovered that pain sensitivity in rats was inherited. Devor injured nerves in the legs of rats and observed that, while some rats scratched and nibbled at their toes, suggesting that they were in pain, others hardly reacted to the injury. Devor then allowed the rats to breed and have pups. When the pups matured, they in turn were subjected to the same nerve-injury test. Devor found that the offspring of rats that had been most sensitive to pain were pain-sensitive themselves, while pups of the pain-insensitive rats weren't bothered by the pain test either.

Genetics influences human pain sensitivity, too. Although congenital insensitivity to pain is extremely rare, Steven Pete's brother also suffers from it. Even among individuals who don't have CIP, multiple studies have demonstrated that approximately one-third of pain sensitivity is due to inherited factors. To identify the specific genetic variants that contribute to differences in human pain sensitivity, scientists test patients with histories of extreme pain sensitivity, or healthy volunteers who are subjected to mildly painful stimuli, in a laboratory setting. The researchers then search for genetic variants consistently found in individuals reporting unusually high or low pain sensitivity. Since people (other than identical twins) have millions of differences in their DNA sequences, determining which genetic variations are consistently linked to pain perception requires testing large numbers of people. Until recently, such large-scale genetic testing was prohibitively expensive, so earlier investigations included small numbers of subjects. As a result, the findings were often inconsistent, and many genetic associations with pain perception could not be replicated. Nevertheless, a few genetic variants were consistently linked with pain sensitivity. In most cases, these variants were found in genes encoding one specific class of proteins, the ion channel proteins.

ION CHANNELS, NEURONS AND ELECTRICAL SIGNALING

To understand ion channel proteins and how they influence the experience of pain, we need to look at how the nervous system transmits pain signals. Pain starts with some irritant (a flame, a knife, perhaps a virus or bacteria) making contact with our body. Next, nerve cells specialized for sensing such negative, noxious stimuli send signals to the brain. In response to these signals, the brain generates the feelings that we experience as pain.

For the sensation of pain to enable us to remove our hand from a fire before it gets burned the noxious signal must be transmitted to the brain quickly. Consequently, rather than use the relatively slow chemical signaling of hormones and neurotransmitters, our nervous system evolved large nerve cells, which can be over 3 feet long, that transmit signals electrically. In contrast to conventional electrical signaling, though, which uses electrons, neuronal signaling uses charged atoms. Such charged atoms are called ions, and those most commonly used by the nervous system include sodium, potassium, calcium and chloride ions.

Using ions rather than electrons has advantages for nervous system signaling. Different ions can be used to transmit separate signals. Combining multiple types of ions enables transmission of more complex signals. Since ions are much larger than electrons, nerve cells can regulate signal transmission using microscopic pores in their membranes. By successively opening and closing such pores – called ion channels – the nerve cell can regulate whether the ions can pass through the membrane and thereby control the movement of electrical signals. Figures 21.1 and 21.2 illustrate ion channels and how neurons use them to propagate electrical signals.

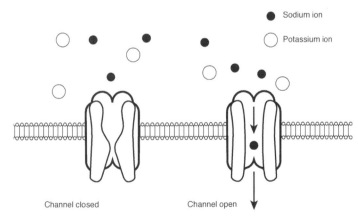

Figure 21.1. Signal transmission using ion channels. Two sodium ion channels are show; the one on the left is closed, while the one on the right has been

opened (possibly because of binding to an activating protein, which is not shown). In a channel's open state, sodium ions can cross the membrane, while potassium ions are not able to pass through.

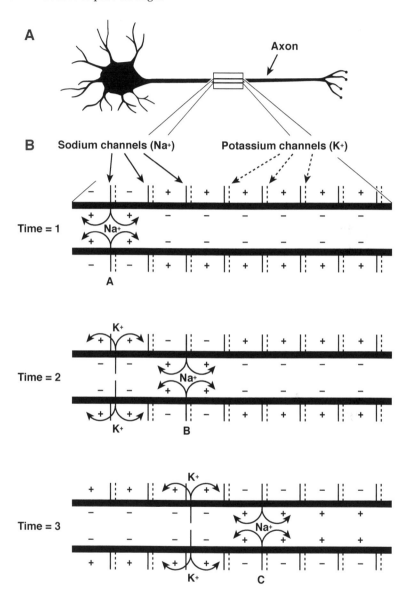

Figure 21.2 Electrical signal propagation along a neuron. A: Schematic of a nerve cell. The globular body on the left is the cell body, while the long protrusion to the right, the axon, is the part of the cell along which electrical signals propagate

toward the next neuron in the signaling pathway. The open rectangle indicates a small section of the axon, which is shown at three successive times in B: At time = 1, the leftmost sodium channel (at position A) has opened, allowing sodium ions to enter. This ion movement triggers a change in the local electric fields, causing an electrical signal to move along the membrane of the axon. This electrical variation causes the sodium channels at position B to open at time = 2. At the same time, the altered electrical fields at position A have caused the potassium channels to open and the sodium channels to close. Finally, at time = 3, the sodium ion signal has propagated to position C, while position A has returned to its quiescent state.

ION CHANNELS AND GENES

Like other cell structures, ion channels are made of proteins, and the recipes for making these proteins are encoded in our DNA. Consequently, DNA sequence variations may cause some people to produce more ion channel protein and others to make less. DNA variations may also lead to ion channel proteins with slightly different shapes, as a result of which ions may enter a cell with a bit less difficulty, or perhaps a little more. If a person has a genetic variant that causes their nerve cells to make more of certain ion channels, or to have channels that transmit ions more easily, then their nervous system may transmit signals more efficiently. If the signals being transmitted are pain signals, then the individual may well be more sensitive to pain.

The most dramatic example of a gene that influences the experience of pain has the innocuous name of *SCN9A*. The *SCN9A* gene encodes the SCN9A protein,* which is an essential component of one type of sodium ion channel. Neurons use this sodium channel to signal the presence of noxious stimuli to the brain. As a result, variations in the *SCN9A* gene can influence our perception of pain.

No one understands the importance of *SCN9A* variants more than Steven Pete and Paul Waters. Pete and Waters have rare mutations in both copies of their *SCN9A* genes. As a result, their neurons can't produce functional SCN9A protein, and because of this single mutation in one ion channel gene, they are unable to feel pain. In contrast, some people have *SCN9A* variants that produce SCN9A protein that is not only functional but that make the sodium ion channel work extremely efficiently. Not surprisingly, people with these SCN9A variants may experience extreme pain from physical stimuli that other people find only slightly unpleasant.

---

* In the scientific literature, this protein is known as Nav1.7.

While *SCN9A* genetic variants that completely disable the SCN9A protein are very rare, other variants of the *SCN9A* are quite common. One such variation is a substitution of a single genetic base (corresponding to position 1150 in the SCN9A protein's amino acid sequence): while 90% of us have a G in our DNA sequence at this location, 10% of us have an A.* Does having an A or a G at this location in the *SCN9A* gene influence our sensitivity to pain? Current evidence suggests the answer is yes. The A-to-G nucleotide variation changes one amino acid in the SCN9A protein from an arginine amino acid to a tryptophan, which slightly alters the structure of the SCN9A protein. Electrical measurements inside neurons have shown that sodium ion channels built from these two types of SCN9A proteins have different electrical properties.

To determine whether such subtle ion channel changes affect pain sensitivity, scientists tested whether *SCN9A* gene variants are correlated with differences in pain sensitivity. Over a thousand people were tested, including both healthy volunteers and patients seeking medical help for recurrent pain. Among both patients and healthy volunteers, those with the A variant in the *SCN9A* gene consistently reported higher levels of pain than individuals with the G variant. Just how the A variant changes the sodium ion channel to increase pain sensitivity is still unknown, but finding out could lead to better pain-killing drugs for those people with *SCN9A* A gene variants.

PAIN SENSITIVITY BEYOND SCN9A

SCN9A is just one component of the nervous system's pain signaling pathways. Other proteins besides SCN9A regulate sodium-ion transport, while potassium ions and calcium ions also contribute to pain signaling. Consequently, scientists weren't surprised to discover that genetic variants in other ion-channel proteins, including the potassium-channel protein KCSN1 and the calcium-channel protein CACNAD2D3, also affect pain sensitivity. In each case, the protein's link to pain was demonstrated both through animal experiments and from studies of reported pain sensitivity among people.

Some pain-signaling proteins aren't ion channel proteins at all. One example is OPRM1, a type of brain cell receptor known as an opioid receptor. Opioid receptors detect morphine and related opium-derived

---

* This is an oversimplification. Since we have two copies of the SCN9A gene, the more accurate statement is that approximately 81% of us have two G's at this position in the gene (.9 * .9 = .81), about 1% have two A's (.1 * .1 = .01) and approximately 18% of the population have one G and one A (1.0 − .81 − .01 = .18).

chemical compounds (called opiates), as well as a much larger class of synthetic and natural substances (opioids) that have chemical structures similar to those of opiates. OPRM1's role in pain sensation was demonstrated in experiments with *OPRM1* knockout mice. In these experiments, mice were subjected to standard mouse pain tests, such as exposing a mouse's paw to a hot surface and measuring the time before the mouse starts licking its paw. The scientists observed that, on average, the *OPRM1* knockout mice withdrew their paws from the hot surface more rapidly than normal mice. To confirm that the difference in pain sensitivity was in fact linked to morphine-receptor signaling, the researchers then gave the mice morphine during the pain test. While the morphine helped the normal mice tolerate the pain, it didn't help the *OPRM1* knockout mice.

Monkeys also use the OPRM1 receptor in their response to pain. The evidence comes from a 2008 study of emotional pain; specifically the pain that infant monkeys experience when separated from their mothers. (Admittedly, physical and emotional pain are different; we'll explore these differences in the next chapter. For now, suffice it to say that some proteins, such as OPRM1, play an important role in both forms of pain.) The OPRM1 monkey study was carried out by researchers from the NIH who relied on a natural variation in the rhesus monkey *OPRM1* gene. At position 77 in the *OPRM1* gene, some monkey genes have a C, while others have a G. Because this variation alters the resulting OPRM1 protein, some scientists hypothesized that the variant would affect the monkey's perception of pain. Since infant monkeys react to maternal separation by emitting specific distress vocalizations, the researchers could measure emotional pain in the monkeys from those vocalizations, and they in fact found that monkeys with the G *OPRM1* variant consistently emitted more distress vocalizations than those with the C variant.

Some people have an *OPRM1* gene variant similar to that found in monkeys. In humans, the variation is between an A and a G rather than a C and a G, but the resulting change in the OPRM1 protein is similar. Two studies have reported associations between the *OPRM1* gene variant and human pain sensitivity, though the change in pain threshold was small. Since OPRM1 is an opioid receptor, scientists also investigated whether the human *OPRM1* gene variation affects morphine's ability to treat pain. Although a few early studies did suggest that a person's OPRM1 gene variants affected the ability of morphine to counteract pain, subsequent reviews of multiple experiments found no significant effect of *OPRM1* gene variation on pain sensitivity or morphine efficacy.

OPRM1 AND PLEASURE

Since OPRM1 is a receptor for morphine and other opium-related drugs, one might guess that OPRM1 influences our feelings of pleasure just as it affects our sensitivity to pain. This idea is supported by the observation that *OPRM1* knockout mice have decreased sensitivity to pleasure along with their lower pain thresholds. To assess pleasure in a mouse, scientists generally use a method called the place preference test. In this test, mice are alternately injected with morphine while in a cage with one form of decoration (for example, vertical stripes), or else are injected with an inactive drug while in a differently decorated cage (say, with horizontal stripes). Once this training phase is completed, the mice are placed in a larger enclosure from which they can enter either cage (see figure 21.3). When the place preference test is performed with normal mice, the mice generally spend most of the time in the cage in which they had initially received morphine. In contrast, *OPRM1* knockout mice spend equal amounts of time in each cage, suggesting that the morphine doesn't affect them.

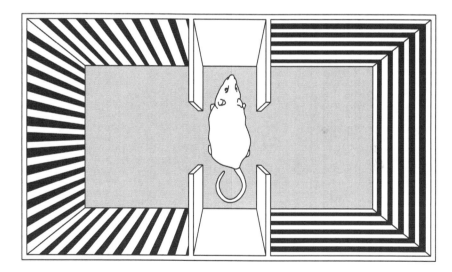

Figure 21.3. Apparatus for the mouse place preference test. The enclosure on the left has vertical stripes and is where the mouse was given the active drug. In the enclosure on right, which has horizontal stripes, the mouse had been injected with an inert drug. During the testing phase, the mouse is placed in center section, as shown, and is allowed to enter either chamber. The relative amount of time the mouse spends in each of the chambers is then used as a measure of the mouse's association of the chamber with the pleasurable memory of the drug.

The influence of OPRM1 receptors on both pleasure and pain continues to be intensively investigated. Such balanced focus on both pain and pleasure is actually relatively uncommon in the world of biomedical research. Typically, researchers focus far more on the biology of pain than the biology of pleasure. If you search PubMed, the NIH's research citation database, with the keyword *pleasure,* you'll find references to approximately 4500 articles. In contrast, searching for *pain* returns over 500,000 references! The world at large doesn't share the same emphasis on pain over pleasure. If you repeat these searches for *pain* and *pleasure* on Google, you'll find a much smaller relative difference (940 million hits for *pain* and 500 million for *pleasure*).

The biomedical research community's focus on pain over pleasure is even starker than these numbers indicate. Many of the scientific papers retrieved with the keyword *pleasure* relate to alcohol, cocaine and other addictive drugs, which, though enjoyable in the short run, are much less pleasant when consumed habitually. This negative emphasis is partly because biomedical science traditionally focuses on "fixing our problems," and pain is nearly always more of a problem than pleasure. Although its name is the National Institutes of *Health,* the NIH primarily supports research studying illness and disease, which seems a shame.

OPIOID RECEPTORS AND THE ENDORPHINS

Despite this imbalance, scientists have learned a considerable amount about how we experience pleasure. Much of that knowledge has come from studying the opiates. Although people have used opiates to reduce pain and increase pleasure for at least 6,000 years, not until the early 1970s did scientists identify the first opioid receptors in the brains of mammals. Since then, five different families of opioid receptors have been discovered, of which the ones most closely linked to our experience of pleasure are those that bind morphine.

Of course, the opioid receptors don't just respond to morphine and other drugs; they originally evolved to detect the brain's natural pleasure-signaling molecules, of which the best known are the endorphins. Many pleasurable activities, including sex, eating and exercise, are believed to be at least partly mediated by endorphins. Until recently, though, verifying this hypothesis in people has been challenging. With experimental animals, scientists can directly measure endorphin binding in the brain, either by making direct brain cell measurements after the animal has been sacrificed or from measurements using implanted brain electrodes. Researchers can even measure how endorphin binding changes during pleasurable activities, such as eating or sex. (By saying an activity is "pleasurable" again I just mean that animals choose it over other activities.) In contrast, measuring endorphin binding in the human brain has in the past been possible only in postmortem studies. Measuring blood endorphin levels is possible, but how closely blood measurements reflect brain endorphin levels is not known.

Recently, though, a technique called positron emission tomography (PET) has enabled scientists to assess endorphin levels in the human brain by measuring the emissions of radioactive compounds. In 2008, a team from Munich Technical University used PET to measure the brain uptake of a compound that binds exclusively to opioid receptors. The goal was to observe the hypothesized release of endorphins during exercise, a phenomenon often referred to as runner's high. The researchers injected the radioactive compound into the bloodstreams of long-distance runners and measured the radioactivity emitted from the runner's brain. Since the compound only bound to opioid receptors, they were able to estimate the concentration of available opioid receptors, that is, those *not* bound by endorphins or other opioids. The runners' brains were scanned 30 minutes after completion of a 2-hour run, as well as after 24 hours of not exercising. The researchers found that after running, the athlete's brains consistently showed less radioactive compound, presumably because more opioid receptors were bound by endorphins and fewer receptors were available to the radioactive compound. In other words, the experiment confirmed that more of the brain's receptors were bound with endorphins after exercising than at other times.

THE BRAIN'S "PLEASURE CENTER"

Remarkably, the brain can experience pleasure without any environmental stimulus, such as food, drugs, sex or exercise. Sixty years ago, James Olds and Peter Milner showed that direct electrical stimulation of the brain can elicit feelings of pleasure in rats. Olds and Milner implanted electrodes in the brains of living rats and, by activating these electrodes, were able to electrically stimulate localized regions in a rat's brain. They also constructed a device with which the rats themselves could activate their brain electrodes by pressing a bar. Olds and Milner discovered that electrically stimulating one place in the rat's brain, called the septal area,* appeared to give the rat pleasure. The rats would continually press the bar to stimulate this brain region, to the extent that they would stop eating and even starve, rather than stop pressing the bar.

Does such a pleasure center also exist in the human brain? In the 1970s, Robert Heath at Tulane University carried out a series of convincing, albeit controversial, experiments that showed that the answer is yes. In his most dramatic experiments, Heath implanted electrodes into several brain regions of two human subjects: one, an intellectually disabled woman with epilepsy; the other, a chronically depressed man who was also epileptic. In part, Heath utilized the electrodes to treat the patients' neurological diseases, but he also used them to perform psychological experiments.

In one set of experiments, Heath activated electrodes at various locations within the patient's brain and found that stimulating one site, the human equivalent of the rat's septal area, resulted in the subject reporting pleasurable sensations. In other experiments, Heath simply monitored the electrical output of the electrodes when the patients were engaged in various activities, observing that the electrical output from the septal area pleasure center increased during activities the subjects described as pleasurable. In one set of experiments, Heath even measured the electrode output from the male patient's brain during both masturbation and intercourse, observing dramatic peaks from the pleasure-center electrodes at the time of orgasm.

Measuring brain responses during sex, from electrodes implanted in intellectually disabled or severely depressed individuals, raises all sorts of ethical issues. The Tulane experiments were, in fact, even more disturbing than I've described; the male patient was being treated as much for his "history of homosexuality" as for his epilepsy and depression. It seems highly unlikely that a medical research ethics panel would approve such a study today.

Subsequent experiments involving electrical stimulation of the human brain have been much less ethically questionable. Within 20 years of the Tulane experiments, brain stimulation had become an effective treatment for Parkinson's disease, chronic pain and dystonia, a neuromuscular movement disorder. Brain stimulation has helped over 80,000 people with

---

* As more precise tools for probing the brain were developed, various substructures within the pleasure center were identified, with ever more arcane names, such as the nucleus accumbens, the ventral pallidum, and the ventral tegmental area. For simplicity, we'll ignore these details of brain anatomy and just refer to this part of the brain as the pleasure center.

diseases that had been resistant to other treatments, and recently, medical researchers have found that brain stimulation may even help people with what is arguably the most devastating lack of pleasure: severe depression.

DOPAMINE AND PLEASURE

Only a few unsuspecting experimental subjects have experienced pleasure from direct electrical brain stimulation. Usually our feelings of pleasure are the result of a complex biochemical pathway that ultimately leads to the brain. The opioids are important players in this signaling pathway, but they are not the only ones. Two other key signaling molecules in pleasure are dopamine and serotonin.

Dopamine and serotonin belong to a class of structurally related neurotransmitters and hormones called the monoamines, which also include adrenaline and noradrenaline. Two-dimensional chemical structures for these monoamines are depicted in figure 21.4 (The details of the figure aren't important; they're provided simply to suggest that these four important molecules have similar structures.)

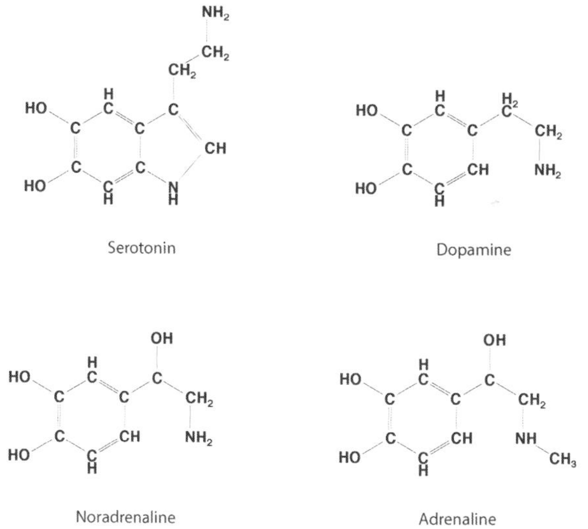

Serotonin

Dopamine

Noradrenaline

Adrenaline

Figure 21.4. Two-dimensional models of the chemical structures of serotonin, dopamine, noradrenaline and adrenaline. C = carbon, H = hydrogen, N =nitrogen, O = oxygen. Single lines indicate chemical bonds involving one electron; double lines indicate bonds involving two electrons. Because of their similar structures (each has a hexagonal substructure of alternating one-electron and two-electron bonds) they are susceptible to degradation by the same enzyme, MAO-A.

Dopamine plays a key role in a wide range of signaling processes of our nervous system. For example, the uncontrolled movement that is the hallmark of Parkinson's disease is caused by the lack of dopamine in a region of the brain called the substantia nigra. Treatment with L-dopa (short for levodopa), a chemical the brain can convert into dopamine, is still a major treatment of Parkinson's disease. Dopamine signaling is involved in learning, breastfeeding, sleep and many other important biological functions. In addition, dopamine is an important link in the "pleasure pathway" within the brain.

The evidence linking dopamine signaling to pleasure again comes largely from mouse experiments. Using a brain-stimulation apparatus similar to that developed by Olds and Milner, scientists have found that mice with their gene for DR2 (a dopamine receptor) knocked out require twice the current to keep them pressing a bar for electrical stimulation for pleasure as normal mice. Dopamine also plays an important role in the experience of pleasure in people. Many pleasurable addictive drugs target the dopamine-signaling pathway. For example, opioids, caffeine and alcohol directly bind to dopamine-secreting neurons, while drugs like cocaine, amphetamines, and MDMA (ecstasy) bind to and block the functioning of GABA neurons, which normally limit the secretion of dopamine.

DOPAMINE AND MOTIVATION

Although the model of dopamine as the "pleasure neurotransmitter" was an appealing and simple one that explained many experimental results, a series of ingenious experiments carried out at the University of Fribourg in Switzerland, showed that this model was a bit too simple. Rather than directly mediating pleasure, dopamine appears to be more closely linked to the *anticipation* of pleasure.

For their experiments, the Fribourg researchers implanted small electrodes in dopamine-secreting neurons of long-tail macaque monkeys. The electrodes were not used to stimulate the secretion of dopamine but simply to monitor the neurons' activity. The researchers then trained the monkeys to expect a food reward by using methods originally developed by Ivan Pavlov at the beginning of the 20th century. The scientists displayed a visual image and two seconds later gave the monkey some food. Meanwhile, the electrodes would measure the activity of the monkey's dopamine neurons both upon receiving the food and in anticipation of receiving the food (that is, when the image was displayed). What the Fribourg scientists discovered was that the dopamine-neuron response was stronger when the image was displayed (that is, when the monkey learned it was about to get food) than when the food was actually delivered.

Subsequent experiments showed that the monkey's dopamine response was more related to the *uncertainty* of receiving pleasure than to the pleasure itself. In these experiments, each monkey was taught to associate visual cues with a *probability* that it would receive food. For example, a green light might be followed by food 100% of the time, while a blue light would be followed by food 50% of the time. The Swiss team discovered that the dopamine signaling was weaker if the monkey knew for sure that it would receive food than if it learned that it had only a 50-50 chance of being rewarded. Apparently uncertainty about whether it would get rewarded was what most stimulated the monkey's dopamine signaling, not unlike the way that the anticipatory excitement of a new adventure or romantic liaison may sometimes eclipse the pleasure of the event itself.

SEROTONIN AND FEELING GOOD

Pleasure, like pain, is a complex phenomenon. It begins with an environmental trigger, but also involves intricate neuronal signaling pathways that can be affected by inherited genetic variations. Opioid detection plays an important role in this neurological pathway; so does dopamine signaling. Another key pleasure-signaling molecule is serotonin. Serotonin is a versatile molecule; we've already seen its influence on sexual orientation and gentleness in mice. Serotonin signaling is also used in regulating the gastrointestinal system. In later chapters, we'll see how serotonin modulates feelings of fear and anger. Serotonin receptors are also important targets of hallucinogenic drugs, such as LSD. Arguably, though, none of serotonin's functions are more important than its influence on pleasure. Serotonin affects our experience of pleasure, in part, by stimulating the brain's dopamine-releasing cells, but serotonin's influence goes well beyond regulating dopamine signaling. As we'll soon see, serotonin affects not only our immediate experience of physical pleasure but also our deeper feelings of contentment and well-being when our lives are going well – as well as the darker experiences of depression during times of stress.

22

WHY ARE SOME PEOPLE HAPPY ... AND OTHERS DEPRESSED?

In April 2012, the United Nation's published its first-ever *World Happiness Report*. The 170-page UN report described Britain's Measuring National Wellbeing program as well as Bhutan's Gross National Happiness index. It detailed environmental factors that can make a person happier, such as secure employment, good health and satisfying relationships. Deep within the report, one could also find a few paragraphs suggesting that our genes might be relevant to how happy we are, as well. In one of those paragraphs, one could even learn of a twin study, which reported that in the United States "a third to a half of within-country variance of happiness can be explained by genetic differences." The UN authors indicated that, rather than consider such genetic factors, they would just focus on "environmental factors affecting happiness, since these are what can be changed." As the UN report was intended for policymakers, this was fair enough. That said, learning how the environment and our genes interact to affect our sense of well-being* is an important topic, and one that may even help us become happier.

Just like our experiences of physical pain and pleasure, our feelings of emotional well-being originate from events in the outside world but are

---

* When scientists study happiness or life satisfaction they often refer to them as feelings of well-being. In this chapter, we'll use these terms interchangeably.

also affected by our brains and nervous system. So we might guess that both genetic and epigenetic factors might influence our feelings of happiness. Is there any evidence that this model of how we experience happiness is correct? Are our feelings of happiness influenced by the biology of our cells and not solely by external events in our lives? In 1978, Philip Brickman and Dan Coates from Northwestern University and Ronnie Janoff-Bulman of the University of Massachusetts (UMass) carried out a classic experiment that began to address these questions. Their approach was to compare the level of happiness among three groups of individuals: people who had won a large sum in the lottery in the previous year, individuals who had recently suffered an accident that left them paralyzed, and people who had neither experienced a dramatically positive or negative life event.

The researchers needed a way to measure happiness, and their approach was a simple one; they just asked. For example, a research subject might be asked to rate how much she enjoyed everyday activities, such as talking with a friend or hearing a funny joke, or simply, "On a 1–10 scale, how would you rate your life overall?" When conducting such surveys, researchers need to take into account fluctuations resulting from catching a participant on a bad day. That said, scientists have found that when such questions are asked over an extended period of time, people's answers are quite consistent.

The Northwestern-UMass researchers were surprised to find that although people did rate winning the lottery as a "highly positive event" they didn't generally rate their *current* level of happiness (which was six months to a year after having won the lottery) higher than other people. People who had suffered crippling accidents did rate their current level of happiness lower than the other two groups, typically approximately 3 on a 0–6 scale, but not *that* much lower – people from the other groups generally rated their happiness as about 4.

Studies similar to this one have been often repeated, and the results consistently indicate that people have a "set point" of subjective well-being that is relatively immune to outside events. Twin studies have even shown that genetics contributes to this set point. For example, in a Dutch study, 1800 identical twin pairs and 1200 same-sex nonidentical twin pairs, ranging in age from adolescents to the elderly, filled out questionnaires asking them to rate on a 1–7 scale how much they agreed or disagreed with statements like "On the whole I'm a happy person." The results indicated a significant genetic component to people's feelings of well-being, with responses from the identical twins having twice as high a correlation as did those of the nonidentical twins.

PHYSICAL FEELINGS AND PSYCHOLOGICAL EMOTIONS

Although the Dutch and Northwestern-UMass studies suggested that biology influences our emotional well-being, they provided little insight into how this occurs. To identify such underlying biological mechanisms, one might hypothesize that our feelings of psychological well-being are related to the superficially similar experiences of *physical* pleasure or pain. If this hypothesis is correct, then our knowledge of the biological pathways influencing physical feelings might yield clues about the brain processes contributing to their psychological counterparts.

But is there any reason to believe that the biological networks underlying our physical feelings and psychological emotions are related? We know that there are at least some differences between physical and psychological feelings. People with congenital insensitivity to pain are quite susceptible to psychological pain. They suffer emotionally just like the rest of us. Yet, we also know physical and psychological feelings aren't completely different. In fact, sometimes it's not even clear whether a feeling should be labeled as physical or psychological. For example, some people even say they can feel pain just by *watching* another person in physical pain, a phenomenon called empathetic pain.

Comparing the biology of physical and psychological feelings is possible with brain imaging. Brain imaging studies are based on the observation that when we have feelings, certain parts of our brain become more active biochemically. This increased activity can be detected in a magnetic resonance imaging (MRI) scan. If physical and psychological feelings are related, then one would expect them to activate related brain structures. In 2011, University of Michigan researchers tested this hypothesis by recruiting 40 volunteers who had recently been rejected in a romantic relationship for a two-part experiment. The researchers first placed a heated object on the forearm of each volunteer, creating a mildly painful sensation. Then they had participants view a picture of their former romantic partner. During both parts of the experiment, the volunteer was inside an MRI machine, which was acquiring images of the subject's brain. Remarkably, the same parts of the brain were activated when the volunteers touched the hot object as when they looked at the picture of their former lover.

MRI scans have even been able to detect the brain's response to empathetic pain. In a 2010 experiment at the University of Birmingham, scientists showed research subjects photographs of people being injured in sports or accidents. Thirty-one of 108 subjects said that while they were looking at the photographs they felt pain in the same part of their body as the injured person. The subjects were then placed in an MRI scanner and shown a similar set of photographs. The researchers found that when the subjects reported feeling pain in response to a photograph, their MRIs also showed increased activity in brain regions associated with pain perception.

Since the MRI experiments suggested that feelings of physical and psychological pain were related, perhaps similar biological pathways mediated both types of experiences. Scientists have tested this hypothesis with two basic strategies. One has been to study people with severe and long-term psychological pain, that is, individuals suffering from depression. The second approach has been to use research subjects whose environment and genetics could be tightly regulated: laboratory animals. In fact some of the most striking advances have come from combining these approaches and studying depression in animals.

How can one study depression in animals? Since one can't ask a mouse how happy it feels or if it still enjoys spending time with other mice,* scientists use indirect methods to assess an animal's feeling of well-being or depression. One approach is to measure physiological changes normally linked to emotional state. Scientists may measure changes in sleep patterns or appetite, or they may monitor brain activity using MRI or electrodes implanted in the animal's brain.

Animal depression can also be measured by observing changes in the animal's behavior. For example, we've learned that infant monkeys vocalize when separated from their mothers. These sounds appear to reflect unhappiness, and one might hypothesize that the loudness of the sounds indicates the intensity of the monkey's unhappiness. To assess mouse depression, scientists use other behavioral tests, such as the forced swim test. In the forced swim test, a mouse is placed into a water tank from which it can't escape. Since being immersed in the water tank is a frightening and stressful experience, the first time a mouse is placed in such an apparatus it swims feverishly in an (unsuccessful) search for a dry surface. After several exposures to the forced swim apparatus, mice typically give up and stop swimming, knowing that their efforts won't be successful anyway. With a bit of an anthropomorphic leap, scientists consider this giving up as a sign of mouse depression. If one accepts this interpretation of the mouse's behavior,

---

* Of course, you *can* ask. You're just not likely to get a useful response.

one can then compare levels of depression by measuring how long it takes a mouse to reach this state of hopelessness.*

THE SEROTONIN-DEFICIENCY MODEL OF DEPRESSION

Through studies of human depression and experiments with animals, scientists are beginning to identify the biological pathways that underlie feelings of well-being. In many cases, the evidence has pointed to the serotonin-signaling network. The initial suggestion that serotonin was involved in feelings of well-being came from fortuitous observations of drug side effects. Already in the 1950s, doctors noticed that iproniazid, a tuberculosis drug, and imipramine, which was used to treat schizophrenia, had an unexpected side effect. Tuberculosis and schizophrenia patients who also happened to suffer from depression sometimes reported that their symptoms of depression improved after taking these drugs.

Iproniazid and imipramine elevate brain concentrations of both serotonin and noradrenaline. Consequently, medical researchers began to investigate whether changes in serotonin and noradrenaline concentrations were alleviating the depression. Eventually they found that other drugs, called SSRIs, or selective serotonin reuptake inhibitors, which specifically target the serotonin generally treat depression just as well as those that target both serotonin and noradrenaline. This led to what became known as the serotonin deficiency model of depression.

SSRIs, such as Prozac, increase serotonin levels by targeting a protein called the serotonin *transporter*. As illustrated in figure 22.1, the serotonin transporter is a membrane protein located on neurons that secrete serotonin. The transporter recycles serotonin before it is detected by a receiving neuron. Proper recycling ensures that the serotonin concentration in the space between neurons is maintained at the optimal level. If the brain's serotonin transporters are overactive, or if too many serotonin transporters are synthesized, then too much serotonin is recycled and insufficient serotonin is available for intercellular signaling.

The development of SSRIs has led to a multibillion dollar drug market. Numerous studies have demonstrated the effectiveness of SSRIs in treating severe depression, though the efficacy of SSRIs in treating mild to moderate depression is not universally accepted. Some studies have even concluded

---

* Experiments that stress monkeys and other animals just to learn how their brains react to negative experiences bring up difficult moral issues. Yet such research does exist, and ignoring what has been learned from it doesn't seem helpful.

that SSRIs are no more effective in treating mild or moderate depression than placebos. That said, since SSRIs are effective in treating severe depression, serotonin signaling must play an important role in our feelings of well-being.

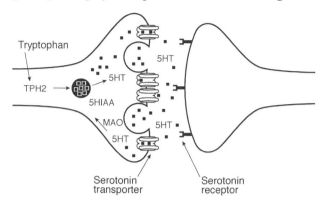

Figure 22.1. Signal transmission between neurons. Serotonin, indicated by small squares and labeled 5HT (the scientific abbreviation of serotonin) is secreted by the neuron on the left and diffuses less than a micron (1/1000th of a millimeter) to the neuron on the right. There, the serotonin is detected by binding to serotonin receptors. Other serotonin molecules are bound by serotonin transporters in the transmitting neuron and then degraded into a related chemical, 5HIAA, by enzymes (labeled MAO). The diagram also shows the initial synthesis of serotonin from tryptophan by the enzyme TPH2.

Animal experiments have also indicated a role for serotonin in depression. SSRIs increase a mouse's resilience against depression, for example, as measured by the forced swim test. Experiments involving *TPH2* transgenic mice have also shown that serotonin mediates feelings of well-being. We've met mice with modified *TPH2* genes before. In chapter 20, we learned that *TPH2* knockout mice, which produce no serotonin at all, are equally sexually attracted to male and female mice. To study serotonin's effects on depression, scientists needed a less-extreme transgenic mouse, one that synthesized *some* serotonin, though not as much as in a normal mouse. When looking to engineer such mice, scientists from Duke University noticed that certain elderly human patients who suffered from depression shared a rare mutation in the *TPH2* gene. This mutation causes a single amino acid change in the TPH2 protein, a histidine amino acid is substituted for an arginine, as a result of which TPH2 synthesizes 80% less serotonin than normal. Based on this observation, the Duke scientists replaced the *TPH2* gene in laboratory mice with the mutated gene found in the human patients. These *TPH2* transgenic mice produced less serotonin in their brains and showed symptoms of depression in standard mouse-depression tests, suggesting that, at least in mice, the lack of serotonin does increase susceptibility to depression.

SEROTONIN TRANSPORTER VARIANTS AND HUMAN FEELINGS OF WELL-BEING

If the serotonin deficiency model of depression is correct, variants in the human serotonin transporter gene, called *SLC6A4*, might influence feelings of well-being in people. Although many such natural variations in the human *SLC6A4* gene exist, most neither change the structure of the serotonin transporter protein nor the amount of transporter protein produced. These *SLC6A4* variants don't seem to make any difference to the neuron. One common genetic variant within the human *SLC6A4* gene does seem to make a difference, though. Some of us have *SLC6A4* genes in which a certain 44–base pair sequence is repeated 12 times, while other people's DNA has only 11 repetitions of the 44–base pair segment, and people with two copies of the longer (12-repeat) variant typically make three times as much serotonin transporter as individuals with two copies of the shorter variant.

Since serotonin transporter knockout mice, which produce either no serotonin transporter or half the normal amount (depending on whether both or only one serotonin transporter gene is knocked out), are more prone to depression than normal mice, perhaps people with *SLC6A4* variants producing less SLC6A4 protein are also more vulnerable to depression? Dozens of investigations have tested this idea, and several early studies reported that individuals with one or two copies of the shorter *SLC6A4* variant scored higher on tests of depression than people with two copies of the longer variant. Later studies also found associations between the *SLC6A4* length variants and other psychological or personality traits, such as anxiety or "neuroticism." One report even claimed that healthy, nondepressed individuals with the longer *SLC6A4* variant rated the "satisfaction with their lives" higher than people with only the shorter variant. Unfortunately, later studies have found little or no correlation between emotional well-being and serotonin transporter variants, and the hypothesis that *SLC6A4* genetic variants affect depression is still unproven.

MATERNAL NURTURING AND EMOTIONAL RESILIENCE

The lack of consistent evidence linking genetic variation with feelings of well-being led some scientists to look for epigenetic mechanisms by which a person's environment might affect their feelings of well-being. Perhaps trauma or other adverse life experiences can trigger epigenetic changes in one's brain cells, making them more vulnerable to depression. What kinds of life experiences might affect one's resilience to adversity? Numerous studies point to childhood abuse or trauma. For example, one study of survivors of child abuse found that almost 80% of child abuse victims suffered from some form of psychiatric disorder as adults, in contrast to 26% of adults from the general population.

That childhood experiences influence resilience to adversity has been dramatically demonstrated in animal experiments. In 2004, researchers at McGill University showed that baby rats reared by mothers that lick and groom them are more resilient to adversity than mice raised by non-nurturing mothers. The experiments also identified some of the underlying biological mechanisms. The researchers found that licking and grooming altered hormone and neurotransmitter signaling, leading to increased brain serotonin. These changes, in turn, resulted in lower levels of the stress hormone, corticosterone.

Their most remarkable discovery was that the different stress responses between nurtured and neglected rats remained long after rat infancy. How did the rats' brains remember whether the rats had been licked and groomed as infants? The experiments suggested that the mechanism was the modification of their brain cells' chromosomes. In chapter 13, we saw how modifying the chromosomes in a mouse's brain cells was a key step in how a mouse learns and remembers the solution to a maze. The McGill experiments suggested that similar brain-cell chromosomal modifications contribute to the forming of memories of traumatic experiences. The adult rats that had been licked and groomed in infancy had different chromosomal modifications (specifically, different patterns of DNA methylation) in the cells of the hippocampus, a part of the brain involved in memory formation. The researchers also found that the chromosomal alterations associated with childhood neglect could be reversed. Giving the neglected rats a drug that suppresses chromosomal modification not only altered the DNA-methylation patterns to resemble those found in the brains of nurtured rats, but also made the neglected rats as emotionally resilient as the nurtured ones.

Might childhood trauma affect brain cell chromosomes in humans as well? Since brain-cell chromosomal modifications can only be measured postmortem, the McGill group decided to compare differences in chromosomal modifications between suicide victims (many of whom were likely to have been depressed) and individuals who had died from accidents. The results suggested that people might not be that different from rodents in the way our brains preserve emotional memories. The researchers found that the pattern of hippocampal chromosomal modifications of the suicide victims *who had been abused as children* was similar to that of rats that hadn't been licked or groomed in infancy. In contrast, the chromosomes in the brains of the accident victims, as well as of suicide victims who had *not* suffered child abuse, were similar to those found in the nurtured rats. The experiment suggested that human and rodent brains remember early life experiences in a similar manner. As a result, many researchers are now testing drugs that inhibit chromosomal modification, similar to those used with the rats, as potential antidepressant drugs for people.

GENES *AND* ENVIRONMENT, NOT GENES *OR* ENVIRONMENT

Although childhood trauma appears to affect emotional resilience in humans, it is clearly not the only factor. Epidemiological studies show that many victims of childhood abuse grow up to become psychologically healthy individuals, suggesting that other factors, possibly including genetic ones, also contribute to human resilience to adversity. So in the early 2000s, scientists again began searching for genetic associations with emotional well-being. This time, though, rather than look for direct links between genetic variants and happiness or depression, the researchers were seeking genetic associations with emotional resilience against stress.

Among the earliest studies linking resilience to adversity to a combination of inherited genetic variants and childhood trauma was a 2002 experiment on rhesus monkeys. In this experiment, conducted by researchers from the NIH's primate unit, 132 infant monkeys were divided into two groups: 60 monkeys were raised by their mothers, while 72 were raised by their peers, without their mothers. For each monkey, the researchers determined the animal's nervous-system serotonin concentration* and its serotonin-transporter variants (monkeys, like humans, have different *SLC6A4* length variants).

---

* Actually, the researchers measured a compound called 5-HIAA in the monkey's cerebral spinal fluid. Spinal fluid 5-HIAA levels are correlated with brain serotonin concentration, while being easier to measure.

The researchers found that all the monkeys raised by their mothers had similar serotonin levels independent of which *SLC6A4* gene variants they had. In contrast, among monkeys raised without their mothers, those with two copies of the longer *SLC6A4* variant had significantly higher serotonin concentrations than those with at least one shorter variant. In other words, the *SLC6A4* gene variant was correlated to a monkey's serotonin level *only* if the monkey had had the stressful experience of being reared without its mother. Although the NIH researchers did not report on the emotional states of the monkeys in this study, it seems plausible that the monkeys with the lower serotonin levels – that is, those raised without their mothers *and* with shorter serotonin-transporter gene variants – would be more vulnerable to depression.

Could a similar combination of stress and a "bad" genetic variant also lead to decreased serotonin and depression in people? A research team from King's College London decided to test this hypothesis with data from a 1000 26 year-olds from Dunedin, New Zealand. These individuals had participated in an epidemiological study in which their health, development and behavior had been followed closely since birth. Each subject was given a blood test to ascertain their *SLC6A4* variants. In addition, the researchers interviewed every participant, to determine whether they had been mistreated as a child, had experienced any stressful life events within the previous five years (such as divorce, lost job) and whether they had suffered from depression.

The study's results were dramatic. Among the young adults who had experienced more than three stressful life events, individuals with two copies of the shorter *SLC6A4* variant were twice as likely to have had an episode of depression as people with two copies of the longer *SLC6A4* variant. Similarly, individuals who had suffered from severe child maltreatment were twice as likely to have been depressed in the last year if they had two copies of the shorter *SLC6A4* variant. On the other hand, young adults without stressful or abusive life experiences had the same rate for depression, independent of which serotonin-transporter variants they had, and individuals with the protective genetic variant had the same incidence of depression, whether or not they had abusive histories. In other words, only people with *both* a stressful history *and* the "bad" *SLC6A4* variant were at higher risk for depression. (Figure 22.2 shows data from this experiment.)

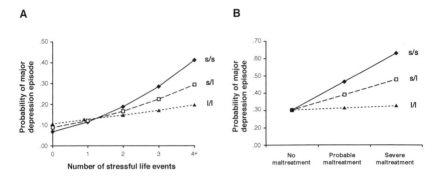

Figure 22.2.   Effect of combined influences of genetic variants, childhood abuse and stressful life events on adult depression. A: Genetic variants and life stresses. This graph shows the probability that a subject had an episode of depression in the previous year as a result of stressful life events in the preceding five years. Individuals who had no stressful life event, or only one, had approximately a 10% probability of a depressive episode, independent of which *SLC6A4* variants they had. In contrast, for people experiencing four or more stressful events, the probability of an episode of depression varied from less than 20% for individuals with two copies of the longer variant to 40% for those with the shorter variant. B: Genetic variants and childhood abuse. The classification of individuals as having had no childhood maltreatment, probable maltreatment or severe maltreatment was based on interviews of the research subjects at age 26. In both graphs the l/l curve displays data from subjects with two copies of the long variant of the *SLC6A4* gene (also called the *5-HTT* gene). s/s indicates those with two copies of the shorter gene variant, while s/l presents the data for those with one copy of each variant. Source: "Influence of Life Stress on Depression: Moderation by a Polymorphism in the 5-HTT Gene," by A. Caspi et al., *Science*, July 18, 2003: vol. 301, no. 5631, pp. 386–389.

The King's College study made a major impact on the biomedical research community. Over the subsequent decade, the 2003 paper was cited over 2000 times in the scientific literature, and scores of research teams sought to replicate and extend its findings. Initially, a number of studies successfully replicated the King's College results; however, several later failed. In 2009, two respected research groups carried out meta-analyses of 19 of these studies and each concluded that the overall data failed to prove that the effects couldn't have been caused by chance. Yet this was not the end of the story. A larger meta-analysis of 54 different studies, published in 2011, concluded that the association between *SLC6A4* variants and vulnerability to depression was statistically significant (that is, unlikely to have been caused by chance), although the magnitude of the effect was smaller than that reported in the original King's College paper. Clearly, the last word on the effect of life stress and serotonin-transporter variants on vulnerability to depression has not yet been written. That said, the results seem to indicate that combined genetic and environmental effects influence our vulnerability to depression. As we'll soon see, they also affect our other emotions, including how we experience fear and anger.

23

WHAT IS FEAR?

In his widely quoted first inaugural address, Franklin D. Roosevelt declared "the only thing we have to fear is fear itself." Taken out of context,* as it sometimes is, this assertion is of course absurd. If you're walking alone in the African bush and an angry elephant charges, being afraid is a healthy response. Fear is an important way that our brain helps protect us from danger. For as long as animals have walked the earth, they have been confronted by danger. If an animal didn't respond to danger, it was eaten. Animals needed to develop finely tuned abilities to sense danger and respond appropriately. They evolved to become aggressive and fight back or to become frightened and run away. Sometimes, though, these self-protective instincts can go awry. Our fear responses, which protected our ancestors so well in the wild, can paralyze us if triggered at the wrong time.

Extreme or inappropriate fear is not rare. According to the U.S. National Institute of Mental Health, each year approximately 40 million Americans adults suffer from some form of fear disorder. Such people are found in all walks of life and include countless talented and successful individuals. Kim Basinger, the Academy Award–winning actress, has been paralyzed by the

---

* Roosevelt's entire statement was "The only thing we have to fear is fear itself – nameless, unreasoning, unjustified terror which paralyzes needed efforts to convert retreat into advance."

fear of strangers and of open places for much of her life. Earl Campbell, star running back for the Houston Oilers football team, had his life thrown into turmoil when he suddenly believed that he was going to have a heart attack, even though his heart was completely healthy. Millions of other people have irrational fears of everything from enclosed places and heights to snakes and insects. Sometimes people suffer from anxiety or panic attacks* without even being able to say what it is that frightens them.

SEROTONIN, ADRENALINE AND ANIMAL FEAR

Like our other emotions, fear results from external events in our environment triggering neuronal pathways in the brain. The functioning of these signaling networks, in turn, is influenced both by genetic variations in the networks' proteins and by epigenetic modifications that alter the brain -cell concentrations of those proteins. For example, a 2001 review of multiple family studies found that identical twins are significantly more likely to both be affected by panic disorder than nonidentical twins, and that relatives of individuals suffering from panic attacks are five times more likely to suffer from a similar disorder themselves than other people are. From these observations, the author concluded that genetics was the principal contributing factor to fear sensitivity and that shared family environment was only of limited importance.[†]

Although family studies can show that fear sensitivity has a genetic component, they do little to identify what the underlying biological factors are. To study these mechanisms, scientists generally rely on experiments with animals. Consequently, researchers need ways to infer an animal's fear from its behavior. One commonly used approach is to measure the amount of time a mouse explores open spaces. Mice generally like to explore new environments, especially if they are dark and not exposed. Mice explore open and well-lit places less frequently, presumably because of fear. Consequently, scientists infer that a mouse that explores relatively exposed spaces more than other mice has a higher fear threshold. Another method is to observe an animal's freezing reaction to a fearful stimulus, as described

---

* Psychologists use several terms to describe human fear. *Fear* and *phobia* are typically used to denote the response to a specific danger, while *anxiety* is used for the emotional reaction in the absence of any clear danger and *panic* is used to refer to sudden or extreme fear or anxiety.

† The review also noted the lack of research studies comparing anxiety and fear thresholds between biological and adopted siblings, which could help differentiate the genetic and environmental factors contributing to fear sensitivity.

in chapter 13 in the context of the fear-conditioning test. A third way is to study the animal's pattern of vocalizations. For example, when in danger, mice typically emit ultrasonic vocalizations, which scientists interpret as expressions of fear.

Using such methods, scientists have studied how neurotransmitter signaling affects fear in mice. Some of the most compelling links between brain signaling and fear have been found in the serotonin and noradrenaline pathways. Serotonin's involvement in fear signaling was not a surprise. SSRI drugs, which increase serotonin availability, not only help patients with depression but also those with anxiety disorders. Experiments with knockout and transgenic mice have also demonstrated serotonin's role in mediating fear. Mice with knocked out serotonin-receptor genes display increased fear, while transgenic mice, with modified serotonin-synthesizing or serotonin-transporting genes that decrease serotonin signaling, exhibit elevated "mouse anxiety" as well.

The noradrenaline/adrenaline pathway also plays a key role in our reaction to danger. Noradrenaline and adrenaline are related signaling molecules that initiate the rush of feelings – which may include fear, anger and excitement – that we experience in response to danger. Adrenaline is secreted by the adrenal gland, while the brain releases noradrenaline. Between them, noradrenaline and adrenaline activate both our physiological response to danger, including increased pulse rate and faster breathing, and our psychological and emotional responses. In other words, noradrenaline and adrenaline are central to our reactions of "fight or flight," so we shouldn't be surprised that abnormal noradrenaline levels in the brain can contribute to extreme emotional responses of fear.

Since the brain's serotonin and adrenaline pathways are involved in fear-signaling, proteins that alter the brain's concentrations of these neurotransmitters would be expected to affect our feelings of fear. One such regulator of serotonin and adrenaline is monoamine oxidase A, or MAO-A. MAO-A is an important enzyme that degrades noradrenaline, adrenaline, serotonin and dopamine. Evidence that MAO-A affects our experience of fear comes from the fact that MAO-A–blocking drugs can be helpful in treating anxiety. Scientists have also studied MAO-A with *MAO-A* knockout mice, but the results have been inconsistent. *MAO-A* knockout mice definitely do exhibit abnormal behavior (male *MAO-A* knockout mice are very aggressive), yet the evidence regarding *MAO-A* deficiency and fear-related behavior is less convincing. Some experiments have suggested that *MAO-A* knockout mice have higher fear thresholds than normal mice, while other experiments that used different measures for assessing mouse anxiety came to the opposite conclusion.

Despite the inconclusive results of the *MAO-A* knockout mice experiments, scientists have looked for human *MAO-A* variants that might contribute to differences in fear thresholds among people. In particular, a 30–base pair DNA segment is repeated either three or four times within the human *MAO-A* gene. Approximately 35% of *MAO-A* genes have the three-repeat variant, with the remainder having four repeated segments.[*] Unfortunately, attempts to link human *MAO-A* variants with susceptibility to panic or anxiety disorders have not been particularly successful either. Some studies found no association. Other studies did report higher occurrence of the four-repeat variant among female panic disorder patients, but the differences found were small, and each study included only approximately 100 women.

FEAR SIGNALING AND TMEM132D

Until recently, studies that searched for genetic variants linked to fear sensitivity focused on known fear-signaling genes, such as those in the serotonin and noradrenaline pathways. The decision to limit the search for fear-related DNA variants in this way was largely made for reasons of expediency. Searching the entire genome in large numbers of people was simply too expensive. (Such studies need to be carried out on very large numbers of subjects to ensure that any detected associations are not simply caused by chance.) As a result, these studies were limited in terms of the fear-related genetic variations they could detect.

In the last few years, the cost of DNA testing has dropped rapidly. It is now economically feasible to screen the entire genomes of large numbers of people, and in 2011, a German research consortium reported three studies, all of which found that certain DNA variants near a gene called *TMEM132D* were more common in people suffering from panic disorder. Since the same variants were identified in three independent samples, the results were unlikely to be the result of chance. The German team therefore decided to investigate TMEM132D in mice. Mice have a natural genetic variation within the *TMEM132D* gene, and when the researchers separated a group of mice into those displaying either "high-anxiety" or "low-anxiety" behavior, they discovered that the high-anxiety mice tended to have one

---

[*] Two rare variants, consisting of either five copies or three and a half copies of the 30–base pair segment, are found in less than 2% of the population.

TMEM132D gene variant, while the low-anxiety mice generally had the other one. They also found that in a region of the mouse brain associated with fear (called the anterior cingulate cortex), the *TMEM132D* messenger RNA* concentration was 34% higher in the high-anxiety mice than in the low-anxiety mice. Encouraged by these results, the German scientists then measured *TMEM132D* messenger RNA levels in human (postmortem) brain tissue. They found that people with the *TMEM132D* variant that had been associated with panic disorder had higher brain levels of *TMEM132D* messenger RNA than individuals with the other variant.

Just how TMEM132D influences fear sensitivity is still unclear. TMEM132D's biological function is unknown, though its location in the cell membrane suggests involvement in intercellular signaling. Perhaps TMEM132D plays a role in the noradrenaline or serotonin pathway. In any case, the data suggest that *TMEM132D* variants affect the amount of TMEM132D protein in the brain, which in turn leads to varying susceptibility to panic disorder.

INHERITED FEAR AND EPIGENETICS

Although family and twin studies have shown that genetics influences a person's overall threshold for fear and anxiety, they don't necessarily indicate that *specific* fears – such as fear of snakes or the dark – can be inherited by means of one's DNA. In fact, many scientists have believed that inheriting specific fears is impossible. Yet, a recent study by researchers at Emory University suggests that specific fears *can* be inherited, at least among mice.

The Emory University experiments studied a group of genetically identical male mice. Half of the mice were trained to be afraid of the smell of a harmless chemical called acetophenone by repeatedly exposing the mice to its smell while administering a mild electric shock to their feet. The remaining mice were trained in a similar way to fear the smell of a different chemical called propanol. After the trainings were completed, each group of mice displayed stereotypical mouse fear behavior – such as freezing in place – when exposed to the odor they had been trained to fear and no response to the other odor.

---

* Ideally, the scientists would have measured TMEM132D *protein*, but messenger RNA concentrations are easier to measure and often (though not always) correlate closely with protein concentrations.

So far, this was just an example of typical animal fear conditioning, a laboratory procedure that scientists have performed for decades. But what the Emory University scientists did next was different. The mice from each group were allowed to mate with female mice that had not received any fear conditioning. The resulting offspring also were not trained to fear any odors. Nevertheless, the offspring of mice that had been trained to fear the odor of acetophenone also displayed fear responses to acetophenone, while the pups of mice that had been trained to fear propanol also feared propanol.

Of course, one might suspect that the parent mice had somehow "taught" their pups to fear the odor that they had been trained to be afraid of. But this was not the case. The scientists proved this in a different experiment using "cross-fostered" pups. The experiment was similar to the initial one, except that the newborn mouse pups were removed from their mothers* (who had been trained to fear the odor of acetophenone) immediately after birth and given to female mice that had been trained to fear propanol. Despite being raised by propanol-fearing mothers, the pups were observed to be fearful of acetophenone and not of propanol. The only logical conclusion was that the pup's fear was being inherited from its biological parents and not from the behavior of the mouse that raised it.

To understand how learned fear was being transmitted to the pups, the scientists analyzed sperm cells of mice that had received acetophenone-odor training. Specifically, the researchers tested sperm-cell DNA near the gene for an important acetophenone odor–receptor protein. They discovered that the pattern of DNA methylation near this gene was quite different from that found in the sperm cells of other mice. Although it's still unknown how a mouse's brain signals its sperm cells to alter their pattern of DNA methylation, the results suggested that fear training led to sperm-cell DNA demethylation and that this was the mechanism by which the fear of acetophenone was inherited by the mouse's offspring.

FEAR AND EPIGENETICS IN PEOPLE

If epigenetic DNA modification, such as DNA demethylation, can store fearful memories in mice, perhaps epigenetic modifications also underlie how people retain memories of frightening experiences. Evidence supporting this hypothesis came from the same Emory University research team that

---

* Note that the researchers had already repeated their experiment with *female* mice that had been trained to fear the two odors. This experiment demonstrated that fear of specific odors could be transmitted to offspring from the mother as well as from the father.

had discovered the inheritance of specific fears in mice. Their human studies focused on the response to fear found in post-traumatic stress disorder (PTSD). Among the most striking symptoms of sufferers of PTSD is their sensitivity to fear, as they often display anxiety and hypervigilance even in nonthreatening environments.

To better understand PTSD's biological origins, the Emory researchers focused on cortisol, a hormone that is secreted by the adrenal gland during stressful and fearful experiences. In their initial experiments, the team discovered the existence of a common DNA variation in the gene for a cortisol regulatory protein called FKBP5. What was interesting was that individuals with one of these *FKBP5* variants – if they had *also* suffered maltreatment as children – were more vulnerable to PTSD as adults than people who had not suffered child abuse or who had the other *FKBP5* variant. This was similar to the discovery described in the previous chapter that the combination of a specific serotonin-transporter genetic variant and a history of child abuse is associated with a greater risk of adult depression.

The researchers also discovered a possible epigenetic mechanism explaining how a *FKBP5* variant combined with a history of child abuse might lead to vulnerability to PTSD. Specifically, the Emory researchers found a specific pattern of DNA methylation in the *FKBP5* genes of the child-abuse survivors who had a certain *FKBP5* DNA variant that was different from the DNA methylation pattern in people who either had other *FKBP5* variants or had not suffered childhood maltreatment. This was the first time that a change in DNA methylation had been associated with a combination of childhood abuse and genetic vulnerability in people. Besides being a significant step in understanding how people store fearful memories, discovering that childhood trauma can change DNA methylation may help researchers create new drugs that reverse such methylation changes and eventually lead to novel treatments for PTSD.

ANXIETY AND OBSESSIVE-COMPULSIVE DISORDER

Among the various manifestations of aberrant fear-related behavior, arguably the strangest is what psychologists call obsessive-compulsive disorder, or OCD. OCD typically includes irrational thoughts or beliefs (the obsession) combined with bizarre, sometimes even self-destructive behavior (the compulsion). Some people suffering from OCD may think that their environment is "unclean" and that they need to be continually cleaning. Others believe that they will lose their possessions and that they need to hoard objects of little value, such as junk mail, old newspapers or trash. Because these obsessive thoughts are often accompanied by feelings of fear, the medical profession considers OCD an anxiety disorder.

Perhaps the best-known person with OCD was the inventor and businessman Howard Hughes. In his later life, Hughes was haunted by a variety of obsessions, based on his fears of disease-carrying germs. Unfortunately, rather than protect him, Hughes' fearful thoughts drove him into a world of compulsive behavior and isolation.

OCD isn't rare; an estimated 1% to 3% of the world's population suffers from OCD. It also has a strong genetic component. Over 80% of monozygotic twins with OCD have a twin who suffers from the disease, while same-sex, dizygotic twins have only a 50% concordance for OCD. Prenatal or early childhood experiences appear to play roles in the development of OCD as well, since non-twin siblings (who share the same amount of DNA as dizygotic twins, but grew up at different times and therefore presumably in somewhat different environments) have only a 10% concordance for OCD.

As OCD is characterized by *behavior,* and not just by feelings, it can be directly studied in animals. In fact, peculiar behavior similar to that found in people with OCD occurs naturally in some animals. Many of us have observed the strange sight of a dog repeatedly chasing its own tail. Other dogs display a behavior called flank sucking, in which they continually lick the sides of their body. Only certain dog breeds display compulsive behaviors; a dog chasing its tail is likely to be a bull terrier or a closely related breed, such as a Jack Russell or Staffordshire terrier. Flank sucking is found primarily in Doberman pinschers. Since dog breeds have developed relatively recently, finding causal genetic variants is easier in dogs than in humans. Although genetic variants linked to compulsive dog behavior haven't been found yet, with luck scientists will start identifying them soon.

While dogs may eventually aid scientists in identifying OCD-linked genetic variants, rodents are already teaching researchers about OCD. Mice and rats can display behavior that is very reminiscent of human compulsive behavior. Some mice groom themselves so incessantly that they lose much of their fur, a symptom similar to the compulsive hair pulling behavior of some people with OCD, while rats may compulsively hoard marbles or similar items, not unlike human "pack rats" with OCD. Moreover, rodents displaying OCD-like behavior can often be successfully treated with SSRI drugs, just as SSRIs sometimes help people with OCD.

Mice are even helping scientists identify which proteins are malfunctioning in OCD. In 2007, scientists from Duke University found that mice with a gene called *SAPAP3* knocked out groom themselves excessively, leading to fur loss and skin lesions. These *SAPAP3* knockout mice also displayed increased general "mouse anxiety," as measured by how much time they spent exploring open areas. When the Duke scientists administered DNA to the knockout mice to restore their SAPAP3 protein levels, the excessive grooming and anxiety symptoms disappeared. The results of the mouse experiments motivated scientists to search for *SAPAP3* variants that might contribute to human OCD, and a 2009 study reported the discovery of rare *SAPAP3* mutations in some people suffering from OCD.

RISK TAKING AND DRD4

There is now considerable evidence that our brain's signaling networks play an important role in our experience of fear. Scientists have learned that when our brain pathways are impaired, our fear threshold can be too low. But what about people whose fear threshold is too high? Might certain genetic variants predispose a person to be an extreme risk taker, having little or no fear, even in truly dangerous situations?

A 2001 study of risk-taking behavior and attitudes in twins suggested the answer may be yes. Researchers from Virginia Commonwealth University surveyed approximately 700 adolescent twin pairs with questions ranging from "Do you use birth-control?" and "Do you wear a seat-belt?" to whether they agreed or disagreed with statements such as "I like to take risks." By comparing the concordance in responses from monozygotic and dizygotic twins, the researchers concluded that approximately 25% to 30% of the respondents' attitudes to risk taking could be attributed to genetic factors.

Researchers at Herzog Hospital and Ben-Gurion University, in Israel, went a step further in the search for genetic links to risk taking by attempting to identify some of the genetic variants that are involved. In 1996, the team published a provocative paper that claimed to find a genetic variant linked to "novelty seeking." The Israeli team surveyed 124 people using a questionnaire that tested four personality factors commonly used in psychology research: harm avoidance, novelty seeking, persistence and reward dependence. They also tested the subjects for genetic variants near a single gene, *DRD4*. They selected *DRD4* because it codes for a dopamine receptor involved in cognitive and emotional brain signaling. The researchers claimed to find a statistically significant association between novelty seeking and a genetic marker in the *DRD4* gene.

To be sure, novelty seeking and risk taking are not quite same thing, and by testing only a single gene, the researchers were ignoring any genetic contributions to novelty seeking or risk taking that might be the result of variations in other genes. Nevertheless, the Israeli study provided hope that studying *DRD4* variants might lead to better treatments for people engaging in ill-advised risk taking, such as using addictive drugs or practicing high-risk sex. Multiple research teams have sought to replicate and extend the findings of the original 1996 study. Unfortunately, the results have been inconsistent. While 10 subsequent studies also found a link between *DRD4* variants and novelty seeking, a similar number of experiments couldn't find any association. In fact, even the Israeli researchers who first announced the link between *DRD4* variants and novelty seeking have had difficulties in replicating their own results.

Even if the link between *DRD4* variants and novelty seeking or risk taking is ultimately confirmed, determining *how* DRD4 influences such behavior remains challenging. DRD4 is a dopamine receptor, and dopamine signaling plays an important role in our experience of pleasure. If *DRD4* variants contribute to increased skydiving or marijuana use, scientists will still need to determine whether this association is caused by increased risk taking or is simply a consequence of dopamine's association with pleasure. Despite these challenges, dozens of studies over the last decade have attempted to link *DRD4* gene variation to just about every type of risky or abnormal human behavior one can imagine. A few such associations have been successfully replicated. Variations in *DRD4*, as well as in other genes, have been associated with susceptibility to attention deficit hyperactivity disorder and with alcohol and nicotine dependence. For other risky or impulsive behaviors, though, no link to dopamine signaling or the DRD4 receptor has been found. In particular, scientists have been unable to find a link between DRD4 and impulsive aggression or violence. Yet, as we'll soon see, that doesn't mean that variants of *other* genes haven't been linked to such behavior.

24

WHAT IS AGGRESSION?

The young woman we'll call Marieke first came to the Radboud University Genetics Clinic in the Netherlands in 1987. Marieke wasn't sick herself, but she was worried about her three young children. Something seemed to be wrong in her family, or, at least, among the men in her family. Eight of her close male relatives (brothers and first cousins) were mildly retarded. More troubling, they were all impulsive and aggressive, often to the point of being violent. One had been convicted of raping his sister, two were known arsonists, and another had attempted to run over his employer. Recently, Marieke's 10-year-old son was also starting to behave abnormally in ways that reminded her of how her brothers had acted when she was a young girl. Marieke was worried that her son, and perhaps her two young daughters as well, had inherited something terrible that was being genetically transmitted within her family.

AGGRESSION AND BIOLOGY

The idea that intellectual or behavioral traits might be inherited has its roots in the work of Francis Galton, a brilliant 19th-century scientist who was a cousin of Charles Darwin. In 1869, Galton published his theories in a book called *Hereditary Genius* in which he proposed that intelligence and societal success might be hereditary. In 1877, a social reformer named Richard Dugdale extended Galton's ideas with a book entitled *The Jukes: A Study in Crime, Pauperism, Disease and Heredity*. Dugdale's book described a multigenerational family in New York he called the Jukes. The family included multiple violent criminals, mentally retarded individuals and other "social deviants." Although Dugdale concluded that the Jukes family's social problems were mainly linked to their impoverished and crime-ridden childhood environment, he also suggested that inborn factors were contributing to their lives of crime and degradation.

The notion of inherited social deviance might have ended with Dugdale's book were it not for the fact that politicians, and even some scientists, began using the Jukes story to promote a sociological theory called social Darwinism. Social Darwinism claimed that, just like animal species, societies also evolve, or at least *should* evolve, to eliminate their less fit members. In this intellectual environment, only the genetic aspect of the Jukes story survived, and with it came the birth of the "science" of eugenics, which proposed that some individuals are genetically unfit for society and should be sterilized so that they don't produce additional undesirable people. In the early 20th century, social Darwinism and eugenics became the rationale for America's immigration policy, which restricted the entry of "undesirable" immigrants from eastern and southern Europe while encouraging the immigration of more "fit" individuals from northern and western Europe.

Only a few decades later, though, the Nazi Holocaust exposed the potential consequences of eugenics and the racist ideology that it could lead to, and eugenics was rejected by virtually all societies. At that point, historians revisited Dugdale's writings and rediscovered the importance of environmental influences in his work. In fact, subsequent historical research showed that some members of the biological Jukes family had been prominent members of the local society, while many other "family members" had not even been biologically related to one another.

Although he was working before the discoveries of modern genetics and epigenetics, Dugdale realized that both inherited and environmental factors influence our behavior. He understood that although inherited factors, which we now know arise from DNA variations, don't *determine* behavior, they can influence it. Recall, for example, that disabling a gene called *Gtf2ird1* made mice "friendlier" and that the loss of *Gtf2ird1* also appeared to be linked to the unusual friendliness found in people with Williams-Beuren syndrome. We've learned how scientists are on the verge of discovering other genetic variations linked to gentleness in dogs, wolves and foxes. Perhaps different genetic variants may predispose one to be *less* gentle? Maybe some genetic variant or other underlying biological factor might make one more susceptible to the unprovoked, angry and hostile behavior that we associate with aggression.

AGGRESSION, TESTOSTERONE AND ESTROGEN

Probably no biological signaling network has been more clearly linked to aggression than the testosterone pathway. Long before laboratory experimentation, farmers knew that castrating bulls makes them less aggressive and that male animals, with more testosterone, tended to be more aggressive than females. In 1849, the German physiologist Arnold Berthold further demonstrated testosterone's key role in a series of dramatic experiments with castrated chickens. Berthold showed not only that castrated male chickens stopped fighting with other males but that the chickens became aggressive again after being surgically transplanted with replacement testes.

Although in most animal species male aggression is directly linked to testosterone levels, this is not universally true. Castrating prairie voles doesn't decrease their aggressive behavior, while among mice, even male mice, aggressive behavior is primarily regulated using estrogen. Estrogen's link to male mouse aggressiveness has been most clearly observed in aromatase knockout mice. Recall that aromatase converts testosterone into estrogen, so aromatase knockout mice can't produce estrogen. In 2001, scientists from Kochi Medical School in Japan engineered aromatase knockout mice and observed that the males weren't aggressive, even though they had plenty of testosterone. When the researchers gave the mice estradiol, a form of estrogen, their normal aggressive behavior (as measured by the mice attacking and biting other mice) returned. In a separate experiment, the Japanese researchers engineered mice that had their estrogen *receptor* gene disabled. These mice had normal amounts of both testosterone and estrogen, yet, without estrogen receptors, their estrogen-signaling pathway was nonfunctional, and they were again less aggressive than normal male mice.

TESTOSTERONE AND AGGRESSION IN MEN

Unlike mice, most animal species use testosterone to regulate male aggression, and testosterone levels usually indicate how aggressively an animal behaves. But do testosterone levels also correlate with aggressiveness in people? As men have more testosterone and are generally more aggressive than women, one might think so. But experiments searching for such correlations in men have been inconclusive. Some studies have found correlations between human aggression and testosterone levels, but others have not.

If testosterone levels aren't directly correlated with human aggression, perhaps some other element of the testosterone-signaling pathway is. For example, everyone's DNA includes the gene for the human testosterone receptor, called the androgen receptor, or AR. One section of the *AR* gene consists of a repeated three–base pair sequence (C-A-G). In some variants of the *AR* gene the C-A-G sequence is repeated as often as 35 times. In other individuals' *AR* gene the C-A-G sequence is repeated only 8 times, or some other number between 8 and 35. As you may recall from the genetic code described in chapter 2, each additional C-A-G sequence within the *AR* gene will cause one more arginine amino acid to be added to the androgen receptor protein. As a result, the number of C-A-G repeats in the *AR* gene changes the length of the arginine amino acid subsection in the resulting androgen receptor. This causes the AR protein to have a somewhat different protein structure, which in turn affects how tightly the receptor can bind testosterone.

Might a difference in testosterone binding influence one's tendency to be aggressive? Two recent studies have found that violent offenders, including murderers and rapists, were more likely to have short androgen-receptor variants (less than 17 repeats in length) than men in the general population. Another experiment, which tested verbally aggressive behavior in male volunteers, also concluded that men with shorter *AR* variants were more aggressive, though the effect was too small to be statistically significant. *AR* variants have even been studied in Akita dogs, and again the more aggressive dogs tended to have the shorter *AR* variants. Although all these studies were performed with small numbers of subjects and need to be replicated in larger populations, their consistent findings are intriguing.

SEROTONIN AND AGGRESSION

Testosterone is only one component of a larger biological signaling network linked to aggression. Another key player is serotonin. That serotonin signaling is implicated in aggressive behavior shouldn't be surprising. We've seen that serotonin signaling is linked to feelings of well-being, and a strong sense of well-being might well lead to less aggressive behavior.

Direct evidence linking serotonin to aggression again comes from animal experiments. Scientists from the University of Groningen in the Netherlands measured brain serotonin levels in six strains of mice, three of which had been bred to be extremely aggressive and three of which had been bred to be docile. On average, the docile mice strains had 30% higher concentrations of serotonin and serotonin receptors than the aggressive mice. Investigations of male song sparrows (which can be quite aggressive) also suggested that serotonin is involved in aggression. The sparrows' aggressiveness was assessed on the basis of behavior, posture and wing positions in response to recordings of male sparrow calls. The experiments found that giving male sparrows a drug that increases serotonin or serotonin-receptor binding leads to less aggressive behavior. Among monkeys as well, more aggressive individuals were found to have lower levels of compounds derived from serotonin in their spinal fluid.

The data linking serotonin levels to animal aggression suggested that serotonin-pathway gene variants might contribute to human differences in aggressive behavior. One suspected genetic variant was the serotonin-transporter-length variant, which had been linked to vulnerability to depression. Although an early study did show a link between one serotonin-transporter variant and aggression in children, subsequent studies failed to replicate this association, and the connection between serotonin-transporter gene variation and aggressive behavior remains controversial.

More convincing evidence of a genetic link to human aggression was found by studying variants of a serotonin-*receptor* gene called *HTR2B*. A 2010 NIH study that included over 200 Finnish men[*] convicted of violent crimes found that 7.5% of the offenders (17 out of 228) had the same mutation in one copy of their *HTR2B* genes, a mutation that leads to a truncated and nonfunctional HTR2B protein. In contrast, only 2.4% of a group of Finnish nonoffenders (7 out of 295) had *HTR2B* genes with the disabling mutation.

It's worth emphasizing that the NIH study didn't demonstrate that the *HTR2B* mutation *causes* violent or aggressive behavior. Rather, the results suggested that the mutation *increases the vulnerability* to such behavior. After all, seven of the Finnish nonoffenders also had the *HTR2B* variant. In addition, as this mutation has only been observed among a small percentage of Finns (who most likely all descended from a single individual who had

---

[*] Since discovering biologically important genetic variants is often easier in small, relatively inbred populations than in large heterogeneous ones, U.S. agencies such as the NIH sometimes fund genetic research on people from other countries, such as Finland.

the original mutation, centuries ago), and never in someone who was not of Finnish descent, it is clearly not a common factor in violent behavior. That said, the fact that the mutation is found almost three times as frequently among offenders than in the general Finnish population does lend credence to the idea of a link between serotonin signaling and aggression.

MAO-A AND AGGRESSION

Another serotonin-regulating protein linked to aggressive behavior is MAO-A. Recall that MAO-A degrades both serotonin and noradrenaline, so MAO-A malfunctioning might well affect susceptibility to aggressive behavior. Scientists from the University of Southern California tested this idea by engineering knockout mice with nonfunctional *MAO-A* genes. The researchers were able to observe the impact of MAO-A on aggressiveness, as the male *MAO-A* knockout mice consistently displayed increased aggressive behavior toward other males than normal male mice.

Insufficient MAO-A not only increases aggressive behavior in mice; it also contributes to aggressiveness in people. In fact, the violent behavior in the family of Marieke was caused by a rare disabling mutation in *MAO-A*. This mutation affected only one nucleotide in the *MAO-A* gene sequence; a single C base was changed to a T. This variation caused a C-A-G sequence within the gene (which normally results in a glutamine amino acid being added to the MAO-A protein) to be converted to a T-A-G sequence, which you may recall is a stop codon. As a result, the MAO-A protein was truncated and nonfunctional.

The results of genetic testing of Marieke's family (depicted in figure 24.1) showed that the aggressive family members all shared the same *MAO-A* gene mutation. Since the *MAO-A* gene is on the X chromosome and men only have a single X chromosome, the affected men did not produce *any* MAO-A protein. The genetic analyses also showed that the unaffected men in the family all had the common, functional variant. Since the female relatives with the *MAO-A* mutation all had a second X chromosome with a normal *MAO-A* gene, their cells produced at least some MAO-A protein, which might explain why none of these women displayed aggressive behavior.

From the perspective of the serotonin deficiency model of depression, diminished MAO-A activity might be expected to *decrease* susceptibility to aggressive behavior. After all, MAO-A destroys serotonin, so diminished MAO-A activity should result in more serotonin. If serotonin boosts feelings of well-being, elevated serotonin levels would presumably lead to less aggression, but this is not the case. Animals (and people) with less MAO-A have increased tendencies toward aggression.

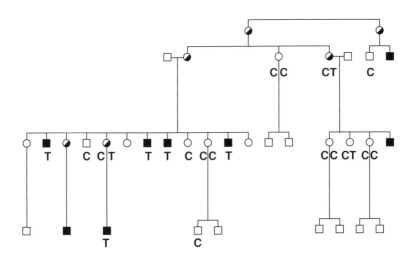

Figure 24.1. Family tree of the Dutch family of Marieke. In the figure, squares indicate males, while circles indicate women. Men showing symptoms of violent aggression are shown as squares that are solid black. The variant (C or T) at the MAO-A mutation is shown for each family member who was available for testing. Men, having only one X chromosome, have only a single variant annotated, while women have two. Half-black circles indicate women who were carriers (they themselves showed no symptoms but could have male offspring who were affected). Note that every affected male who was tested had a T, while the unaffected males all had a C, and the carriers had one C and one T.

The principal reason why *MAO-A* knockout mice are more aggressive is that serotonin acts in other critical signaling pathways besides those leading to enhanced feelings of well-being. In particular, serotonin is a key player in regulating brain development in young animals (and presumably in young people as well). For example, in a series of experiments performed in the 1990s, scientists from the French National Scientific Research Center showed that the primary cause of the aberrant behavior in *MAO-A* knockout mice was excess serotonin in the mouse's brain during early development, leading to the malformation of important substructures in its brain.

A secondary reason that decreased MAO-A activity may lead to increased aggression is that serotonin is not the only target of MAO-A. MAO-

A's targets also include dopamine, adrenaline and noradrenaline. Since noradrenaline is a fundamental component of our fight-or-flight response, finding a link between diminished MAO-A activity and increased aggression is less surprising. Noradrenaline's link to aggressive behavior has been further supported by experiments involving a protein called COMT. Like MAO-A, COMT destroys adrenaline and noradrenaline, but, unlike MAO-A, COMT doesn't degrade serotonin. Consequently, by engineering *COMT* knockout mice, scientists from Rockefeller University could study the effect of increasing noradrenaline without affecting serotonin, and they found that the *COMT* knockout mice were, in fact, more aggressive than normal mice.

MAO-A VARIANTS, CHILDHOOD ABUSE AND ADULT AGGRESSION

The *MAO-A* mutation in Marieke's family is extremely rare; in fact, the variant has only been found in this one family. A different *MAO-A* variant (the repeated 30–base pair variant we encountered as a possible factor in panic disorders) is much more common. Considering the impact of the rare *MAO-A* mutation in the family of Marieke, some scientists wondered whether this more common *MAO-A* variant, which does decrease production of MAO-A protein, might also tend to make men more aggressive.

Although the idea that shorter *MAO-A* variants lead to increased aggression seemed plausible, studies attempting to demonstrate this connection were largely unsuccessful. Unless an *MAO-A* variant completely eliminated MAO-A protein production, as in the family of Marieke, MAO-A protein concentration did not in general appear to be correlated with aggressive male behavior. Yet, perhaps decreased MAO-A levels might be linked to aggressiveness in individuals who had *also* experienced a severe negative life experience, such as childhood trauma. To test this idea, a research team from King's College London (the same group that found the association between serotonin-transporter genetic variants and depression in victims of childhood mistreatment) studied the DNA and behavior patterns of 150 men who had been abused as children. They found that abuse survivors with the shorter *MAO-A* variant, which produces less MAO-A protein, were twice as likely to have been diagnosed with a "conduct disorder" and three times as likely to have been convicted of a violent crime as those with the longer variant. In contrast, among men who had *not* been abused as children, the same level of antisocial behavior was observed, no matter which *MAO-A* gene variant they had.

Several other research groups were also able to find associations between adult violent behavior and the combination of the shorter *MAO-A* variants and childhood abuse, though a few could not. One intriguing report replicated the King's College findings, but only among men of European descent (which was the genetic background of the men in the King's College study). Clearly gene-environment interactions can be complicated. That said, a 2007 meta-analysis of eight studies confirmed that shorter *MAO-A* gene variants are correlated with increased adult antisocial behavior only among survivors of trauma or childhood abuse.

BIOLOGY, VIOLENCE AND THE LAW

If someone has a genetic variant or other biological abnormality that makes them more susceptible to violence and other aggressive behavior, should they still be held responsible for their actions? Or should a genetic propensity for violence, just like insanity, be a mitigating excuse for aggressive criminal behavior? At least in the world of fiction, legal defenses based on such biological considerations are no longer uncommon. Anyone who regularly watched the television series *Law and Order* or has seen similar courtroom dramas will be familiar with the defense attorney's last-ditch argument: He had no choice. He was compelled to commit his violent acts because of his "bad genes" or his brain disease.

Outside of the world of fiction as well, some accused criminals have adopted legal defenses based on their genes. The first was Stephen Mobley, an accused murderer, who in 1994 asked the court to test his *MAO-A* gene. (This was shortly after the publication of the report describing the family of Marieke) Mobley argued that since other members of his family had been violent, he was genetically fated to be violent as well, and that his *MAO-A* gene would prove it. He claimed that he was not responsible for his actions and should be treated more leniently. Unfortunately for Mobley, the Georgia courts disagreed; Mobley's DNA was not tested, and he was executed on March 1, 2005.

Subsequently, other accused or convicted criminals have attempted to introduce genetic evidence in hopes of being found not guilty or of receiving lesser punishment. Most judges have denied these requests. Recently, though, some judges have permitted the consideration of genetic variants in determining a defendant's responsibilities for his actions. One such case was the 2009 murder trial of Bradley Waldroup. Waldroup had brutally murdered a friend of his estranged wife and attempted to kill his wife as well. He was charged with felony murder and attempted first-degree murder.

During the trial, Waldroup's lawyer argued that Waldroup had the *MAO-A* variant linked to vulnerability to violence in childhood-trauma survivors and that Waldroup had, in fact, been abused as a child. Whether this evidence influenced the jury is unclear. One juror was quoted as saying the *MAO-A* data "was just one piece of evidence" in the jury's deliberations. That said, Waldroup was ultimately convicted of voluntary manslaughter and attempted second-degree murder, for which he was sentenced to 32 years in prison, rather than of felony murder, which could have led to the death penalty.

Only six months after the Waldroup trial, another court decision was influenced by genetic evidence. This case involved the sentencing appeal of one Abdelmalek Bayout in a murder trial in Italy. Bayout had confessed to the 2007 murder of Walter Novoa Perez, whom Bayout believed had taunted him and beaten him up. Bayout was sentenced to nine years in prison. Bayout's lawyers appealed the sentence, saying that his genetic variants predisposed him to violence. In September 2009, Pier Reinotti, the judge who reviewed the appeal, reduced Bayout's sentence to eight years, stating that he found the defendant's "*MAO-A* evidence particularly compelling."

Whether increased consideration of genetic variation will make the legal system more just is by no means clear. Many scientists think the opposite is true. A few months after Bayout's resentencing, eight leading geneticists published an editorial in the *European Journal of Human Genetics* deploring the use of genetic variants as grounds for modifying a criminal sentence. No comparable editorial appeared by scientists supporting Judge Reinotti's decision. It's also unclear whether considering genetic variants necessarily helps defendants. Judges or jurors may reason, "Well, he has this genetic variant that makes him vulnerable to impulsive violence. Consequently, he's more likely to commit another violent crime in the future, so we should lock him up for a *longer* time." Indeed, the entire subject of linking genetic variants with legal or moral responsibility is filled with uncertainty. Perhaps the only thing we can be confident of is that as more is learned about the genetics underlying aggression, this topic will play an ever greater role in society's decisions. As members of society making those decisions, it behooves us to understand the science underlying aggression as best we can.

BRAIN ABNORMALITIES, AGGRESSION AND VIOLENT BEHAVIOR

It seems unlikely that DNA variants, such as those in the *MAO-A, HTR2B* or *AR* genes, are the only biological factors influencing human susceptibility to aggression or other antisocial behavior. Considerable evidence also suggests that antisocial behavior may be linked to structural changes in the brain. Although this evidence is often anecdotal,* it is intriguing. Consider the story of Charles Whitman. As a child, Whitman was a good student. He was an Eagle Scout and an excellent pianist. Although Whitman got in trouble for gambling and threatening another soldier while he was a Marine, for most of his life he was a law-abiding individual. Then, on the morning of August 1, 1966, after murdering his mother and his wife, Whitman climbed to the observation deck of the Tower at the University of Texas, in Austin, and started shooting, killing another 14 people before being shot to death by police.

What makes Whitman's story different from that of other mass murderers is that Whitman left a note, which was found after his death. In the note, Whitman complained of headaches and requested that his body be autopsied. He wrote that he hoped the autopsy might shed light on his awful acts, which Whitman claimed he, himself, couldn't understand. On August 2, his body was autopsied, and Whitman was found to have had a malignant brain tumor. The tumor was located in the hypothalamus, adjacent to the amygdala, a part of the brain linked to aggression. A medical commission was organized in an attempt to understand the cause of this tragedy. Although the commissioners stated they lacked sufficient evidence for a definitive conclusion, they speculated that Whitman's brain tumor contributed to his actions. Of course, most murderers don't have brain tumors, and people with brain tumors rarely become violent. Whitman's tumor might well have been a coincidence, but it's an intriguing one.

The report of an even more remarkable possible link between a brain tumor and aberrant behavior appeared in the journal *Archives of Neurology* in 2003. The article described an apparently healthy, married schoolteacher I'll call Mr. Jones, who for the first 40 years of his life had lived a normal, unexceptional life. At the age of 40, though, Jones suddenly found himself uncontrollably attracted to young children. Initially, this sexual obsession was limited to secretly viewing child pornography. Later, he began making veiled sexual advances to children, and Jones was arrested and convicted of child molestation.

So far Jones' story, though sad, was not unusual. What made his story different was that shortly after his conviction, Jones began to complain of headaches. An MRI scan revealed that he had a brain tumor. The tumor

---

* The problem with anecdotal evidence is that it comes from just a single association between behavior and an alleged cause. As a result, the association may well be coincidental, rather than an example of cause and effect.

was removed and Jones' headaches stopped. Remarkably, so did his sexual obsession with children. He was released and resumed his life. One year later, though, Jones realized that his abnormal sexual urges had returned. He was given another brain scan, and the brain images showed that his tumor had recurred. Shortly thereafter, the second tumor was also surgically removed, and Jones' sexual obsessions disappeared once again.

Was this a highly unusual story? Absolutely. Were the simultaneous occurrences of Jones' brain tumor and his sexual obsessions simply a coincidence? Perhaps, but, if so, it was a remarkable coincidence. Nonanecdotal evidence also suggests that brain abnormalities may contribute to violent and antisocial behavior. The data comes from brain images of hundreds of violent criminal offenders. A recent review of 43 brain-imaging studies reported consistent abnormalities in a region of the brain called the prefrontal cortex. The studies found that the criminal offender tended to have a less fully developed prefrontal cortex, with less biochemical activity. Since the prefrontal cortex is associated with guiding our thoughts and inhibiting inappropriate behavior, the abnormal brain images again suggest that antisocial and violent tendencies may have partly biological origins. As we'll explore in the following chapter, the evidence that biology affects our behavior raises challenging moral and philosophical questions. If biological factors can alter who we are and what we do, should we still be held responsible for our actions? In fact, can we still even consider ourselves autonomous beings with free will – or are we just biological automatons?

25

WHAT ARE SCIENCE'S LIMITS?

Molecular biology is beginning to provide answers to fundamental questions about what it means to be human. Yet science can't answer every question. It is vital for us to understand science's limits, lest we become frustrated and impatient and tempted to reject science altogether. In this last chapter, we'll look at questions science hasn't answered. Some are scientific questions with answers still beyond our grasp. Others are moral and philosophical questions, which lie outside the realm of science but involve aspects of the human condition that can be studied scientifically. These moral and philosophical questions include some of the most challenging society faces. Although this chapter won't be able to answer these questions, hopefully, it will provide guidelines showing how science can help us address them.

SCIENTIFIC PROGRESS IS NEVER FAST ENOUGH

The last half century has seen explosive growth in our understanding of human biology and disease. To appreciate this, one needs to look no further than the leading research database of inherited diseases and conditions, the Online Mendelian Inheritance in Man (OMIM) database.* When OMIM was started in started, in 1966, it included 1400 entries, almost none of which contained the description of an underlying DNA variant or molecular mechanism. In fact, in 1966, the techniques for reading DNA sequences hadn't even been developed yet. Less than 50 years later, OMIM contains 22,000 genetic conditions, and over 18,000 have been linked to a specific gene or DNA variant. Many of these conditions are devastating and life threatening. Some, such as cystic fibrosis, Huntington's disease and sickle cell anemia, are well known. Others are extremely rare conditions about which very little is known.

Despite this astounding progress, the origin of over 3500 inherited conditions remains unknown. Meanwhile, more complex diseases, such as cancer, heart disease or diabetes, which involve multiple interacting genes or environmental factors, are even less well understood. And even where a disease's molecular roots have been identified, translating that knowledge into effective medical treatment has often been slow.

As a result, while our knowledge has grown rapidly, scientific progress is always too slow for those afflicted with disease. They may well want to push science to move faster, by speeding new drugs through regulatory approval or by relying on evidence solely from animal experiments or limited human studies. Although such strategies may accelerate scientific discovery and the introduction of desperately needed drugs, they may also lead to the treatments that are ineffective or even harmful.

As we've seen in examples ranging from the difficulty replicating caloric restriction experiments in primates to the inability to reproduce the link between prenatal drug use and sexual orientation, hypotheses in biology need to be carefully and repeatedly confirmed. Rushing science without consistent verification can have disastrous consequences, as occurred, for example, when the drug thalidomide was introduced without adequate testing, leading to the birth of thousands of infants with terrible deformities. Of course, for a family faced with a devastating disease, waiting for a multiyear experiment or clinical trial may not seem a viable option. Nevertheless, we must be careful not to push science too fast and rush to treatments that may have more risks than benefits.

Those facing diseases with no known treatment are not the only ones at risk of trying to push science too fast. Ironically, the very speed of recent discoveries has led us to expect that researchers can already answer every

---

* Named in honor of Gregor Mendel, the 19th-century European monk who discovered the fundamental laws of genetic inheritance.

scientific question. We expect science to predict the future of our bodies and brains from our DNA and to bring us pills that make us smarter and keep us from aging. *Someday* such scientific feats may be possible, but expecting them now will only lead us to disappointment as well as ill-advised and dangerous decisions.

Modern society provides us with some protection from such unrealistic expectations. We have laws that prevent pharmaceutical companies from marketing drugs that haven't gone through extensive long-term testing and regulations that stop consumer DNA-testing companies from making unsubstantiated claims about the medical significance of our genetic variants. Beyond government protection, it is equally important that we protect ourselves by learning how scientific research works and what its limits are. By doing so, we'll be able to accept that the rate at which science unravels nature's mysteries depends solely on the skill, ingenuity and perseverance of the scientific community – and not on how much we may want or need to know the answer.

MORAL CHALLENGES CAN POSE AS SCIENTIFIC QUESTIONS

Although its pace of discovery may not be as fast as we like, the scientific method eventually does lead to answers to even the most challenging biological problems. Some questions, though, will never be answered by scientific research. These include many important moral and ethical questions that our new knowledge of biology is thrusting upon us. For example:

- When should criminal offenders not be held responsible for their actions? Only if they are mentally incompetent? Or should genetic variations or brain abnormalities be considered excuses for violent or antisocial behavior?

- Under what circumstances should society restrict prospective parents from terminating a pregnancy? Should abortion be permitted if there is only a *possibility* that the child will get a disease as an adult? What if the unborn child simply doesn't have the gender desired by the parents? What if a future genetic test enables parents to determine that the child isn't likely to excel in athletic or intellectual pursuits, or that the child's future sexual orientation isn't that desired by the parents?

- How do we protect personal privacy when we can learn about someone simply by reading their DNA? Should researchers, the government or insurance companies be allowed to test our DNA? How should we balance genetic privacy with the needs of researchers attempting to cure deadly diseases? Should we be allowed to keep our DNA sequence secret even if that information might solve a crime or exonerate an unjustly convicted individual?

- Should genetic research be restricted to prevent its findings from being exploited by those with racist or otherwise bigoted agendas?
- Should companies be allowed to patent information encoded in our DNA?
- What types of animal experiments are morally acceptable? When are scientific advances valuable enough to justify deliberately hurting a nonhuman primate or other animal? Are the potential benefits of experiments on transgenic primates worth the scientific risks of creating a "Frankenstein" and the moral risks of creating a new form of life?

These are not scientific questions, and we should not expect science to resolve them for us. They are moral and ethical questions. Scientific questions address the way the world *is*, while moral questions help us identify the values we believe the world *should* have. Yet moral and scientific questions are often linked. Even if science can't resolve moral or ethical questions, it is important for us to be able to isolate the scientific aspects of such questions so that we can understand the scientific implications of our moral choices.

BIOLOGY AND BEING HUMAN

Beyond its impact on ethics and morality, molecular biology is altering our very concept of what it means to be human. As we've seen, almost every aspect of our lives is influenced by interactions of proteins and nucleic acids in our cells. We need to be prepared for the possibility that our biology has greater impact on who we are than we had believed. It is becoming ever more difficult to believe that people are unique and unlike the other creatures on this planet. If humans are, in fact, created to mirror the "image of God," then it seems likely that chimpanzees, chickens and worms are created in that image as well.

Though some people may find this image of humanity attractive, others may well find it unsettling. Its implications have been noted before, perhaps nowhere more eloquently than in Steven Pinker's 2002 book, *The Blank Slate*. Pinker suggested that our increasing knowledge of biology is feeding some of our deepest fears regarding the essence of human nature. We are learning that humans are not created equal, at least not biologically speaking.

Each of us is born with a unique genetic recipe in our DNA and an epigenetic pattern in our chromosomes. This causes some people to be predisposed to disease while others are not. Even normal, healthy individuals do not appear to be born with the same genetic and epigenetic gifts, for example, in athletic or intellectual potential. At the same time that we are learning how much humans differ from one another we are also discovering how biologically similar humans are to other animal species. And we are finding out that many of these animal species that we resemble, at least at the level of molecular biology, exhibit behavior that is less appealing than we had imagined, including infanticide, rape and cannibalism.

Some of the lessons we might choose to draw from these observations are disturbing. We might decide that if humans are not all created the same, then prejudice and discrimination are justified. We might conclude that if we are not so different from other animals, then cruel and violent human behavior is inevitable. If we are simply the products of our biology, we might even find it hard to believe that we have free will and are responsible for our actions. In fact, if we are shaped just by our genes and our brain biochemistry, can we still say that we are the masters of our own destinies and that our lives can have meaning and purpose? Can we claim that we make real choices in our lives, or are we compelled to see ourselves as biochemical robots, in which our feelings of love, happiness or fear are the result of molecular interactions in our brain?

These disturbing arguments have led some people to reject the discoveries of the biological nature of humans. In effect, they say that if science presents such a dismal view of the human condition, then we should simply reject science. Fortunately, this grim picture of human nature is extreme and unjustified, and we do not need to abandon science to continue to believe that people can control their lives. Molecular biology doesn't teach that our actions or feelings are *determined* by genetic and epigenetic events in our cells. Rather, we are learning that biological phenomena *influence* who we are and how we act. Examples of extreme biological conditions that overwhelmingly affect our behavior, like those of Mr. Jones and Marieke's relatives described in the last chapter, are *very* rare. Yes, we are affected by genetic and other biological factors outside our control, but science doesn't dispute that we do make choices in our lives. We are responsible for the consequences of our choices, and we can still make them in ways that give our lives meaning.

Rejecting science is not the way to address the uncomfortable implications of its discoveries. Instead, we must learn how to separate the factual findings of science from the moral issues raised by those findings. We need to accept that we are biological animals and that our lives are influenced by microscopic events within our bodies and brains. This understanding should not lead us to a fatalistic view of our lives or a desire to remain ignorant of how our biology affects us. That we are born with different genetic potentials does not imply that some people are better than others, and it certainly does not justify discrimination, oppression or violence against those who are different. Genetic diversity does not give any group of people the moral right to take advantage of others who simply were born with a different set of genetic or epigenetic variations in their cells. In fact, biology teaches us that diversity is one of nature's most effective defenses against environmental change. Hopefully, rather than seeking the illusion that we are unaffected by our biology or that we are all biologically the same, our scientific knowledge will provide us with a more profound understanding of ourselves. If biology has taught us anything, it is that we should celebrate the entire diverse spectrum of humanity that nature has created, and treat all of its members with kindness, tolerance and respect.

ACKNOWLEDGEMENTS

I am grateful to the many people who have guided and assisted me on the path that ultimately has led to this book. First of all, I'd like to thank all the talented scientists whose hard work and ingenuity led to the dramatic discoveries described in the book. For reasons of space I have chosen to note researchers' names only in those cases where it seemed to be an important part of the story. In many cases, scientists' names will be found in the Chapter Notes and Further Readings, at the end of the book, though there, as well, for reasons of lack of space I have annotated only a small fraction of the relevant research papers. My apologies to all the researchers whose names and papers are not noted in the book. I hope they will understand that the omissions in no way reflect a lack of profound respect for the importance of their work.

I'd like to express my appreciation to all of my teachers, from Steven Weinberg and Tony Duncan who years ago introduced me to the extraordinary world of particle physics to Todd Lowe and Sean Eddy, my mentors as I transitioned into becoming a biologist. I am thankful to the many biologists, including Han Brunner, John Crabbe, Simon Fisher, Richard Holt, Peter Koopman, Susan Lessin, Farah Lubin, Nina Riehs, Antoine Roux, Karl Skorecki, Moshe Szyf, Joe Tsien and Kelli Vaughn, who reviewed parts of the book and whose feedback helped me eliminate errors that crept into the manuscript. Of course, any errors that may remain are solely my responsibility. I am also indebted to my nonbiologist readers: Eileen Feldman, Marc Fine, Bob Henninge, Larry Rosenberg, Ralph Small and especially Ferd Wulkan. Their insightful comments have made the book clearer and easier to understand. I am particularly appreciative of Jay Chesavage, Scott and Ellen Fowler and Beverly Littlejohn for reading the entire manuscript and offering countless helpful suggestions and improvements. I would like to thank Lee Hotz, Sandra Blakeslee and the other teachers and participants from the 2013 Santa Fe Science Writing Workshop for the helpful feedback I received on an early version of the book.

I am grateful for the outstanding help I have received in the book's production. Leslie Tilley has been everything one could want in a top-notch editor and more – consistently enabling me express my thoughts more clearly and gracefully and ensuring that the facts and statistics I quoted were accurate. Paul Brittenham's excellent illustrations contribute greatly to making the book clear and understandable. His cheerful willingness to redraw illustrations (and, if necessary, redraw them again) until they were exactly what I had envisioned has been greatly appreciated. I have also been fortunate to find someone as talented as Glenn Edelstein for envisioning and executing such an attractive cover for the book. I'd also like to thank the team at Reality Information Systems and Mary Harper for preparing the book's interior layout and index, respectively.

I would like to acknowledge the following individuals and institutions for granting permission to reproduce artwork:

Cover and title page photographs: Thinkstock by Getty Images for use of the photographs of the bicyclist, the boy blowing bubbles, and the mother and infant.

Cover and title page photograph: Shutterstock.com for use of the photograph of the elderly couple.

Figure 3.2: Rabson Wuriga and Modreck Maeresera. Photograph by Sandy Leeder, courtesy of Kulanu, Inc.

Figure 4.2: Public Domain. Originally from /Alexander Palace Forums/, Boasson and Eggler, St. Petersburg, Nevsky 24 (1913).

Figure 9.1: Elsevier for use of figure 1 of L. McGraw and L. Young, "The prairie vole: an emerging model organism for understanding the social brain," *Trends in Neurosciences* 33 (2): 103–109, ©2010, and Dr. Lloyd Glenn Ingles and The California Academy of Sciences, ©1999 for use of the montane vole image.

Figure 11.1: The National Academy of Sciences for use of figure 1 of R. Agate et al., "Neural, Not Gonadal, Origin of Brain Sex Differences in a Gynandromorphic Finch," *PNAS* 100 (8): 4873–4878, ©2003.

Figure 15.1: Dr. Antoine Roux.

Figures 16.1 and 17.3: John Wiley and Sons for use of figures 2 and 4 from R. Holt "Is Human Growth Hormone an Ergogenic Aid?" *Drug Testing and Analysis* 1: 412–418 (2009).

Figures 16.2 and 16.3: The Public Library of Science for use of figure 3 from S-J. Lee "Quadrupling Muscle Mass in Mice by Targeting TGF-ß Signaling Pathways" *PLoS ONE* 2(8): e789 (2007); figure 1 from D. Mosher et al. "A Mutation in the Myostatin Gene Increases Muscle Mass and Enhances Racing Performance in Heterozygote Dogs" *PLoS Genetics* 3 (5): e79 (2007).

Figure 17.1: The Sports Museum of Finland.

Figures 20.2, 22.2 and 24.1: The American Academy of Arts and Sciences for use of figure 3 from D. Hamer et al. "A Linkage Between DNA Markers on the X Chromosome and Male Sexual Orientation" *Science* 261 (5119): 321–327 (1993); figures 1 and 2 from A. Caspi et al. "Influence of Life Stress on Depression: Moderation by a Polymorphism in the 5-HTT Gene" *Science* 301: 386 (2003); figure 1 from H. Brunner et al. "Abnormal Behavior Associated with a Point Mutation in the Structural Gene for Monoamine Oxidase A" *Science* 262: 578 (1993).

I've saved my biggest thank-yous for last. To Susan Lessin, my wife and life partner, I owe an enormous debt for providing the necessary encouragement and emotional support when the task of creating this book seemed overwhelming. And last but not least, I want to thank you, dear reader, for having stayed the course with me in reading this book. I hope you have enjoyed reading about the amazing new world of molecular biology as much as I've enjoyed writing about it. And I have one more thank you to express, this time in advance. If you've enjoyed this book and found it useful to you, I'd be grateful if you'd take a few moments to share your thoughts about it on Amazon, or whatever book review site you use, to help me spread the word about it. Thanks again.

CHAPTER NOTES AND FURTHER READING

All the stories in *Sex, Love and DNA* were originally reported in the scientific literature. Many have been written about in the popular press as well. Most of the sources cited here are research review articles, which integrate the results of multiple scientific investigations. Although primarily written for scientists, review articles are often understandable by the nonspecialist. When reviews are unavailable, references to the original research articles are included. Such papers are harder for the nonspecialist to follow, but one can often get a sense of the experimental results and their significance by just reading the introduction and discussion sections.

Abstracts and links to most of the papers can be found using PubMed, the search engine of the National Library of Medicine website (www.ncbi. nlm.nih.gov/pubmed). Often they can be found simply by doing a web search on the article's title as well.

Unfortunately, obtaining copies of entire articles can become expensive; in fact, downloading one or two articles may cost more than buying this book. Although this situation is improving, with fewer journals charging fees, reading scientific journals is still expensive unless one has access to a university or research library.

I have also included some references to popular retellings of the scientific results. I will also be periodically updating the stories and science described in *Sex, Love and DNA* on the book's website at PeterSchattner. com. Popular accounts have the obvious advantage of being easier to read. Their disadvantage is that they also often oversimplify the scientific concepts and can be very misleading – or simply wrong. Consequently, popular accounts of scientific advances (including this book, of course) should be read with caution and a healthy dose of skepticism..

On the other hand, there are many excellent textbooks and books for the general public available that present the fundamentals of molecular biology. Among the textbooks, B. Alberts, *Molecular Biology of the Cell*, (2008), is especially well written. D.P. Clark, *Molecular Biology Made Simple and*

Fun, (1997), and R.F. Kratz, *Molecular and Cell Biology for Dummies*, (2009), are good introductions that are less technical and more oriented to the general reader. Wikipedia is another excellent resource on many of the topics described in this book (though the caveat about maintaining skepticism while reading is relevant here as well.)

Many popular books also recount stories in genetics, including M. Ridley, *Genome*, (1999), P. Reilly, *The Strongest Boy in the World*, (2006), and S. Kean, *The Violinist's Thumb*, (2012).

F.S. Collins, *The Language of Life*, (2010), is an excellent general introduction to genetics focused primarily on health and disease prevention. P. Reilly, *Abraham Lincoln's DNA*, (2000), is another good, though somewhat dated, introduction to genetics that emphasizes ethical and legal issues.

Chapter 2: Can a Protein Save You from AIDS?

To learn more about the basics of proteins and the genetic code see chapters 3–5 of Alberts, *Molecular Biology of the Cell*, (2008), and chapters 1 and 6 of Kratz, *Molecular and Cell Biology for Dummies*, (2009). Chapter numbers in Alberts' *Molecular Biology* refer to the 5th edition, though earlier editions generally contain very similar material. In particular, you can search (but not read) the 4th edition for free on the National Library of Medicine Website at www.ncbi.nlm.nih.gov/books/NBK21054/.

For more about Steve Crohn, the AIDS survivor with the *CCR5* gene variant, see the *Nova* documentary "Surviving AIDS" and E. Woo, "Obituaries; Stephen Crohn, 1946 - 2013; Immune to HIV but Not Its Tragedy," *Los Angeles Times* (2013) (which also notes Crohn's subsequent suicide).

The experimental data describing how *CCR5* gene variants provide HIV immunity is reviewed in R. Liu, "Homozygous Defect in HIV-1 Coreceptor Accounts for Resistance of Some Multiply-Exposed Individuals to HIV-1 Infection," *Cell* 86 (1996), and M. Samson, "Resistance to HIV-1 Infection in Caucasian Individuals Bearing Mutant Alleles of the CCR-5 Chemokine Receptor Gene," *Nature* 382 (1996).

G. Hutter, "Eradication of HIV by Transplantation of CCR5-Deficient Hematopoietic Stem Cells," *Scientific World Journal* 11 (2011), describes the successful bone-marrow-transplant treatment of Timothy Ray Brown.

See W.G. Glass, "CCR5 Deficiency Increases Risk of Symptomatic West

Nile Virus Infection," *The Journal of Experimental Medicine* 203 (2006), for more information on the link between CCR5, HIV and West Nile virus.

Chapter 3: Who Are Our Fathers?

More detailed descriptions of the genetics of the Y chromosome can be found in chapter 21 of Alberts, *Molecular Biology of the Cell*, (2008), and chapter 16 of Kratz, *Molecular and Cell Biology for Dummies*, (2009).

The early Y chromosome research indicating the Middle Eastern origins of the Jewish paternal ancestry is described in M.F. Hammer, "Jewish and Middle Eastern Non-Jewish Populations Share a Common Pool of Y-Chromosome Biallelic Haplotypes," *Proceedings of the National Academy of Sciences of the United States of America* 97 (2000).

The later papers that extended the research using data from all 46 human chromosomes are G. Atzmon, "Abraham's Children in the Genome Era: Major Jewish Diaspora Populations Comprise Distinct Genetic Clusters with Shared Middle Eastern Ancestry," *American Journal of Human Genetics* 86 (2010), and D.M. Behar, "The Genome-Wide Structure of the Jewish People," *Nature* 466 (2010).

One recent study, E. Elhaik, "The Missing Link of Jewish European Ancestry: Contrasting the Rhineland and the Khazarian Hypotheses," *Genome Biology and Evolution* 5 (2013), still finds support for the Khazar theory of Ashkenazi Jewish origins, but its methodology and findings have been widely challenged within the scientific community.

The Cohan modal haplotype is described in M.G. Thomas, "Origins of Old Testament Priests," *Nature* 394 (1998).

The DNA evidence supporting the connection of the Lemba tribe to the Jews was reported in M.G. Thomas, "Y Chromosomes Traveling South: The Cohen Modal Haplotype and the Origins of the Lemba--the "Black Jews of Southern Africa"," *American Journal of Human Genetics* 66 (2000). The Lemba story is recounted for the general public in the books H. Ostrer, *Legacy*, (2012), and M. Le Roux, *The Lemba*, (2003). For a recent scientific review, see M.F. Hammer, "Extended Y Chromosome Haplotypes Resolve Multiple and Unique Lineages of the Jewish Priesthood," *Human Genetics* 126 (2009).

T. Zerjal, "Y-Chromosomal Insights into the Genetic Impact of the Caste System in India," *Human Genetics* 121 (2007), and D.R. Carvalho-Silva, "The Grandest Genetic Experiment Ever Performed on Man? - a

Y-Chromosomal Perspective on Genetic Variation in India," *International Journal of Human Genetics* 8 (2008), present the genetic investigations of the history of the Indian caste system.

The research behind the Genghis Khan Y chromosome is described in T. Zerjal, "The Genetic Legacy of the Mongols," *American Journal of Human Genetics* 72 (2003).

Chapter 4: Who Are Our Mothers?

For more information on mitochondrial DNA, see chapter 14 of Alberts, *Molecular Biology of the Cell*, (2008), or chapter 2 of Kratz, *Molecular and Cell Biology for Dummies*, (2009).

Vy Higginsen's story originally appeared in a segment on the television show *60 Minutes* and can be found on the CBS website at www.cbsnews.com/stories/2007/10/05/60minutes/main3334427.shtml.

The study that attempted to determine African American maternal ancestry using mitochondrial DNA analysis is B. Ely, "African-American Mitochondrial DNAs Often Match mtDNAs Found in Multiple African Ethnic Groups," *BMC Biology* 4 (2006). For more information on whole-genome methods to trace African American ancestry, see O. Lao, "Evaluating Self-Declared Ancestry of U.S. Americans with Autosomal, Y-Chromosomal and Mitochondrial DNA," *Human Mutation* 31 (2010).

The history of Anastasia is told in G. King, *The Resurrection of the Romanovs*, (2011). The scientific papers identifying the Romanoff remains and proving that Anna Anderson was not a Romanoff are E.I. Rogaev, "Genomic Identification in the Historical Case of the Nicholas II Royal Family," *Proceedings of the National Academy of Sciences of the United States of America* 106 (2009), and M.D. Coble, "The Identification of the Romanovs: Can We (Finally) Put the Controversies to Rest?," *Investigative Genetics* 2 (2011).

The mtDNA study finding mainly European origins in extreme maternal Jewish ancestry is M.D. Costa, "A Substantial Prehistoric European Ancestry Amongst Ashkenazi Maternal Lineages," *Nature Communications* 4 (2013).

Chapter 21 of Alberts, *Molecular Biology of the Cell*, (2008), and chapters 14 and 16 of Kratz, *Molecular and Cell Biology for Dummies*, (2009), present a more detailed introduction to genetic recombination.

The study applying genetic recombination to the history of the Puerto

Rican slave trade is A. Moreno-Estrada, "Reconstructing the Population Genetic History of the Caribbean," *PLoS Genetics* 9 (2013).

Chapter 5: Can We Raise the Dead?

The story of the brief resurrection of the Pyrenean ibex is from J. Folch, "First Birth of an Animal from an Extinct Subspecies (Capra Pyrenaica Pyrenaica) by Cloning," *Theriogenology* 71 (2009).

The identification of dinosaur opsins is described in B.S. Chang, "Recreating a Functional Ancestral Archosaur Visual Pigment," *Molecular Biology and Evolution* 19 (2002).

For more on discovering ancient proteins and using them to develop novel anti-toxins, see M. Goldsmith, "Enzyme Engineering by Targeted Libraries," *Methods in Enzymology* 523 (2013), and R.D. Gupta, "Directed Evolution of Hydrolases for Prevention of G-Type Nerve Agent Intoxication," *Nature Chemical Biology* 7 (2011).

The resurrection of an extinct human virus was reported in Y.N. Lee, "Reconstitution of an Infectious Human Endogenous Retrovirus," *PLoS Pathogens* 3 (2007), and M. Dewannieux, "Identification of an Infectious Progenitor for the Multiple-Copy HERV-K Human Endogenous Retroelements," *Genome Research* 16 (2006).

For a review of the methods of cloning animals, see N.V. Thuan, "How to Improve the Success Rate of Mouse Cloning Technology," *The Journal of Reproduction and Development* 56 (2010).

The attempt to clone the gaur is described in R.P. Lanza, "Cloning of an Endangered Species (Bos Gaurus) Using Interspecies Nuclear Transfer," *Cloning* 2 (2000).

General information on the genetics of the wooly mammoth can be found in A.B.P.G. Lister, *Mammoths*, (2007). The sequencing of the mammoth genome was first published in W. Miller, "Sequencing the Nuclear Genome of the Extinct Woolly Mammoth," *Nature* 456 (2008).

An overview of the field of cryonics is given in D. Shaw, "Cryoethics: Seeking Life after Death," *Bioethics* 23 (2009). For details on the controversies surrounding cryonics and the preservation of Ted Williams' body, see the Wikipedia entry for "Larry Johnson (author)" and references therein.

Chapter 6: Who Owns Our Past?

The ill-fated collaboration between the Havasupai tribe and Arizona State University is recounted in A. Harmon, "Indian Tribe Wins Fight to Limit Research of Its DNA," *New York Times* (2010). The original scientific report on diabetes and the Havasupai is K. Zuerlein, "NIDDM: Basic Research Plus Education," *Lancet* 338 (1991).

The general topics of informed consent and the implications it has on the genomic world are described in E.C. Hayden, "Informed Consent: A Broken Contract," *Nature* 486 (2012).

The court papers describing the attempts to force Duke University to release its research records in the John Trosper Bradley case can be found at http://caselaw.findlaw.com/nc-court-of-appeals/1000946.html.

The discovery of the genetic origins of phenylketonuria is told in H. Bickel, "The First Treatment of Phenylketonuria," *European Journal of Pediatrics* 155 Suppl 1 (1996).

L.F. Ross, "Mandatory Versus Voluntary Consent for Newborn Screening?," *Kennedy Institute of Ethics Journal* 20 (2010), describes the controversies surrounding mandatory newborn screening for sickle cell anemia and other diseases.

For the savior sibling stories of the "Agostinos" and "Molly N" see G.R. Burgio, "Conceiving a Hematopoietic Stem Cell Donor: Twenty-Five Years after Our Decision to Save a Child," *Haematologica* 97 (2012), and Y. Verlinsky, "Preimplantation Diagnosis for Fanconi Anemia Combined with HLA Matching," *Journal of the American Medical Association* 285 (2001), respectively.

P. Jha, "Trends in Selective Abortions of Girls in India: Analysis of Nationally Representative Birth Histories from 1990 to 2005 and Census Data from 1991 to 2011," *Lancet* 377 (2011) presents the data on sex-selective abortions and sex-birth ratios in India.

Chapter 7: Can a Gene Keep You from Thinking Clearly?

For more on messenger RNA, transcription and translation, and regulatory DNA, see any molecular biology textbook, for example, chapters 6 and 7 of Alberts, *Molecular Biology of the Cell*, (2008), or chapters 7 and 18 of Kratz, *Molecular and Cell Biology for Dummies*, (2009).

A.L. Bhakar, "The Pathophysiology of Fragile X (and What It Teaches Us About Synapses)," *Annual Review of Neuroscience* 35 (2012), gives an overview of fragile X disease and the discovery of its underlying mutation.

Martin and Bell's original paper was J.P. Martin, "A Pedigree of Mental Defect Showing Sex-Linkage," *Journal of Neurology and Psychiatry* 6 (1943). The mouse experiments detailing the promising new treatments for fragile X are reported in A. Michalon, "Chronic Pharmacological mGlu5 Inhibition Corrects Fragile X in Adult Mice," *Neuron* 74 (2012).

Chapter 8: How Much Sleep Do You Need?

The discovery of the DEC2 mutation in human short sleepers is described in Y. He, "The Transcriptional Repressor DEC2 Regulates Sleep Length in Mammals," *Science* 325 (2009).

The original paper reporting the golden hamsters with 20-hour sleep cycles is M.R. Ralph, "A Mutation of the Circadian System in Golden Hamsters," *Science* 241 (1988).

Chapter 9: What Is Love?

Z.R. Donaldson, "Oxytocin, Vasopressin, and the Neurogenetics of Sociality," *Science* 322 (2008), presents a good scientific review of the experiments describing the connections between genetic variants and vole behavior.

A description of this work for the general reader can be found in G. Sinha, "You Dirty Love," *Popular Science* 261 (2002).

Studies linking the protein CD38 with prosocial behavior are described in D. Jin, "CD38 Is Critical for Social Behaviour by Regulating Oxytocin Secretion," *Nature* 446 (2007).

For information on the impact of oxytocin and vasopressin in rats and other rodents, see the review by T.R. Insel, "The Challenge of Translation in Social Neuroscience: A Review of Oxytocin, Vasopressin, and Affiliative Behavior," *Neuron* 65 (2010).

The association between oxytocin and autism is described in E. Andari, "Promoting Social Behavior with Oxytocin in High-Functioning Autism Spectrum Disorders," *Proceedings of the National Academy of Sciences of the United States of America* 107 (2010), and H. Kosaka, "Long-Term Oxytocin Administration Improves Social Behaviors in a Girl with Autistic Disorder," *BMC Psychiatry* 12 (2012).

R.P. Ebstein, "Arginine Vasopressin and Oxytocin Modulate Human Social Behavior," *Annals of the New York Academy of Sciences* 1167 (2009), reviews the attempts to link variations in the *AVPR1A* gene with human social behavior.

The studies linking oxytocin with romantic love and relationship communication are I. Schneiderman, "Oxytocin During the Initial Stages of Romantic Attachment: Relations to Couples' Interactive Reciprocity," *Psychoneuroendocrinology* 37 (2012), and B. Ditzen, "Intranasal Oxytocin Increases Positive Communication and Reduces Cortisol Levels During Couple Conflict," *Biological Psychiatry* 65 (2009).

D. Scheele, "Oxytocin Modulates Social Distance between Males and Females," *The Journal of Neuroscience* 32 (2012) describes the experiment showing oxytocin's influence on monogamous men.

The association between *AVPR1A* variants and human pair-bonding is reported in H. Walum, "Genetic Variation in the Vasopressin Receptor 1a Gene (AVPR1A) Associates with Pair-Bonding Behavior in Humans," *Proceedings of the National Academy of Sciences of the United States of America* 105 (2008).

See C. Dulac, "Neural Mechanisms Underlying Sex-Specific Behaviors in Vertebrates," *Current Opinion in Neurobiology* 17 (2007), for a review of behavior in *TRPC2* knockout mice and E.R. Liman, "TRPC2 and the Molecular Biology of Pheromone Detection in Mammals," in *TRP Ion Channel Function in Sensory Transduction and Cellular Signaling Cascades* (2007), for more information on the TRPC2 protein in humans.

A recent critical perspective on the menstrual cycle–synchrony controversy can be found in J.C. Schank, "Do Human Menstrual-Cycle Pheromones Exist?," *Human Nature* 17 (2006).

For a review of the experiments on MHC molecules, pheromones and "smell tests" for human romantic attraction, see J. Havlicek, "MHC-Correlated Mate Choice in Humans: A Review," *Psychoneuroendocrinology* 34 (2009).

The paper claiming that couples with differing MHC profiles are more sexually compatible is C.E. Garver-Apgar, "Major Histocompatibility Complex Alleles, Sexual Responsivity, and Unfaithfulness in Romantic Couples," *Psychological Science* 17 (2006).

Chapter 10: What Is Kindness?

The experiments comparing dog and wolf DNA in an attempt to track down behavior-related genes are described in B.M. Vonholdt, "Genome-Wide SNP and Haplotype Analyses Reveal a Rich History Underlying Dog Domestication," *Nature* 464 (2010). For a more detailed introduction to the phenomenon of positive selection in genetics, see P.C. Sabeti, "Positive Natural Selection in the Human Lineage," *Science* 312 (2006), or the Wikipedia entry for "directional selection."

The experiments on mice with Williams syndrome–like behavior are detailed in E.J. Young, "Reduced Fear and Aggression and Altered Serotonin Metabolism in Gtf2ird1-Targeted Mice," *Genes, Brain, and Behavior* 7 (2008).

L. Edelmann, "An Atypical Deletion of the Williams-Beuren Syndrome Interval Implicates Genes Associated with Defective Visuospatial Processing and Autism," *Journal of Medical Genetics* 44 (2007), describes the human patient with "inappropriately friendly" behavior resulting from a deletion of the *Gtf2ird1* and *GTF2I* genes.

The report about the girl whose WBS deletion did not include *GTF2I* is L. Dai, "Is It Williams Syndrome? GTF2IRD1 Implicated in Visual-Spatial Construction and GTF2I in Sociability Revealed by High Resolution Arrays," *American Journal of Medical Genetics. Part A* 149A (2009).

A.V. Kukekova, "Genetics of Behavior in the Silver Fox," *Mammalian Genome* 23 (2012), is a good scientific review of the silver fox experiments. E. Ratliff, "Taming the Wild," *National Geographic* 219 (2011), is a well-written description of the project for the general reader.

A.V. Kukekova, "Sequence Comparison of Prefrontal Cortical Brain Transcriptome from a Tame and an Aggressive Silver Fox (Vulpes Vulpes)," *BMC Genomics* 12 (2011), reports the early results of the search for genetic variants that correlate with tameness in these foxes.

The paper describing the experiments of "rat empathy" is I. Ben-Ami Bartal, "Empathy and Pro-Social Behavior in Rats," *Science* 334 (2011). The criticism of the interpretation of the rat empathy experiment is found in M. Vasconcelos, "Pro-Sociality without Empathy," *Biology Letters* 8 (2012). The "ant empathy" experiment is reported in E. Nowbahari, "Ants, Cataglyphis Cursor, Use Precisely Directed Rescue Behavior to Free Entrapped Relatives," *PLoS One* 4 (2009).

Chapter 11: What Is Sex?

For a fascinating, if rather opinionated, journey through the diverse world of sex determination in animals see J. Roughgarden, *Evolution's Rainbow*,

(2004). For a more scientific introduction, see J.A. Marshall Graves, "Weird Animal Genomes and the Evolution of Vertebrate Sex and Sex Chromosomes," *Annual Review of Genetics* 42 (2008).

D. Zhao, "Somatic Sex Identity Is Cell Autonomous in the Chicken," *Nature* 464 (2010), presents the research explaining gynandromorphy in birds.

The discovery of sex reversal in wrasse was first reported in D.R. Robertson, "Social Control of Sex Reversal in a Coral-Reef Fish," *Science* 177 (1972).

To learn more about the early research of Painter and Jost, see L. DiNapoli, "SRY and the Standoff in Sex Determination," *Molecular Endocrinology* 22 (2008).

The experiments that identified SRY as the principal sex-determining factor in mammals are reviewed in K. Kashimada, "SRY: The Master Switch in Mammalian Sex Determination," *Development* 137 (2010).

For more details on the overall sex-determining pathways, see P. Koopman, "The Delicate Balance between Male and Female Sex Determining Pathways: Potential for Disruption of Early Steps in Sexual Development," *International Journal of Andrology* 33 (2010).

L. DiNapoli, "SRY and the Standoff in Sex Determination," *Molecular Endocrinology* 22 (2008), reviews the impact of variations in the *SOX9* gene on sex reversal.

A. Biason-Lauber, "Control of Sex Development," *Best Practice & Rsearch. Clinical Endocrinology & Metabolism* 24 (2010), presents further details on other important sex-determining proteins, including SF-1, M-33 and CBX-2.

The stories of Mario and Gino, the men with two X chromosomes, were reported in A. Vetro, "XX Males SRY Negative: A Confirmed Cause of Infertility," *Journal of Medical Genetics* 48 (2011), and in G. Micali, "Association of Palmoplantar Keratoderma, Cutaneous Squamous Cell Carcinoma, Dental Anomalies, and Hypogenitalism in Four Siblings with 46,XX Karyotype: A New Syndrome," *Journal of the American Academy of Dermatology* 53 (2005), and B. Capel, "R-Spondin1 Tips the Balance in Sex Determination," *Nature Genetics* 38 (2006), respectively.

Background on sex reversal and 5-AR and 17β-HSD3 deficiency can be found in P.T. Cohen-Kettenis, "Gender Change in 46,XY Persons with 5alpha-Reductase-2 Deficiency and 17beta-Hydroxysteroid Dehydrogenase-3 Deficiency," *Archives of Sexual Behavior* 34 (2005), and J. Imperato-McGinley, "Androgens and Male Physiology the Syndrome of 5alpha-Reductase-2 Deficiency," *Molecular and Cellular Endocrinology* 198 (2002).

For more on complete androgen insensitivity syndrome, see I.A. Hughes, "Androgen Insensitivity Syndrome," *Lancet* 380 (2012).

The relationship between congenital adrenal hyperplasia and 21-hydroxylase is described in M.I. New, "Diagnosis and Management of Congenital Adrenal Hyperplasia," *Annual Review of Medicine* 49 (1998), and M. Hines, "Prenatal Endocrine Influences on Sexual Orientation and on Sexually Differentiated Childhood Behavior," *Frontiers in Neuroendocrinology* 32 (2011).

Chapter 12: Can You Get Cancer from a Gene?

For a more detailed introduction to genetics and inheritance, including the concepts of genotype and phenotypes, see any modern textbook on genetics or molecular biology, such as chapter 8 of Alberts, *Molecular Biology of the Cell*, (2008), or chapter 15 of Kratz, *Molecular and Cell Biology for Dummies*, (2009).

Angelina Jolie's story is from A. Jolie, "My Medical Choice," *New York Times* (2013).

The animal experiments leading to the discovery of the causes of narcolepsy are described in M. Hungs, "Hypocretin/Orexin, Sleep and Narcolepsy," *Bioessays* 23 (2001).

To learn more about inherited breast cancer and the discovery of *BRCA1*, see P.L. Welcsh, "BRCA1 and BRCA2 and the Genetics of Breast and Ovarian Cancer," *Human Molecular Genetics* 10 (2001), and C.I. Szabo, "Inherited Breast and Ovarian Cancer," *Human Molecular Genetics* 4 Spec No (1995).

For a review of the investigations that have been carried out on transgenic mice in attempts to uncover the biological function of *BRCA1* and other cancer-related genes, see P. Taneja, "Transgenic and Knockout Mice Models to Reveal the Functions of Tumor Suppressor Genes," *Clinical Medicine Insights. Oncology* 5 (2011).

The studies suggesting a link between breast cancer and dioxin exposure are U. Manuwald, "Mortality Study of Chemical Workers Exposed to Dioxins: Follow-up 23 Years after Chemical Plant Closure," *Occupational and Environmental Medicine* 69 (2012), and A.C. Pesatori, "Cancer Incidence in the Population Exposed to Dioxin after the "Seveso Accident": Twenty Years of Follow-Up," *Environmental Health* 8 (2009).

Chapter 13: What Is Learning? What Is Memory?

Susannah Cahalan's story is recounted in S. Cahalan, Brain on Fire, (2012).

N. Carey, *The Epigenetics Revolution*, (2012) provides a comprehensive overview of epigenetics intended for the general reader.

For descriptions of the connections between methylation and learning, see J.J. Day, "DNA Methylation and Memory Formation," *Nature Neuroscience* 13 (2010), and J. Feng, "Dnmt1 and Dnmt3a Maintain DNA Methylation and Regulate Synaptic Function in Adult Forebrain Neurons," *Nature Neuroscience* 13 (2010).

The experiments of epigenetic changes in aging mice are reported in S. Peleg, "Altered Histone Acetylation Is Associated with Age-Dependent Memory Impairment in Mice," *Science* 328 (2010).

The discovery of anti-NMDA-receptor encephalitis is described in G.S. Day, "Anti-NMDA-receptor Encephalitis: Case Report and Literature Review of an under-Recognized Condition," *Journal of General Internal Medicine* 26 (2011).

Chapter 14: Can We Live to 120 by Eating Less?

For general information on lifespan extension by dietary restriction see the review articles C.J. Kenyon, "The Genetics of Ageing," *Nature* 464 (2010), and H.E. Wheeler, "Genetics and Genomics of Human Ageing," *Philosophical Transactions of the Royal Society of London. Series B, Biological Sciences* 366 (2011).

For more on the original rat experiments see C.M. McCay, "Prolonging the Life Span," *The Scientific Monthly* 39 (1934).

R.J. Colman, "Caloric Restriction Delays Disease Onset and Mortality in Rhesus Monkeys," *Science* 325 (2009), and J.A. Mattison, "Impact of Caloric Restriction on Health and Survival in Rhesus Monkeys from the NIA Study," *Nature* 489 (2012), present the results of the caloric-restriction experiments in monkeys.

The comparisons of caloric restriction in laboratory mice and wild mice were reported in J.M. Harper, "Does Caloric Restriction Extend Life in Wild Mice?," *Aging Cell* 5 (2006).

For more information on the Okinawan centenarians, see B.J. Willcox, "Caloric Restriction, the Traditional Okinawan Diet, and Healthy Aging: The Diet of the World's Longest-Lived People and Its Potential Impact on Morbidity and Life Span," *Annals of the New York Academy of Sciences* 1114 (2007).

The physiological tests of the members of the Caloric Restriction Society were reported in L. Fontana, "Long-Term Calorie Restriction Is Highly Effective in Reducing the Risk for Atherosclerosis in Humans," *Proceedings of the National Academy of Sciences of the United States of America* 101 (2004).

The CALERIE Project was described in the scientific literature in A.D. Rickman, "The CALERIE Study: Design and Methods of an Innovative 25% Caloric Restriction Intervention," *Contemporary Clinical Trials* 32 (2011), and in the general press in J. Gertner, "The Calorie-Restriction Experiment," *New York Times Magazine* (2009).

The experiments performed on methionine-deprived rats were originally reported in N. Orentreich, "Low Methionine Ingestion by Rats Extends Life Span," *The Journal of Nutrition* 123 (1993).

The 2014 study linking low-protein diets to decreased human mortality rates is M.E. Levine, "Low Protein Intake Is Associated with a Major Reduction in IGF-1, Cancer, and Overall Mortality in the 65 and Younger but Not Older Population," *Cell Metabolism* 19 (2014).

For experiments describing the impact on animal lifespan of oxidative damage, zero gravity and changes in reproductive status, see J.R. Cypser, "Multiple Stressors in Caenorhabditis Elegans Induce Stress Hormesis and Extended Longevity," *The Journals of Gerontology. Series A, Biological Sciences and Medical Sciences* 57 (2002), Y. Honda, "Genes Down-Regulated in Spaceflight Are Involved in the Control of Longevity in Caenorhabditis Elegans," *Scientific Reports* 2 (2012), and H. Hsin, "Signals from the Reproductive System Regulate the Lifespan of C. Elegans," *Nature* 399 (1999), respectively.

J.B. Mason, "Transplantation of Young Ovaries Restored Cardioprotective Influence in Postreproductive-Aged Mice," *Aging Cell* 10 (2011), and S.L. Cargill, "Age of Ovary Determines Remaining Life Expectancy in Old Ovariectomized Mice," *Aging Cell* 2 (2003) described the research on female mice with transplanted ovaries.

The comparison of long-term survival between women with and without ovaries is reported in W.H. Parker, "Ovarian Conservation at the Time of Hysterectomy and Long-Term Health Outcomes in the Nurses' Health

Study," *Obstetrics and Gynecology* 113 (2009), and W.H. Parker, "Effect of Bilateral Oophorectomy on Women's Long-Term Health," *Womens Health* 5 (2009).

The analysis of oral contraceptive use and lifespan in women is M. Vessey, "Factors Affecting Mortality in a Large Cohort Study with Special Reference to Oral Contraceptive Use," *Contraception* 82 (2010).

The study of longevity in Korean eunuchs is K.J. Min, "The Lifespan of Korean Eunuchs," *Current Biology* 22 (2012).

Chapter 15: Can We Live to 120 – And Still Eat Cake?

The story of the Kahn family is from J. Green, "What Do a Bunch of Old Jews Know About Living Forever?," *New York Times Magazine* (2011).

A good review of the genetic experiments in animals that demonstrated lifespan extension without dietary restriction is C.J. Kenyon, "The Genetics of Ageing," *Nature* 464 (2010).

The study of longevity in the siblings of centenarians is T.T. Perls, "Life-Long Sustained Mortality Advantage of Siblings of Centenarians," *Proceedings of the National Academy of Sciences of the United States of America* 99 (2002).

H.E. Wheeler, "Genetics and Genomics of Human Ageing," *Philosophical Transactions of the Royal Society of London. Series B, Biological Sciences* 366 (2011), reviews the genetic searches for DNA variants associated with human longevity.

The 2013 Boston University study of genetic variation in centenarians is P. Sebastiani, "Meta-Analysis of Genetic Variants Associated with Human Exceptional Longevity," *Aging* 5 (2013).

For general background on longevity in naked mole rats see R. Buffenstein, "The Naked Mole-Rat: A New Long-Living Model for Human Aging Research," *The Journals of Gerontology. Series A, Biological Sciences and Medical Sciences* 60 (2005).

The recent discoveries of molecular mechanisms that appear to underlie mole rat longevity are reported in J. Azpurua, "Naked Mole-Rat Has Increased Translational Fidelity Compared with the Mouse, as Well as a Unique 28s Ribosomal RNA Cleavage," *Proceedings of the National Academy of Sciences of the United States of America* 110 (2013).

The molecular pathways through which dietary restriction and other environmental stresses lead to longer life are described in C.J. Kenyon, "The Genetics of Ageing," *Nature* 464 (2010).

The transgenic worm with the 10 times longer lifespan was reported in S. Ayyadevara, "Remarkable Longevity and Stress Resistance of Nematode Pi3k-Null Mutants," *Aging Cell* 7 (2008).

For more information on cellular aging mechanisms see E. Sahin, "Axis of Ageing: Telomeres, P53 and Mitochondria," *Nature Reviews. Molecular Cell Biology* 13 (2012).

The research linking telomere length with health and aging is reviewed in I. Shalev, "Stress and Telomere Biology: A Lifespan Perspective," *Psycho-neuroendocrinology* 38 (2013).

J. Guevara-Aguirre, "Growth Hormone Receptor Deficiency Is Associated with a Major Reduction in Pro-Aging Signaling, Cancer, and Diabetes in Humans," *Science Translational Medicine* 3 (2011) described the anti-aging effects of growth hormone receptor deficiency, and Y. Suh, "Functionally Significant Insulin-Like Growth Factor I Receptor Mutations in Centenarians," *Proceedings of the National Academy of Sciences of the United States of America* 105 (2008), reported on the *IGF1*-receptor mutations in centenarians.

The experiments identifying the effects on aging from senescent cell signaling can be found in D.J. Baker, "Clearance of P16ink4a-Positive Senescent Cells Delays Ageing-Associated Disorders," *Nature* 479 (2011).

D.E. Harrison, "Rapamycin Fed Late in Life Extends Lifespan in Genetically Heterogeneous Mice," *Nature* 460 (2009), describes the effects of rapamycin on longevity in mice.

See R.A. Miller, "Rapamycin, but Not Resveratrol or Simvastatin, Extends Life Span of Genetically Heterogeneous Mice," *The Journals of Gerontology. Series A, Biological Sciences and Medical Sciences* 66 (2011), for the report on rapamycin and resveratrol as anti-aging drugs.

V.N. Anisimov, "If Started Early in Life, Metformin Treatment Increases Life Span and Postpones Tumors in Female SHR Mice," *Aging* 3 (2011), describes the experiments using metformin to increase mouse lifespan.

The review of the effects of antioxidants and hormones to combat human aging is N.S. Kamel, "Antioxidants and Hormones as Antiaging Therapies: High Hopes, Disappointing Results," *Cleveland Clinic Journal of Medicine* 73 (2006).

The story of Brooke Greenberg was reported in the scientific literature in R.F. Walker, "A Case Study of "Disorganized Development" and Its Possible Relevance to Genetic Determinants of Aging," *Mechanisms of Ageing and Development* 130 (2009).

Chapter 16: What Makes a Super-athlete? Part I

D.J. Epstein, *The Sports Gene*, (2013), is a comprehensive recent introduction to genetics and athletic performance for the general reader.

Scientific reviews of the impact of genetics and epigenetics on athletic ability include E.A. Ostrander, "Genetics of Athletic Performance," *Annual Review of Genomics and Human Genetics* 10 (2009), and T. Ehlert, "Epigenetics in Sports," *Sports Medicine* 43 (2013).

See Y.M. Hur, "Genetic Influences on the Difference in Variability of Height, Weight and Body Mass Index between Caucasian and East Asian Adolescent Twins," *International Journal of Obesity* 32 (2008), for studies of height variation in twins.

H.S. Chahal, "AIP Mutation in Pituitary Adenomas in the 18th Century and Today," *The New England Journal of Medicine* 364 (2011), describes gigantism, the AIP gene and the giant Charles Byrne.

For additional information on Marfan syndrome, achrondroplasia and the *SHOX* gene see H.C. Dietz, "Marfan Syndrome Caused by a Recurrent De Novo Missense Mutation in the Fibrillin Gene," *Nature* 352 (1991), W.A. Horton, "Achondroplasia," *Lancet* 370 (2007), and G. Binder, "Short Stature Due to SHOX Deficiency: Genotype, Phenotype, and Therapy," *Hormone Research in Paediatrics* 75 (2011), respectively.

M.I. McCarthy, "Genome-Wide Association Studies for Complex Traits: Consensus, Uncertainty and Challenges," *Nature Reviews. Genetics* 9 (2008), presents a general overview of genome-wide association studies and their applications.

The large GWAS study that attempted to identify the genetic origins of human height variations is M.N. Weedon, "A Common Variant of HMGA2 Is Associated with Adult and Childhood Height in the General Population," *Nature Genetics* 39 (2007).

The identification of the 180 gene variants that contribute to human height variation was originally reported in H. Lango Allen, "Hundreds of Variants Clustered in Genomic Loci and Biological Pathways Affect Human Height," *Nature* 467 (2010).

To learn how dogs aid in the search for genes that cause size variation, as well as other traits, see H.G. Parker, "Man's Best Friend Becomes Biology's Best in Show: Genome Analyses in the Domestic Dog," *Annual Review of Genetics* 44 (2010).

The story of the super-strong German boy was initially reported in M. Schuelke, "Myostatin Mutation Associated with Gross Muscle Hypertrophy in a Child," *The New England Journal of Medicine* 350 (2004), and is recounted in P. Reilly, *The Strongest Boy in the World*, (2006).

High-performing myostatin-deficient dogs are reviewed in D.S. Mosher, "A Mutation in the Myostatin Gene Increases Muscle Mass and Enhances Racing Performance in Heterozygote Dogs," *PLoS Genetics* 3 (2007).

Myostatin knockout mice were first reported in A.C. McPherron, "Regulation of Skeletal Muscle Mass in Mice by a New TGF-beta Superfamily Member," *Nature* 387 (1997), and are reviewed in S.J. Lee, "Myostatin and the Control of Skeletal Muscle Mass," *Current Opinion in Genetics & Development* 9 (1999).

The experiments testing endurance in myostatin knockout mice were reported in K.J. Savage, "Endurance Exercise Training in Myostatin Null Mice," *Muscle and Nerve* 42 (2010).

S.J. Lee, "Quadrupling Muscle Mass in Mice by Targeting TGF-beta Signaling Pathways," *PLoS One* 2 (2007), describes the discovery of the myostatin inhibitor follistatin.

Chapter 17: What Makes a Super-athlete? Part II

The genetic studies of Eero Mäntyranta and his relatives were reported in A. de la Chapelle, "Truncated Erythropoietin Receptor Causes Dominantly Inherited Benign Human Erythrocytosis," *Proceedings of the National Academy of Sciences of the United States of America* 90 (1993).

The *ACTN3* knockout mouse experiments are reviewed in D.G. MacArthur, "An Actn3 Knockout Mouse Provides Mechanistic Insights into the Association between Alpha-Actinin-3 Deficiency and Human Athletic Performance," *Human Molecular Genetics* 17 (2008).

N. Yang, "ACTN3 Genotype Is Associated with Human Elite Athletic Performance," *American Journal of Human Genetics* 73 (2003), describes the link between *ACTN3* gene variants and human athletic performance.

See A.G. Williams, "Bradykinin Receptor Gene Variant and Human Physical Performance," *Journal of Applied Physiology* 96 (2004), for the data linking variation in the genes for the bradykinin receptors and cardiovascular performance.

For more information on angiotensin-converting enzyme and athletic performance, see Z. Puthucheary, "The ACE Gene and Human Performance: 12 Years On," *Sports Medicine* 41 (2011).

The unusual genetic variants found in the hypoxia gene family in the mountain dwellers of Tibet are described in A. Bigham, "Identifying Signatures of Natural Selection in Tibetan and Andean Populations Using Dense Genome Scan Data," *PLoS Genetics* 6 (2010), and X. Yi, "Sequencing of 50 Human Exomes Reveals Adaptation to High Altitude," *Science* 329 (2010).

The connection between adaptation to high-altitude hypoxia and hemoglobin levels is presented in C.M. Beall, "Andean, Tibetan, and Ethiopian Patterns of Adaptation to High-Altitude Hypoxia," *Integr Comp Biol* 46 (2006).

R.W. Hanson, "Born to Run; the Story of the PEPCK-Cmus Mouse," *Biochimie* 90 (2008), reviews the PEPCK "super-mouse."

See Y.X. Wang, "Regulation of Muscle Fiber Type and Running Endurance by PPARdelta," *PLoS Biology* 2 (2004), and B. Egan, "Exercise Metabolism and the Molecular Regulation of Skeletal Muscle Adaptation," *Cell Metabolism* 17 (2013), for more information describing the interactions among exercise, PPARδ protein, and fast and slow twitch muscles.

The experiments linking exercise with epigenetic DNA methylation were reported in R. Barres, "Acute Exercise Remodels Promoter Methylation in Human Skeletal Muscle," *Cell Metabolism* 15 (2012), and T. Ronn, "A Six Months Exercise Intervention Influences the Genome-Wide DNA Methylation Pattern in Human Adipose Tissue," *PLoS Genetics* 9 (2013).

L. DeFrancesco, "The Faking of Champions," *Nature Biotechnology* 22 (2004), provides an overview of the use of drugs to enhance athletic performance.

The data indicating that drugs improve athletic performance is from R.I. Holt, "Is Human Growth Hormone an Ergogenic Aid?," *Drug Testing and Analysis* 1 (2009).

The tests of the drugs GW1516 and AICAR on mice are presented in V.A. Narkar, "AMPK and PPARdelta Agonists Are Exercise Mimetics," *Cell* 134 (2008).

Resveratrol's impact on mouse endurance was reported in the scientific literature in M. Lagouge, "Resveratrol Improves Mitochondrial Function and Protects against Metabolic Disease by Activating SIRT1 and PGC-1alpha," *Cell* 127 (2006), and in the popular press in N. Wade, "Drug Doubles Endurance, Study Says," *New York Times* (2006).

For commercial genetic tests claiming to tell whether one is more suited for marathons or sprints, see the Atlas Sports Genetics website (www. atlasgene.com).

Chapter 18: Why Are Some People So Smart?

The *NR2B* transgenic mouse experiments were originally reported in Y.P. Tang, "Genetic Enhancement of Learning and Memory in Mice," *Nature* 401 (1999). J.Z. Tsien, "Building a Brainier Mouse," *Scientific American* 282 (2000), is an excellent review for the general reader.

Solving mazes and learned responses to external stimuli, the principal methods by which scientists measure learning, memory and other forms of "intelligence" in mice, are described in S. Sharma, "Assessment of Spatial Memory in Mice," *Life Sciences* 87 (2010), and J.N. Crawley, "Behavioral Phenotyping Strategies for Mutant Mice," *Neuron* 57 (2008).

The experiments demonstrating KIF-17's and CDK5's influence on mouse intelligence are described in R.W. Wong, "Overexpression of Motor Protein KIF17 Enhances Spatial and Working Memory in Transgenic Mice," *Proceedings of the National Academy of Sciences of the United States of America* 99 (2002), and A.H. Hawasli, "Cyclin-Dependent Kinase 5 Governs Learning and Synaptic Plasticity Via Control of NMDAR Degradation," *Nature Neuroscience* 10 (2007).

For the experiments that uncovered de novo mutations underlying human intellectual disability, see L.E. Vissers, "A De Novo Paradigm for Mental Retardation," *Nature Genetics* 42 (2010), and F.F. Hamdan, "Excess of De Novo Deleterious Mutations in Genes Associated with Glutamatergic Systems in Nonsyndromic Intellectual Disability," *American Journal of Human Genetics* 88 (2011).

Studies comparing intelligence between twins include S.A. Petrill, "The Genetic and Environmental Etiology of High Math Performance in 10-Year-Old Twins," *Behavior Genetics* 39 (2009), Y. Kovas, "'Generalist Genes' and Mathematics in 7-Year-Old Twins," *Intelligence* 33 (2005), and T.J. Polderman, "A Longitudinal Twin Study on IQ, Executive Functioning, and Attention Problems During Childhood and Early Adolescence," *Acta Neurologica Belgica* 106 (2006).

The study on intellectual ability in twins raised apart is T.J. Bouchard, Jr., "Sources of Human Psychological Differences: The Minnesota Study of Twins Reared Apart," *Science* 250 (1990).

For more on the searches for specific genetic variants linked to intelligence and brain size, see G. Davies, "Genome-Wide Association Studies Establish That Human Intelligence Is Highly Heritable and Polygenic," *Molecular Psychiatry* 16 (2011), and J.L. Stein, "Identification of Common Variants Associated with Human Hippocampal and Intracranial Volumes," *Nature Genetics* 44 (2012).

The DUF1220 region and its link to brain size were reported in L.J. Dumas, "DUF1220-Domain Copy Number Implicated in Human Brain-Size Pathology and Evolution," *American Journal of Human Genetics* 91 (2012).

See C.I. Ragan, "What Should We Do About Student Use of Cognitive Enhancers? An Analysis of Current Evidence," *Neuropharmacology* 64 (2013), for a review of smart drugs and their implications for society.

S.E. McCabe, "Non-Medical Use of Prescription Stimulants among Us College Students: Prevalence and Correlates from a National Survey," *Addiction* 100 (2005), and A.D. DeSantis, "Illicit Use of Prescription ADHD Medications on a College Campus: A Multimethodological Approach," *Journal of American College Health* 57 (2008), assessed student use of smart drugs.

See S.E. Ward, "Challenges for and Current Status of Research into Positive Modulators of AMPA Receptors," *British Journal of Pharmacology* 160 (2010), for a review of the effectiveness of ampakine drugs on learning and memory.

The three meta-analyses of cognitive enhancing drugs are D. Repantis, "Acetylcholinesterase Inhibitors and Memantine for Neuroenhancement in Healthy Individuals: A Systematic Review," *Pharmacological Research* 61 (2010), D. Repantis, "Modafinil and Methylphenidate for Neuroenhancement in Healthy Individuals: A Systematic Review," *Pharmacological Research* 62 (2010), and M.E. Smith, "Are Prescription Stimulants "Smart Pills"? The Epidemiology and Cognitive Neuroscience of Prescription Stimulant Use by Normal Healthy Individuals," *Psychological Bulletin* 137 (2011).

The review of nonpharmaceutical interventions for cognitive enhancement is M. Dresler, "Non-Pharmacological Cognitive Enhancement," *Neuropharmacology* 64 (2013).

Chapter 19: Where Does Language Come From?

For general information on verbal dyspraxia, specific language impairment and dyslexia, see the corresponding articles in Wikipedia.

The original paper describing the K. E. family is J.A. Hurst, "An Extended Family with a Dominantly Inherited Speech Disorder," *Developmental Medicine and Child Neurology* 32 (1990).

O. Feher, "De Novo Establishment of Wild-Type Song Culture in the Zebra Finch," *Nature* 459 (2009), describes the experiments that were performed to create the new songbird culture.

For more on the origins of Nicaraguan Sign Language, see A. Senghas,

"Children Creating Core Properties of Language: Evidence from an Emerging Sign Language in Nicaragua," *Science* 305 (2004).

The report identifying FOXP2 as the protein causing the K. E. family's verbal dyspraxia is C.S. Lai, "A Forkhead-Domain Gene Is Mutated in a Severe Speech and Language Disorder," *Nature* 413 (2001).

For an account of the FOXP2 story for the general reader, see E. Yong, "One Gene Speaks Volumes About Evolution," *New Scientist* 199 (2008).

For additional information on the influence of FOXP2 in communication in mice, songbirds and humans, see S.E. Fisher, "FOXP2 as a Molecular Window into Speech and Language," *Trends in Genetics* 25 (2009).

Studies assessing SRPX2's involvement in language ability include P. Roll, "SRPX2 Mutations in Disorders of Language Cortex and Cognition," *Human Molecular Genetics* 15 (2006), and G.M. Sia, "The Human Language-Associated Gene SRPX2 Regulates Synapse Formation and Vocalization in Mice," *Science* 342 (2013).

The associations between *CNTNAP2* variants and language impairment and normal language abilities are presented in M. Alarcon, "Linkage, Association, and Gene-Expression Analyses Identify CNTNAP2 as an Autism-Susceptibility Gene," *American Journal of Human Genetics* 82 (2008), and A.J. Whitehouse, "CNTNAP2 Variants Affect Early Language Development in the General Population," *Genes, Brain, and Behavior* 10 (2011), respectively.

See chapter 6 of Alberts, *Molecular Biology of the Cell*, (2008), or chapters 18 and 19 of Kratz, *Molecular and Cell Biology for Dummies*, (2009), for further explanations of RNA splicing.

D.F. Newbury, "Recent Advances in the Genetics of Language Impairment," *Genome Medicine* 2 (2010), reviews multiple studies describing the impact of CMIP and ATP2C2, as well as CNTNAP2, on human language development.

Research on DYX1C1 and other proteins linked to dyslexia is reviewed in G. Poelmans, "A Theoretical Molecular Network for Dyslexia: Integrating Available Genetic Findings," *Molecular Psychiatry* 16 (2011), and S.E. Fisher, "Genes, Cognition and Dyslexia: Learning to Read the Genome," *Trends in Cognitive Sciences* 10 (2006).

The Finnish family with inherited dyslexia was reported in J. Nopola-Hemmi, "Two Translocations of Chromosome 15q Associated with Dyslexia," *Journal of Medical Genetics* 37 (2000).

The study of dyslexia in twins is in C. Kang, "Genetics of Speech and Language Disorders," *Annual Review of Genomics and Human Genetics* 12 (2011).

Chapter 20: What Is Gender?

The experiments that showed how sexual gender preference in laboratory mice and rats is influenced by testosterone and estrogen are reviewed in M.V. Wu, "Control of Masculinization of the Brain and Behavior," *Current Opinion in Neurobiology* 21 (2011), and M. Hines, "Prenatal Endocrine Influences on Sexual Orientation and on Sexually Differentiated Childhood Behavior," *Frontiers in Neuroendocrinology* 32 (2011), respectively.

The research that demonstrated how manipulating the proteins fucose mutarotase and TPH2 influences sex preference in mice was reported in D. Park, "Male-Like Sexual Behavior of Female Mouse Lacking Fucose Mutarotase," *BMC Genetics* 11 (2010), and Y. Liu, "Molecular Regulation of Sexual Preference Revealed by Genetic Studies of 5-HT in the Brains of Male Mice," *Nature* 472 (2011), respectively.

The study showing that, in mice, changes in gender specific behavior are linked to chromosomal modifications is J.R. Kurian, "Sex Differences in Epigenetic Regulation of the Estrogen Receptor-Alpha Promoter within the Developing Preoptic Area," *Endocrinology* 151 (2010).

For a lively description of the variety of forms of sexual orientation among animals in the wild, see J. Roughgarden, *Evolution's Rainbow*, (2004). B. Bagemihl, *Biological Exuberance*, (1999), provides an exhaustive account of the diversity of animal sexual orientation.

The study reporting the results of the "sexual-orientation conversion program" is S.L. Jones, "A Longitudinal Study of Attempted Religiously Mediated Sexual Orientation Change," *Journal of Sex & Marital Therapy* 37 (2011).

The data indicating the ages at which people report feelings of same-gender sexual attraction comes from R.B.K. Crooks, *Our Sexuality*, (1990).

The   gender-specific differences in infant behavior were reported in G.M. Alexander, "Sex Differences in Infants' Visual Interest in Toys," *Archives of Sexual Behavior* 38 (2009), A.M. Bao, "Sexual Differentiation of the Human Brain: Relation to Gender Identity, Sexual Orientation and Neuropsychiatric Disorders," *Frontiers in Neuroendocrinology* 32 (2011), and J. Connellan, "Sex Differences in Human Neonatal Social Perception," *Infant Behavior and Development* 23 (2000).

The similar experiments in infant monkeys are from G.M. Alexander, "Sex Differences in Response to Children's Toys in Nonhuman Primates (Cercopithecus Aethiops Sabaeus)," *Evolution and Human Behavior* 23 (2002).

Statistics on the incidence of concordant homosexuality in twins is from N. Langstrom, "Genetic and Environmental Effects on Same-Sex Sexual Behavior: A Population Study of Twins in Sweden," *Archives of Sexual Behavior* 39 (2010), and J.M. Bailey, "Genetic and Environmental Influences on Sexual Orientation and Its Correlates in an Australian Twin Sample," *Journal of Personality and Social Psychology* 78 (2000).

The 1993 paper identifying maternal inheritance of male homosexuality and claiming to find a link between homosexual preference and a region on the X chromosome was D.H. Hamer, "A Linkage between DNA Markers on the X Chromosome and Male Sexual Orientation," *Science* 261 (1993).

The 2005 paper by the same group, in which they extend their search for genes associated with homosexuality across the entire genome and acknowledge their lack of success at replicating their findings is B.S. Mustanski, "A Genomewide Scan of Male Sexual Orientation," *Human Genetics* 116 (2005).

The idea that male homosexuality might be linked to maternal X chromosome modifications was presented in S. Bocklandt, "Extreme Skewing of X Chromosome Inactivation in Mothers of Homosexual Men," *Human Genetics* 118 (2006).

For more on the hypotheses that sexual orientation might be linked to prenatal stress or hormone use, see G. Schmidt, "Does Peace Prevent Homosexuality?," *Archives of Sexual Behavior* 19 (1990), or M. Hines, "Prenatal Endocrine Influences on Sexual Orientation and on Sexually Differentiated Childhood Behavior," *Frontiers in Neuroendocrinology* 32 (2011), respectively.

The study refuting the idea that male human sexual orientation is linked to birth order is A.M. Francis, "Family and Sexual Orientation: The Family-Demographic Correlates of Homosexuality in Men and Women," *Journal of Sex Research* 45 (2008).

A.L. Roberts, "Does Maltreatment in Childhood Affect Sexual Orientation in Adulthood?," *Archives of Sexual Behavior* 42 (2013), is the recent study that suggested that childhood abuse might be linked to some cases of atypical sexual orientation.

For more general information on transgenderism and transsexuality see the entries for "transgender" and "transsexual" in Wikipedia or the North American Intersex Society website (www.isna.org/faq).

The story of Lynn Conway was reported in M.A. Hiltzik, "Through the Gender Labyrinth; How a Bright Boy with a Penchant for Tinkering Grew up to Be One of the Top Women in Her High- Tech Field," *Los Angeles Times* (2000).

David Reimer's life is recounted in J. Colapinto, *As Nature Made Him*, (2006).

For data on the other XY children raised as girls because of atypical penile anatomy see W.G. Reiner, "Discordant Sexual Identity in Some Genetic Males with Cloacal Exstrophy Assigned to Female Sex at Birth," *The New England Journal of Medicine* 350 (2004).

Attempts to link transgenderism with genetic and environmental factors are reviewed in D.F. Swaab, "Sexual Differentiation of the Human Brain: Relevance for Gender Identity, Transsexualism and Sexual Orientation," *Gynecological Endocrinology* 19 (2004).

The transgender community in Thailand is described in S. Winter, "Thai Transgenders in Focus: Their Beliefs About Attitudes Towards and Origins of Transgender," *International Journal of Transgenderism* 9 (2006).

D.F. Swaab, *We Are Our Brains*, (2014), is a book for the general reader covering many of topics discussed in the chapter, written by one of the pioneers in the field of applying brain research to understanding human sexuality.

Chapter 21: What Is Pain? What Is Pleasure?

The stories of Steven Pete and Paul Waters are recounted on their website www.thefactsofpainlesspeople.com. For a general review of congenital insensitivity to pain see A.I. Basbaum, "Cellular and Molecular Mechanisms of Pain," *Cell* 139 (2009).

The early experiment demonstrating inheritance of sensitivity to pain in rats is M. Devor, "Heritability of Symptoms in an Experimental Model of Neuropathic Pain," *Pain* 42 (1990).

Twin studies on human sensitivity to pain are described in C.S. Nielsen, "What Can We Learn from Twin Studies of Pain and Analgesia?," *Pain* 153 (2012).

To learn more about ion channels and neuronal signaling, see chapter 11 of Alberts, *Molecular Biology of the Cell*, (2008) .

To learn more details about SCN9A and how it modulates pain, see F. Reimann, "Pain Perception Is Altered by a Nucleotide Polymorphism in SCN9A," *Proceedings of the National Academy of Sciences of the United States of America* 107 (2010).

For descriptions of other proteins in the pain-perception pathway, such as KCSN1 and CACNAD2D3, see M. Costigan, "Multiple Chronic Pain States Are Associated with a Common Amino Acid-Changing Allele in KCNS1,"

Brain 133 (2010), and G.G. Neely, "A Genome-Wide Drosophila Screen for Heat Nociception Identifies Alpha2delta3 as an Evolutionarily Conserved Pain Gene," *Cell* 143 (2010).

OPRM1's role in pain perception in mice is reported in I. Sora, "Opiate Receptor Knockout Mice Define Mu Receptor Roles in Endogenous Nociceptive Responses and Morphine-Induced Analgesia," *Proceedings of the National Academy of Sciences of the United States of America* 94 (1997). For OPRM1's involvement in pain in monkeys, see C.S. Barr, "Variation at the Mu-Opioid Receptor Gene (OPRM1) Influences Attachment Behavior in Infant Primates," *Proceedings of the National Academy of Sciences of the United States of America* 105 (2008).

See C. Walter, "Meta-Analysis of the Relevance of the OPRM1 118A>G Genetic Variant for Pain Treatment," *Pain* 146 (2009), and C. Walter, "Micro-Opioid Receptor Gene Variant OPRM1 118 A>G: A Summary of Its Molecular and Clinical Consequences for Pain," *Pharmacogenomics* 14 (2013), for more on *OPRM1* gene variants and pain.

The PET study of endorphin binding and the runner's high is H. Boecker, "The Runner's High: Opioidergic Mechanisms in the Human Brain," *Cerebral Cortex* 18 (2008).

Olds' and Mills' original study of rats pressing a bar to receive electrical brain stimulation is J. Olds, "Positive Reinforcement Produced by Electrical Stimulation of Septal Area and Other Regions of Rat Brain," *Journal of Comparative and Physiological Psychology* 47 (1954).

R.G. Heath, "Pleasure and Brain Activity in Man. Deep and Surface Electroencephalograms During Orgasm," *The Journal of Nervous and Mental Disease* 154 (1972), is a report of Heath's early experiments with electronic stimulation and measurement of the human brain in pleasure.

For a description of the opioids and their impact on the experience of pleasure see P.W. Kalivas, "Cocaine and Amphetamine-Like Psychostimulants: Neurocircuitry and Glutamate Neuroplasticity," *Dialogues in Clinical Neuroscience* 9 (2007).

Pleasure sensitivity in dopamine-receptor knockout mice was studied in G.I. Elmer, "Brain Stimulation and Morphine Reward Deficits in Dopamine D2 Receptor-Deficient Mice," *Psychopharmacology* 182 (2005).

The monkey study linking dopamine to the anticipation of pleasure was C.D. Fiorillo, "Discrete Coding of Reward Probability and Uncertainty by Dopamine Neurons," *Science* 299 (2003).

Chapter 22: Why Are Some People Happy?

The World Happiness Report is J. Helliwell, World Happiness Report, (2012).

Brickman et al.'s early study of happiness is P. Brickman, "Lottery Winners and Accident Victims: Is Happiness Relative?," *Journal of Personality and Social Psychology* 36 (1978).

The Dutch report of happiness assessment in twins is M. Bartels, "Heritability and Genome-Wide Linkage Scan of Subjective Happiness," *Twin Research and Human Genetics* 13 (2010).

The MRI study comparing brain responses to physical and emotional pain is E. Kross, "Social Rejection Shares Somatosensory Representations with Physical Pain," *Proceedings of the National Academy of Sciences of the United States of America* 108 (2011).

See J. Osborn, "Pain Sensation Evoked by Observing Injury in Others," *Pain* 148 (2010), for the MRI experiment on empathetic pain.

Approaches for measuring depression and other emotional states in animals are described in V. Krishnan, "Animal Models of Depression: Molecular Perspectives," *Current Topics in Behavioral Neuroscience* 7 (2011), and E.J. Nestler, "Animal Models of Neuropsychiatric Disorders," *Nature Neuroscience* 13 (2010).

To learn more about the overall effects of serotonin on happiness and depression, see M.J. Owens, "Role of Serotonin in the Pathophysiology of Depression: Focus on the Serotonin Transporter," *Clinical Chemistry* 40 (1994), and C.B. Nemeroff, "The Role of Serotonin in the Pathophysiology of Depression: As Important as Ever," *Clinical Chemistry* 55 (2009).

J.C. Fournier, "Antidepressant Drug Effects and Depression Severity: A Patient-Level Meta-Analysis," *Journal of the American Medical Association* 303 (2010), and R. Pies, "Are Antidepressants Effective in the Acute and Long-Term Treatment of Depression? Sic et Non," *Innovations in Clinical Neuroscience* 9 (2012), assess the effectiveness of SSRIs in treating human depression.

G.V. Carr, "The Role of Serotonin Receptor Subtypes in Treating Depression: A Review of Animal Studies," *Psychopharmacology* 213 (2011), reviews experiments studying serotonin transmission and SSRIs in animal depression.

The study measuring depression in TPH2 transgenic mice is J.P. Jacobsen, "The 5-HT Deficiency Theory of Depression: Perspectives from a Naturalistic 5-HT Deficiency Model, the Tryptophan Hydroxylase 2Arg439His Knockin Mouse," *Philosophical Transactions of the Royal Society of Lon-*

don. Series B, Biological Sciences 367 (2012).

H. Clarke, "Association of the 5- HTTLPR Genotype and Unipolar Depression: A Meta-Analysis," *Psychological Medicine* 40 (2010) presents a meta-analysis of studies measuring the association of *SLC6A4* variants and human depression.

An overview of the long-term behavioral consequences of child abuse can be found in A.B. Silverman, "The Long-Term Sequelae of Child and Adolescent Abuse: A Longitudinal Community Study," *Child Abuse & Neglect* 20 (1996).

The experiments demonstrating epigenetic changes resulting from maltreatment of infant mice is reported in I.C. Weaver, "Epigenetic Programming by Maternal Behavior," *Nature Neuroscience* 7 (2004), and T.Y. Zhang, "Epigenetic Mechanisms for the Early Environmental Regulation of Hippocampal Glucocorticoid Receptor Gene Expression in Rodents and Humans," *Neuropsychopharmacology* 38 (2013).

The studies linking epigenetic changes, suicide and child abuse are P.O. McGowan, "Epigenetic Regulation of the Glucocorticoid Receptor in Human Brain Associates with Childhood Abuse," *Nature Neuroscience* 12 (2009), and P.O. McGowan, "Promoter-Wide Hypermethylation of the Ribosomal RNA Gene Promoter in the Suicide Brain," *PLoS One* 3 (2008).

The interaction of genetic variation and environmental influences on brain functioning in monkeys is described in A.J. Bennett, "Early Experience and Serotonin Transporter Gene Variation Interact to Influence Primate Cns Function," *Molecular Psychiatry* 7 (2002).

The classic experiments of Caspi and Moffitt on genetic variation and childhood abuse on adult depression are reported in A. Caspi, "Influence of Life Stress on Depression: Moderation by a Polymorphism in the 5-HTT Gene," *Science* 301 (2003). For meta-analyses of the studies that have attempted to replicate the Caspi and Moffitt findings, see N. Risch, "Interaction between the Serotonin Transporter Gene (5-HTTLPR), Stressful Life Events, and Risk of Depression: A Meta-Analysis," *Journal of the American Medical Association* 301 (2009), and K. Karg, "The Serotonin Transporter Promoter Variant (5-HTTLPR), Stress, and Depression Meta-Analysis Revisited: Evidence of Genetic Moderation," *Archives of General Psychiatry* 68 (2011).

Chapter 23: What Is Fear?

For a review of twin studies of anxiety disorder see J.M. Hettema, "A Review and Meta-Analysis of the Genetic Epidemiology of Anxiety Disorders," *The American Journal of Psychiatry* 158 (2001).

Transgenic serotonin-transporter mouse behavior is described in D.A. Finn, "Genetic Animal Models of Anxiety," *Neurogenetics* 4 (2003), and J.P. Jacobsen, "The 5-HT Deficiency Theory of Depression: Perspectives from a Naturalistic 5-HT Deficiency Model, the Tryptophan Hydroxylase 2Arg439His Knockin Mouse," *Philosophical Transactions of the Royal Society of London. Series B, Biological Sciences* 367 (2012).

The effects of MAO-A deficiency on anxiety and other behavior in mice was reported in O. Cases, "Aggressive Behavior and Altered Amounts of Brain Serotonin and Norepinephrine in Mice Lacking MAOA," *Science* 268 (1995), and J.J. Kim, "Selective Enhancement of Emotional, but Not Motor, Learning in Monoamine Oxidase a-Deficient Mice," *Proceedings of the National Academy of Sciences of the United States of America* 94 (1997).

Attempts to associate differences in human anxiety levels with *MAO-A* gene variants are described in J. Samochowiec, "Association Studies of MAO-A, COMT, and 5-HTT Genes Polymorphisms in Patients with Anxiety Disorders of the Phobic Spectrum," *Psychiatry Research* 128 (2004).

A. Erhardt, "TMEM132D, a New Candidate for Anxiety Phenotypes: Evidence from Human and Mouse Studies," *Molecular Psychiatry* 16 (2011), presents the human and mouse studies indicating a link between fear sensitivity and the protein TMEM.

B.G. Dias, "Parental Olfactory Experience Influences Behavior and Neural Structure in Subsequent Generations," *Nature Neuroscience* 17 (2014) describes the experiments that demonstrated the inheritance of fear of specific odors in mice.

The impact of *FKBP5* genetic variants and childhood abuse on PTSD and DNA methylation states can be found in E.B. Binder, "Association of FKBP5 Polymorphisms and Childhood Abuse with Risk of Posttraumatic Stress Disorder Symptoms in Adults," *Journal of the American Medical Association* 299 (2008), and T. Klengel, "Allele-Specific FKBP5 DNA Demethylation Mediates Gene-Childhood Trauma Interactions," *Nature Neuroscience* 16 (2013).

For an overview of how obsessive-compulsive behavior is studied in mice see H.R. Robertson, "Annual Research Review: Transgenic Mouse Models of Childhood-Onset Psychiatric Disorders," *Journal of Child Psy-*

chology and Psychiatry 52 (2011).

The studies of obsessive-compulsive behavior and the SAPAP3 protein are reported in J.M. Welch, "Cortico-Striatal Synaptic Defects and OCD-Like Behaviours in Sapap3-Mutant Mice," *Nature* 448 (2007), and S. Zuchner, "Multiple Rare SAPAP3 Missense Variants in Trichotillomania and OCD," *Molecular Psychiatry* 14 (2009).

The survey comparing risk-taking behavior between twins is D.R. Miles, "A Twin Study on Sensation Seeking, Risk Taking Behavior and Marijuana Use," *Drug and Alcohol Dependence* 62 (2001).

R.P. Ebstein, "Dopamine D4 Receptor (D4DR) Exon III Polymorphism Associated with the Human Personality Trait of Novelty Seeking," *Nature Genetics* 12 (1996), is the original report attempting to link *DRD4* genetic variants and novelty-seeking behavior.

For a review of experiments seeking to associate variations in the *DRD4* gene with a variety of behavioral traits see R. Ptacek, "Dopamine D4 Receptor Gene DRD4 and Its Association with Psychiatric Disorders," *Medical Science Monitor* 17 (2011).

Chapter 24: What Is Aggression?

D.M.D. Goldman, *Our Genes, Our Choices*, (2012), is a comprehensive description of one laboratory's efforts to link genetic variants to different types of behavior, including aggression.

For a good scientific review of experiments on genetic variation and aggressive behavior see I.W. Craig, "Genetics of Human Aggressive Behaviour," *Human Genetics* 126 (2009).

The studies of aggression in aromatase- and estrogen-receptor knockout mice are reported in K. Toda, "A Loss of Aggressive Behaviour and Its Reinstatement by Oestrogen in Mice Lacking the Aromatase Gene (Cyp19)," *The Journal of Endocrinology* 168 (2001), and S. Ogawa, "Behavioral Effects of Estrogen Receptor Gene Disruption in Male Mice," *Proceedings of the National Academy of Sciences of the United States of America* 94 (1997), respectively.

The experiments linking serotonin levels and aggressive behavior in mice and songbirds are presented in D. Caramaschi, "Differential Role of the 5-HT1A Receptor in Aggressive and Non-Aggressive Mice: An across-Strain Comparison," *Physiology and Behavior* 90 (2007), and T.S. Sperry, "Effects of Acute Treatment with 8-OH-DPAT and Fluoxetine on Aggressive

Behaviour in Male Song Sparrows (Melospiza Melodia Morphna)," *Journal of Neuroendocrinology* 15 (2003).

O. Cases, "Aggressive Behavior and Altered Amounts of Brain Serotonin and Norepinephrine in Mice Lacking MAOA," *Science* 268 (1995), describes the increased aggression in *MAO-A* knockout mice and the underlying mechanism of elevated brain serotonin levels during development.

The King's College data on the interaction of *MAO-A* genetic variants and stress on adult aggression was published in A. Caspi, "Role of Genotype in the Cycle of Violence in Maltreated Children," *Science* 297 (2002).

The meta-analysis of the attempts to replicate the King's College findings is A. Taylor, "Meta-Analysis of Gene-Environment Interactions in Developmental Psychopathology," *Development and Psychopathology* 19 (2007).

J.A. Gogos, "Catechol-O-methyltransferase-deficient Mice Exhibit Sexually Dimorphic Changes in Catecholamine Levels and Behavior," *Proceedings of the National Academy of Sciences of the United States of America* 95 (1998), presented the data on aggressive behavior in COMT-deficient mice.

Dugdale's book on the Jukes family is R. Dugdale, *The Jukes*, (1877).

S. Christianson, "Bad Seed or Bad Science? The Story of the Notorious Jukes Family," *New York Times* (2003), presents recent data on the Jukes family.

The violent Dutch family with a rare mutation in the *MAO-A* gene was reported initially in H.G. Brunner, "Abnormal Behavior Associated with a Point Mutation in the Structural Gene for Monoamine Oxidase A," *Science* 262 (1993).

The reports describing links between human aggression and variations in serotonin-transporter or -receptor genes are J.H. Beitchman, "Serotonin Transporter Polymorphisms and Persistent, Pervasive Childhood Aggression," *The American Journal of Psychiatry* 163 (2006), and L. Bevilacqua, "A Population-Specific HTR2B Stop Codon Predisposes to Severe Impulsivity," *Nature* 468 (2010).

D. Cheng, "Association Study of Androgen Receptor CAG Repeat Polymorphism and Male Violent Criminal Activity," *Psychoneuroendocrinology* 31 (2006), and S. Rajender, "Reduced CAG Repeats Length in Androgen Receptor Gene Is Associated with Violent Criminal Behavior," *International Journal of Legal Medicine* 122 (2008), present the measurements of different androgen receptor variants among violent criminals.

The study attempting to link androgen-receptor variants and aggressive behavior in normal people is E.G. Jonsson, "Androgen Receptor Trinucleotide Repeat Polymorphism and Personality Traits," *Psychiatric Genetics*

11 (2001). A. Konno, "Androgen Receptor Gene Polymorphisms Are Associated with Aggression in Japanese Akita Inu," *Biology Letters* 7 (2011), is the similar study using Akita dogs.

The Charles Whitman story can be found in G.M. Lavergne, *A Sniper in the Tower*, (1997).

The man whose pedophilia was linked to a brain tumor was originally reported in J.M. Burns, "Right Orbitofrontal Tumor with Pedophilia Symptom and Constructional Apraxia Sign," *Archives of Neurology* 60 (2003).

The court cases of Mobley and Waldroup are described in M. Baum, "The Monoamine Oxidase a (MAOA) Genetic Predisposition to Impulsive Violence: Is It Relevant to Criminal Trials?," *Neuroethics* 6 (2013).

The Bayout trial is reviewed in F. Forzano, "Italian Appeal Court: A Genetic Predisposition to Commit Murder?," *European Journal of Human Genetics* 18 (2010), and E. Feresin, "Lighter Sentence for Murderer with 'Bad Genes'," *Nature* (2009).

Y. Yang, "Prefrontal Structural and Functional Brain Imaging Findings in Antisocial, Violent, and Psychopathic Individuals: A Meta-Analysis," *Psychiatry Research* 174 (2009), presents the data describing the MRI scans of violent offenders.

Chapter 25: What Are Science's Limits?

For discussions of some of the ethical issues raised by genetics, see E.C. Hayden, "Informed Consent: A Broken Contract," *Nature* 486 (2012), L.F. Ross, "Mandatory Versus Voluntary Consent for Newborn Screening?," *Kennedy Institute of Ethics Journal* 20 (2010), K.L. Hudson, "Genomics, Health Care, and Society," *The New England Journal of Medicine* 365 (2011), and P. Reilly, "*Abraham Lincoln's DNA*," (2000).

Pinker's analysis of the philosophical implications of our emerging understanding of biology is S. Pinker, *The Blank Slate*, (2002).

GLOSSARY

adenine. One of the four fundamental nucleotides of DNA.

aerobic. Using oxygen.

allele. See **polymorphism**.

amino acids. A class of small carbon-based molecules that are the building blocks of proteins.

anaerobic. Not using oxygen.

androgen. A sex hormone, such as testosterone or dihydrotestosterone (DHT), produced primarily by the testes, that stimulates male sexual development.

aromatase. Key enzyme in the synthesis of estrogen from testosterone.

association study. See **genome-wide association study**.

ATP. Adenosine triphosphate, the primary energy-storing molecule of the cell.

autosome. Any chromosome other than one of the sex chromosomes.

bacteria. A large family of micro-organisms that consist of a single cell without a cell nucleus.

base. A subunit of a DNA or RNA molecule. The DNA bases are adenine, cytosine, guanine and thymine. The RNA bases are adenine, cytosine, guanine and uracil.

base pair. A nucleic acid base bound together with its complementary base.

bisexual. An individual who is sexually attracted to both males and females.

blind experiment. An experiment in which the research subjects don't know whether they are receiving the actual drug (or other substance) being tested or an inert substance (placebo) with no biological effect.

bone marrow. A type of tissue found within bone that is the source of cells of the blood and immune systems.

bone marrow transplant. A medical procedure in which healthy bone marrow cells are transplanted into an individual suffering from one of certain leukemias or other serious blood diseases.

carrier. A carrier of a recessive genetic disease is a person who has one normal and one disabling variant of some gene. A carrier will not have the disease. However, if two carriers have a child, there is a 1 in 4 chance that the child will inherit the disabling variant from each parent and hence be afflicted with the disease.

catalyst. A molecule that enables a biochemical reaction to occur at a much faster rate than would otherwise be possible.

causal variant. A variation in an individual's genome that causes some phenotype to occur or be more likely to occur.

cell. The structural subunit of all forms of life. Some living organisms, such as bacteria, consist of a single cell, while more complex organisms may consist of trillions of cells.

cellular differentiation. The process by which cells become specialized cells, such as blood cells, muscle cells or neurons.

chromosomal location. A position along a chromosome measured in base pairs from one tip (the "short-arm end") of the chromosome. Since each individual's DNA sequence is slightly different in length from everyone else's, chromosomal location is defined in terms of a specific genome, which is nearly always a reference genome.

chromosome. A DNA molecule together with the proteins that are bound to it.

codon. Sequence of three consecutive nucleotides that is translated by the ribosome into a single amino acid.

complementary base. A base that is able to make a chemical bond with a given base; adenine is the complementary base of thymine, and cytosine is the complementary base of guanine.

complete sex reversal. The condition in which an individual has the normal appearance of one sex while having the normal chromosomal complement of the other sex.

concordance. The occurrence or absence of a trait or phenotype in both members of a pair of twins. In such cases, the twins are said to be concordant; if one twin has the trait and the other does not, they are described as discordant for the trait.

conditional knockout animal. An animal that has been genetically manipulated so that one or more of its genes have been disabled, but only in certain of the animal's cells or at a specific time during the animal's development.

control. In the context of the design of a scientific experiment, a control is a normal or baseline condition against which some unusual or novel condition or intervention is compared. An experiment using a control is called a controlled experiment.

correlation coefficient. Mathematical measure of similarity between continuously varying quantities, such as height or IQ score. Correlation coefficient values range from +1 (exactly similar) to −1 (completely different). Thus a correlation coefficient of zero indicates no relationship between the two quantities.

cytosine. One of the four fundamental nucleotides of DNA.

demethylation. A chemical reaction in which a methyl group is removed from a molecule.

de novo mutation. A mutation that occurs during the production of the sperm or egg cell (in contrast to most mutations, which are inherited from a parent's DNA).

diploid genome. The entire sequence of DNA letters of an organism, including all of the DNA the organism inherited from both of its parents.

discordant. In twin studies, indicating that a pair of twins do not share a trait or phenotype.

dizygotic twins. Twins that develop from two distinct eggs that were fertilized at the same time by different sperm; also called fraternal or nonidentical twins.

DNA. Extremely long nucleic acid molecule that stores the genetic "blueprints" for making our proteins.

DNA strand. Either of the paired nucleotide biopolymers that make up the double-stranded DNA molecule.

dominant variant. A variant that affects an individual's phenotype even if they only have a single copy of the variant.

environmental sex reversal. Change of an organism's sex as a result of a change in the environment, such as a change in ambient temperature or acidity.

enzyme. A protein that acts as a catalyst in a biochemical reaction.

epigenetic modification. A biochemical modification of either a DNA nucleotide or a protein bound to DNA within a chromosome.

epigenome. The complete set of epigenetic modifications to the chromosomes of a cell.

essential amino acid. An amino acid that can't be synthesized by an animal and therefore must be obtained by the animal from its food.

essential gene. A gene that – if disabled in both copies – leads to an organism not being able to live.

estrogen. Any of the related female sex hormones of the estrogen family, including estrone, estradiol and estriol.

eugenics. The idea that individuals with heritable traits considered desirable to society should be encouraged to have more offspring, while those with undesirable traits should be prevented from reproducing.

extreme maternal ancestor. At any level of ancestry, the individual's mother's mother's ... mother. For example, at the level of great grandparents, the extreme maternal ancestor is their mother's mother's mother.

extreme paternal ancestor. At any level of ancestry, the individual's father's father's ... father. For example, at the level of great grandparents, the extreme paternal ancestor is their father's father's father.

fear conditioning. An experimental procedure in which animals first learn to associate a visual or aural cue with an unpleasant experience and subsequently are tested on how well they remember the association (judged by their fearful behavior in response to the cue). May be used to assess anxiety or learning.

gender. Set of characteristics that distinguish males from females. Generally, in this book the term *sex* is used to indicate physical distinctions between males and females, while *gender* is used to specify behavioral differences between males and females.

gender dysphoria or **gender identity reversal.** The condition of having a gender self-image that is not in agreement with one's outward physical characteristics.

gender identity. The gender with which an individual identifies as part of their self-image.

gene. A region of DNA that codes for a single protein.

gene therapy. Modification of an organism's DNA to treat a genetic disease.

genetic. Relating to the specific sequence of nucleotides in an organism's DNA.

genetic code. The set of rules specifying which DNA nucleotide triplet specifies which amino acid in the "blueprint" for a protein, as well as the nucleotide triplets that determine when a protein blueprint in the DNA starts or stops.

genetic marker. A location in the (haploid) genome where frequent variation occurs within a population.

genome. May refer to either the haploid genome or the diploid genome, depending on the context.

genome-wide association study (GWAS). A large study (often tens of thousands of people) comparing genetic markers throughout the genome of people who have a given trait with people who don't have the trait.

genotype. The specific genetic sequence of an individual member of a population. As a verb, *to genotype* means to determine the genetic sequence of a member of a population.

germ cell. A sperm or egg cell.

glucose. A simple form of sugar used as an energy source by most organisms.

gonads. Sex organs. May refer to testes, ovaries or to the initial undifferentiated sex organs.

guanine. One of the four fundamental nucleotides of DNA.

GWAS. See **genome-wide association study**.

gynandromorph. An organism that is male on one side of its body and female on the other side.

haploid genome. The entire sequence of DNA letters that an organism inherits from one of its parents.

haplotype. A set of genetic variants, typically inherited together, that are contained within a single region of a chromosome.

heterosexual. Someone who is primarily or exclusively sexually attracted to individuals of the opposite gender.

heterozygous location. A location in the genome where the genetic sequence that the individual inherited from their mother differs from the sequence they inherited from their father.

homologous chromosomes or **homologous pair.** Two chromosomes of the same type that the individual has inherited, one from their father and one from their mother.

homosexual. A person (or animal) who is primarily or exclusively sexually attracted to individuals of the same sex.

homozygous location. A location in the genome where an individual inherited the same genetic sequence from their mother and their father.

hormone. A ligand that the body uses to send signals over relatively long distances throughout the body, typically via the bloodstream.

immune system. The part of our body that protects us from infection.

inbreeding. The mating of two individuals that are genetically related.

intersex. Having partially male and partially female sexual genotype and phenotype.

ion. An electrically charged atom or molecule.

knockin animal. An animal that has been genetically manipulated so that one or more genes have been added or gained a new function.

knockout animal. An animal that has been genetically manipulated so that one or more genes have been disabled.

ligand. A small molecule – often but not always a protein – that cells secrete to send a signal to other cells.

location. See **chromosomal location**.

lysine. One of the 20 amino acids.

lysine acetylation. A specific epigenetic modification of a lysine amino acid, important in gene regulation by chromosomal protein alteration.

marker. See **genetic marker**.

messenger RNA (mRNA). A transcribed section of RNA that can be translated into a protein by the ribosome.

meta-analysis. An aggregation and review of multiple related scientific reports, typically using sophisticated statistical techniques to combine the results of studies with widely varying numbers of research subjects.

methionine. One of the 20 amino acids. Because the methionine codon is also the start codon of the genetic code, all proteins start with a methionine amino acid.

methylation. A chemical reaction in which a methyl group is attached to a molecule.

methyl group. A chemical ion consisting of one carbon atom and three hydrogen atoms.

mitochondrial DNA. The small amount of DNA that is located in each mitochondrion and is transmitted directly from mother to child (in contrast to the nuclear DNA that is contained in the cell's nucleus and is inherited from both the father and the mother).

mitochondrion (plural mitochondria). Cellular substructure used by the cell to convert food and oxygen into usable chemical energy in the form of the molecule ATP.

monoamine neurotransmitter. A neurotransmitter with a chemical structure consisting of a nitrogen (amine) chemical group linked to a carbon ring chemical group. Examples include serotonin, dopamine, adrenaline and noradrenaline.

monozygotic twins. Twins that develop from a single fertilized egg, also called identical twins.

mRNA. See **messenger RNA**.

mutation. A genetic variant that has appeared more recently in a family or species than other variant(s) found at that location.

neuron. Cell of the nervous system.

neurotransmitter. A ligand used by neurons to signal nearby neurons or other cells.

neutral mutation or **neutral variant.** A genetic variant that makes the organism neither more nor less fit.

NIH. National Institutes of Health, the U.S. government's medical research agency.

NMDA receptor. One of the brain's key nerve cell receptors for the detection of glutamate signals.

normal. Typical or average; not necessarily healthy or desirable.

nucleic acid. A class of biological molecules consisting of a long backbone made of phosphates to which a sequence of nucleotides is attached.

nucleotide. See **base**.

nucleus. Cellular substructure containing a cell's genetic material, that is, its DNA.

ovaries. Female sex organs.

oxidative damage. DNA and protein damage caused by oxygen-carrying ions produced during energy generation in a cell's mitochondria.

partial sex reversal. See **intersex**.

penetrance. The fraction of individuals carrying a genetic variant that actually displays the phenotype associated with that variant.

phenotype. An observable property or trait of an organism or cell.

pheromone. A molecule whose scent is used by animals for sexual signaling.

place preference test. A test to determine whether an animal has a preference for a drug. The animal is initially trained to associate a drug with a specific location, and subsequently tested to see whether it prefers to return to the location where it previously received the drug.

placebo. An inert substance (often a sugar pill) given to someone who is told that the substance may have a beneficial effect. Placebos are often used in scientific studies to separate the physiological effects of a medical treatment from the psychological effects of simply being *told* that the treatment is beneficial.

polymorphism. A genetic variation observed between individuals of the same species.

positive selection. The increased occurrence over time of genetic variants that increase the fitness of an organism.

progeria. A disease in which a principal symptom is premature aging.

protein. Class of biomolecules that make up most of the (nonwater) structures of living organisms. Proteins consist of long strings of amino acids.

receptor. A protein structure that binds ligands and thereby enables cells to detect the ligands.

recessive variant. A variant that affects an individual's phenotype only if both copies of their DNA have the variant.

recombination. The mixing of DNA between pairs of homologous chromosomes during the formation of sperm and egg cells.

reference genome or **reference sequence.** The entire DNA sequence of a single individual of a species or of some "average" individual. Used as a standard to describe how any one individual's genotype differs from the genomes of other members of the species.

review article. A scientific publication that integrates the results of multiple related experiments and studies.

ribosome. Cellular enzyme consisting of proteins and RNA that converts mRNA into protein.

RNA. A nucleic-acid molecule related to DNA that serves numerous functions within cells, including being the messenger molecule (in the form of messenger RNA) that transmits the genetic prescription for making a protein from the DNA in the nucleus to the protein-building ribosomes.

senescent cells. Cells that have matured to the point where they cannot divide under any conditions.

sexual orientation. Characteristic describing an individual's physical and/or emotional attraction to the same and/or opposite sex.

sexual preference. The sex an individual is *most* attracted to when given a choice.

species. Group of organisms with closely related genomes. Among sexually reproducing organisms, the term *species* indicates a class of organisms whose mating can produce healthy, fertile offspring.

start codon. The initial codon of an mRNA sequence that is translated by the ribosome; this is nearly always AUG (corresponding to a DNA sequence of ATG).

statistically significant. Description of a statistical test that has a large enough sample size that one can be confident that the test's result was not caused by chance.

stop codon. Codon that signals the ribosome to terminate mRNA translation; generally either UGA, UAA or UAG (corresponding to a DNA sequences of TGA, TAA or TAG).

strain. A population of genetically identical animals.

strand. See **DNA strand**.

structural variant. A large variation in an individual's DNA that may be hundreds of thousands or even millions of nucleotides in length. Structural variants are often either large deletions or duplications of DNA, or may be transfers of large pieces of DNA from one chromosome to another.

telomerase. An enzyme that repairs the telomeres of chromosomes.

telomere. The end of a chromosome.

testes. Male sex organs.

thymine. One of the four fundamental nucleotides of DNA.

transcription. The copying of a section of DNA into RNA by an enzyme (called RNA polymerase).

transcription factor. A nuclear protein that regulates gene transcription.

transgender or **transsexual.** Someone whose gender self-image is not in agreement with their outward physical characteristics. See **gender dysphoria**.

transgenic animal. An animal that has been genetically manipulated so that one or more of its genes have been modified in some controlled manner.

translation. The synthesis of a sequence of amino acids by the ribosome from an mRNA sequence.

uracil. One of the four fundamental nucleotides of RNA. It is closely related to the DNA nucleotide thymine.

virus. A large class of microorganisms that consist of a small amount of DNA or RNA surrounded by a protein coat. Viruses are unable to survive or replicate unless located within a host organism whose biological machinery they can manipulate.

X chromosome. One of the two chromosomes in mammals that determine sex. Generally females have two X chromosomes and males only one.

Y chromosome. One of the two chromosomes in mammals that determine sex. Generally males have one Y chromosome and females have none.

INDEX

ABOUT THE AUTHOR

Peter Schattner is a scientist, educator and writer with over 30 years of research experience in molecular biology, biomedical instrumentation and physics. He received his PhD in physics from MIT under Nobel Laureate Steven Weinberg and has held research and teaching positions at the University of California, California State University, Stanford Research Institute and Diasonics, Inc., where he contributed to the early development of medical ultrasound and MRI scanners. He is the author of numerous scientific articles and reviews as well as the textbook *Genomes, Browsers and Databases*. He was a Woodrow Wilson Fellow and a winner of the American Institute of Ultrasound in Medicine's Award for Technical Innovation. When not doing science, he enjoys hiking, music and wildlife photography. He lives in the San Francisco Bay Area with his wife Susan and a motley assortment of stuffed animals. *Sex, Love and DNA* is his first book for nonscientists. Visit PeterSchattner.com for current information relating to the author and the science of *Sex, Love and DNA*.